PHILOSOPHY *of* BIOLOGY

SECOND EDITION

PHILOSOPHY *of* BIOLOGY

edited by MICHAEL RUSE

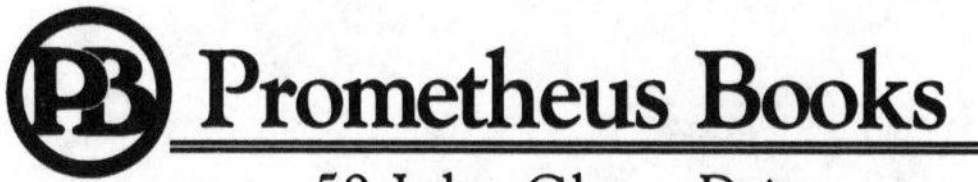

Prometheus Books
59 John Glenn Drive
Amherst, New York 14228-2119

Published 2007 by Prometheus Books

Inquiries should be addressed to
Prometheus Books
59 John Glenn Drive
Amherst, New York 14228–2119
VOICE: 716–691–0133, ext. 210
FAX: 716–691–0137
WWW.PROMETHEUSBOOKS.COM

11 10 09 08 07   5 4 3 2 1

Library of Congress Cataloging-in-Publication Data

Philosophy of biology / ed. by Michael Ruse.
p. cm.
Includes bibliographical references.
ISBN 978–1–59102–527–6
1. Biology—Philosophy. I. Ruse, Michael.

QH331.P468 2007
570.1—dc22   2007011067

Printed in the United States on acid-free paper

CONTENTS

INTRODUCTION

What is philosophy? What is biology? What is the philosophy of biology? Let me start with the second question, which might seem the easiest. Biology, surely, is the science of life, of living things. But, what is a living thing? Examples come readily to mind. Humans are living things, living humans, that is! And so are snakes and oak trees and the bacteria that make food go bad. On the other hand, planets are not living things, and neither are shovels or spades or rocks or mountains or the sea. "Or the sea?" Is it irrational to talk about "the living ocean" and how it ebbs and flows, attacks and protects, helps and hinders? Or, what about a thunderstorm, if you do not believe the sea has the kind of self-maintenance often associated with life? A thunderstorm picks up energy as it goes along, throwing out waste and destruction behind it, just like an elephant, only more so.

Now, at this moment, I do not want to press this particular train of thought. Apart from anything else, we shall be coming back to the notion of "life." Rather, I want to use the discussion to make a point. What have we been doing in the last paragraph? We have certainly not been practicing biology—no dead frogs were laid out on a table. Rather, we have been thinking *about* biology. And this provides a clue to our other questions: to philosophy in general and to the philosophy of biology in particular. Philosophy is a second-order inquiry that looks at other subjects—politics, art, religion, or, in our case, biology—and asks questions about them. The biologist qua biologist dissects and studies frogs and humans and oak trees and bacteria. But, when it comes to questions about what the biologist is doing, then we are in the realm of philosophy.

Is this not all somewhat presumptuous? A nonexpert in the field calmly enters it and tells its practitioners what they are doing and how they should behave. Sometimes, I confess it is presumptuous; unless the philosopher makes some effort to understand the field being studied, grief will not be far behind. However, just as the experienced auto mechanic knows that there are certain principles common to all cars and that a detached general knowledge is not necessarily bad, so the experienced philosopher learns that certain principles apply through all human endeavors, and various problems and concerns keep arising. A detached general knowledge is not necessarily bad.

Moreover, apart from the intrinsic interest of philosophical problems, some outside help can often be of much use to the success of a first-order subject. An example illustrates this point perfectly for biology. In recent years various extreme evangelical Christians, those who insist on taking every word of the Bible absolutely literally (so-called fundamentalists), have been trying hard to have their (Creationist) views inserted into biology curricula. Whatever the ultimate resolution of these efforts, what has become apparent is that there is more here than straight science. Can claims that the earth is only six thousand years old and that human beings were created miraculously by God ever, even in principle, be part of geology and biology, or are they necessarily religious? These, clearly are questions about science rather than questions within science. In other words, they are philosophical questions.

Since even the least reflective of us are bound to encounter philosophical problems sooner or later, we had better get them out on the table and look at them explicitly. But, how should we set about doing this, particularly in biology? My experience is that approaching philosophy is similar to driving a car: There is only so much talking you can do about it; it is far better to get behind the wheel and try it yourself. This is the approach I have taken in this volume, a collection of readings by biologists and philosophers, past and present. I have divided my selection of readings into sections, trying as much as possible to let people of different opinions each have their say; I myself simply try to offer a few guidelines to disentangle the various threads.

There is a reasonable continuity to the sections, but feel free to sample those subjects of particular personal interest. I suspect, however, that the topics with the greatest intellectual appeal, such as human nature, morality, and God, which I have put toward the end, will take on a fuller and richer meaning when some of the earlier sections have been covered. I should add that I have not chosen readings because they are simple. My guide has been to select topics and contributions that are interesting and important. But, I have tried as far as possible to keep matters jargon-free. In addition, at the end I have provided some suggestions for further reading.

The eighteenth-century English lawyer Oliver Edwards, a friend of Dr. Samuel Johnson, is reputed to have said that although he had tried to practice philosophy on several occasions, he could never stay miserable long enough! This image of philosophy persists to this day. I cannot pretend that this collection is a barrel of laughs, but I shall be very disappointed if the reader at no time is gripped by the material. The ideas are important, and there is a real thrill in the cut and thrust of intellectual debate. Philosophy may not be funny, but it can be fun.

BIOLOGICAL EXPLANATION

Ernst Mayr was one of the giants of twentieth-century evolutionary biology. Born in Germany, as a young man he came to and settled in America, dying in 2004 at the great age of one hundred. He was an ornithologist and leading theoretician of taxonomy (the practice of biological classification). His *Systematics and the Origin of Species* (1942), in which he established the overall, gradual nature of evolution, is rightly considered one of the great classics of the field. Mayr was always interested in the methodology and foundations of his science, and this led him naturally into philosophical inquiries. One of his most celebrated papers, reprinted here as a kick-off introduction to this collection, deals with the nature of biological explanation. What kind of question is it that fascinates a biologist? What sorts of information are they after, and what kinds of understanding do they hope to achieve?

The great Greek philosopher Aristotle—you will be meeting him in person in the next section—argued that real scientific understanding involves causes. You have to go beyond describing what happens to trying to find out why and how things happen. You have to go beyond, let us say, hearing a noise, to finding out what made that noise happen. You have to come up with an answer like: The falling rock made the noise, or was the cause of the noise. Mayr agrees with this kind of thinking and (in fact still following Aristotle on this matter) argues that there are basically two kinds of questions that need answering in biology—questions that lead to rather different sorts of causes. On the one hand, there are questions demanding what Mayr calls *proximate* causes. These deal with the mechanics of the situation. They tell you how things happen. In Mayr's example, what is it that triggers a bird's flying south? On the other hand, there are questions dealing with what Mayr calls *ultimate* causes. These tell you why things happen. In Mayr's example, why the bird flies south is to avoid the New England winter and to find food stuffs.

Throughout this collection you will find both kinds of causes being invoked. It would be fair to say, however, that there will be more discussion of ultimate (what earlier thinkers, including Charles Darwin, called final) causes. This is because they are so central to evolutionary thinking and for many reasons—from the apparent distinctiveness of its nature to the all-too-real clashes with religion—evolutionary theory has garnered a great deal of philosophical attention. The important point is to keep in mind Mayr's caution—ask first what kind of question is being asked, and ask second what kind of answer would satisfy.

LIFE AND ITS ORIGINS?

Go back to the beginning discussion. For all the similarities, we probably do not want to say that the ocean or a thunderstorm is a living thing. But, wherein does lie the difference between life and nonlife? It is tempting to suppose that living things must be made of different substances from nonliving things—and this is often true. The body of Abraham Lincoln was not the same substance as the statue of Lincoln in Washington. Yet, we know now that living beings are made, ultimately, from the same minerals as nonliving things. Moreover, since the early nineteenth century it has been possible to synthesize the compounds of living beings from inert substances.

This all leads one to think—it has certainly led many philosophers to think—that living things must have something special *in addition* to make them living, a sort of life force or fluid, akin perhaps to the mind but living in or through all life and separating the quick from the dead. Such a force has been called, naturally, a "vital" force (and its enthusiasts "vitalists"). This view goes back to Aristotle, who was as distinguished a biologist as he was a philosopher.

I can think of no better person than Aristotle himself to begin this collection, and so our first reading is from one of his best known works, *On the Generation of Animals*, in which he wrestles with the way in which the sperm or the seed affects the growing organism. Aristotle thought that the male parent provides the form of the offspring whereas the female parent provides the substance, and he argues that somehow this form gets transmitted through "soul." Precisely what he means by this notion has engaged scholars for over two thousand years. It is certainly not the Christian notion, but it represents some sort of animating force that gives vitality and feeling.

Everyone who reads Aristotle comes away with tremendous respect for his biological acumen. Nevertheless, I suspect that in this doctrine of soul he would find few modern followers. Let me therefore say that there is nothing at all foolish about such a position. After all, there is something different about the living and the nonliving. Nor is it significant that we can never see the soul or vital force. If I open your skull, I doubt I will see your mind; but, I do not doubt you have one. However, this said, the trouble with vital forces is they do not seem to do much. Or, perhaps, the problem is that they do too much. The more biologists delve into living things, the more they find that not only are the substances the same as for the nonliving, but the workings are not so very different either. A bird stays aloft because it exploits the powers of nature—gravity, winds, heat, and the like—no less (and no more) than does a cloud or a handful of sand in a storm.

This is an exaggeration, of course. The bird puts things together much better than does a sandstorm: a point that brings me to our second reading, by J. B. S. Haldane. He was (as we shall learn later) a distinguished and creative biologist who thought and worked in the second quarter of this century. He was also an ardent Marxist, and the piece I have chosen appeared during the Second World War in the *Daily Worker*, the English communist newspaper. (This explains the reference to Engels and perhaps also the enthusiasm for *Alexander Nevsky*, a film by the great Russian director, Serge Eisenstein.) Haldane rejects vital forces, as I have just rejected them. But he suggests that there is more to the organization of life and, in particular, what this organization does, than a trivial and slighting glance would indicate. In essence, he invites us to start shifting our gaze from what life *is* to what life *does*. We must stop thinking in terms of the static and start thinking in terms of the dynamic.

Whether this approach will work is another matter. Certainly, we are going to need much more detail than Haldane's sketch provides. But, surely, Haldane is onto something. After all, what is life if it is not activity, be this eating, drinking, defecating, copulating, or practicing biology or philosophy? For this reason, we should not dismiss Haldane's suggestion out of hand, particularly when, nearly a half century later, we see that much of his science is no less out of date than Aristotle's. Haldane thought that proteins are key substances in life. Now we know that is only part of the story. Yet, the thrust of Haldane's suggestion that "life is a pattern of chemical processes" makes just as much sense now as it did then—indeed, with our deeper understanding of the chemistry of life, it might perhaps make more sense.

The third piece in this section is rather different. Leslie Orgel is today one of the leading researchers in origin of life studies, and this is a recent survey that he wrote about the field. It is obviously a bit more technical than the first two pieces, but don't let that bother you unduly, because the basic ideas are relatively easy to follow. Start with the great discovery in 1953 by James Watson and Francis Crick that the material carrying the information about life—the molecular equivalent of the gene—is a long molecule (deoxyribonucleic acid, DNA) which is twisted in a double helix around a fellow. The information comes not so much in the composition of the molecule but in the order in which sub-molecules occur along the string. The information is read off the DNA molecule by another long molecule (ribonucleic acid, RNA), and this then picks up other smaller molecules in the cell (amino acids) and strings them together to make proteins. These latter are the building blocks of the cell and also (as enzymes) make the cellular processes work.

There was great excitement in the 1950s when it was discovered that one could make amino acids naturally, that is, under conditions that were thought to be the case when life on this planet started. (The universe is about fifteen billion years old, the Earth is about four and one-half billion years old, and the fossil record and other evidence suggests that life started about three and three-quarter billion years ago.) It seemed as if the solution to the origin of life problem was just around the corner. However since then, enthusiasm has cooled significantly. For a start, it is now not at all clear what kind of world we had when life started. Was it a place with an atmosphere with lots of hydrogen (reducing), hence making oxygen-containing products (like amino acids) stable, or not? If not, then are there other alternatives? Orgel points out that the vents in deep sea beds are today favored by some as the place where such products might have been synthesized and retained. For a second, which came first, the chicken or the egg? You need DNA to makes proteins. You need proteins to make DNA. Orgel notes that today many think the RNA molecule had a crucial role to play. In certain circumstances it can not only carry information but synthesize itself.

The theme of Orgel's piece is that we seem to have a detective story with too many clues and too many suggested solutions. "It would be hard to find two investigators who agree on even the broad outline of the events that occurred so long ago and made possible the subsequent evolution of life in all its variety." But note also his ending. We may be very far "from knowing whodunit." Nevertheless: "The only rational certainty is that there will be a solution." Is this the only rational certainty? Orgel and his fellow workers assume the laws that created life are the regular laws of physics and chemistry. Are they right about this? Could it be that we need some special kinds of laws? Could it be that science has to stop at this point and give up the fight? The well-known American philosopher Alvin Plantinga is (somewhat atypically for a Calvinist) deeply hostile to modern science and quite scathing about biology. About the claim that "life arose by naturalistic means"—that life arose through the normal laws of nature—he sneers: "This seems to me for the most part mere arrogant bluster; given our present state of knowledge, I believe it is vastly less probable, on our present evidence, than its denial." Is this really so? Or should we bet that somehow, somewhere, the scientific solution will appear?

EXPLAINING DESIGN

Even if we agree with Haldane that the key question does not concern the composition of organisms but rather how they work and how they are organized, we have to go further to distinguish the quick from the dead or, rather, the living from the inert. What is it about biological organization that makes it special? Continuing a tradition that goes back to the ancient Greeks, many believe today that the distinguishing property of life is that it shows "irrefragible evidence of creative forethought," as the nineteenth-century British anatomist Richard Owen once put it. Unlike shovels or spades or rocks or mountains—or even the sea or thunderstorms—snakes and oak trees and bacteria uniquely show the marks of design. They look as if they were planned and created by a thinking intelligence.

In the past, for most people the appearance of design meant only one thing: Organisms look designed because they *are* designed—by God! This section, therefore, opens with the classic exposition of the Argument for Design (or "Teleological Argument"), first published in 1802 by the English clergyman, Archdeacon William Paley. Yet, for all the authority and support of the very greatest theologians, and for all the brilliance of his own arguments, Paley's position leaves the modern mind dissatisfied. His solution to the mysteries of organic nature takes almost all the important and interesting questions out of the domain of science. Basic questions about organic origins and organic nature can be answered only by religion. And this, in the opinion of many—including sincere believers—is altogether too severe a constriction on biological understanding.

Things have changed greatly since the time of Paley. Biological opinion today is that life neither sprang out of nowhere nor was the result of some instantaneous creation out of nonliving substances. Rather, it is believed that life "evolved," that is, it developed continuously by natural processes, from simple forms (probably, ultimately, from basic minerals), and that such development has produced the complex diversity we see today. The Englishman Charles Robert Darwin (1809–1882) played a key role in advancing the concept of evolution—primarily in his major work, published in 1859, *On the Origin of Species.*

However, even today there is controversy. All sorts of different and conflicting claims are made: about evolution being "only a theory not a fact," about gaps in the fossil record, about Darwin's theory concerning biology being unduly teleological (whatever that might mean), and so forth. In this and the next few sections, I hope we can progress in analyzing some of the problems and answering some of the questions just posed, as well as others.

As a start toward understanding, let me begin by making a three-part division. When people talk about "evolution" they mean several things, and although there are certainly connections, evidence for one is not necessarily evidence for another. First, there is evolution as *fact*, that is, that species are not fixed but developed out of other species. This assertion is of course in conflict with the teachings of the Bible (taken absolutely literally) and of most philosophers and biologists before the nineteenth century. Second, there is evolution as *path*. This involves the actual routes that evolution took. (The technical term for such a route is "phylogeny.") For example, some people think the birds evolved from the dinosaurs. Others are not so sure. These are questions about evolution as path. Third, there is evolution as *mechanism* or *theory*. What was the force or cause or power behind evolutionary change?

Now, relating these three things to Darwin, in the *Origin* he made a strong case for evolution as fact. So strong that this is the case we accept today. How Darwin did it is not often realized by nonbiologists. It is usually thought that evolution stands or falls by the fossils, but although fossils do play an important role in evolutionary theorizing, to establish the fact of evolution Darwin took another route. Realizing that he had to persuade people of the truth of something they will never see, he did what we always do when we are dealing with the unknown—he used circumstantial evidence. Just as in a court of law you establish a suspect's guilt through clues, so Darwin looked through the living world for clues of evolution.

And he found them. Consider the remarkable similarities between the forelimbs of animals, even though those limbs are used for such different purposes (see fig. 1.1. Such similarities are known technically as "homologies"). How could this be? As a result of evolution from a shared ancestor, answered Darwin. He was greatly impressed, for example, by the similarities among tortoises on different islands of the Galapagos archipelago (see figure 1.2.). How could these similarities have come into existence except through evolution from a common ancestor? And so the questions were asked and the answers given. The clues of life point to evolution, and, conversely, belief in the fact of evolution explains the clues of life. (This appeal to all of the evidence is known technically as a "consilience of inductions." In Darwin's day, it was a method of proof much favored by the philosopher William Whewell, and there is good historical evidence that Darwin learned it from him (see fig. 1.3.).

Is evolution just a theory, not a fact? It depends what you mean by "theory" and what you mean by "fact." If by "theory" you mean that a mechanism is pos-

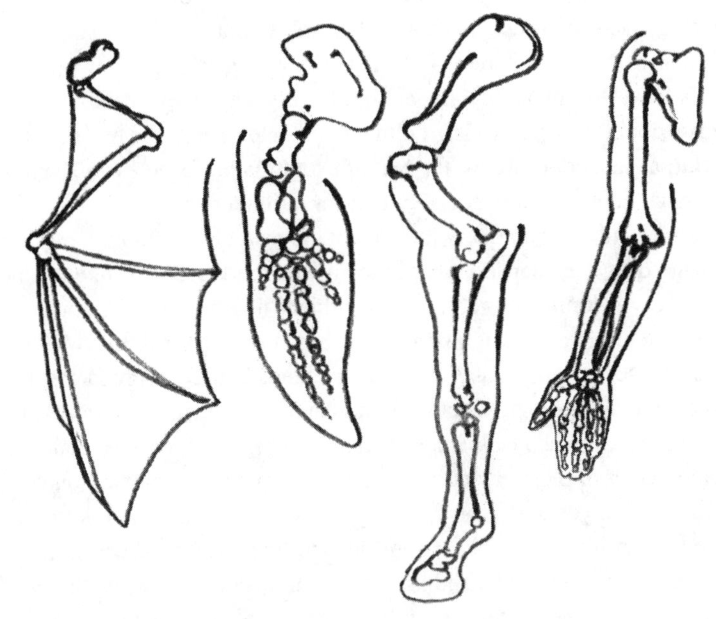

Figure 1.1. Homology between the forelimbs of several vertebrates. Copyright © 2007 Michael Ruse.

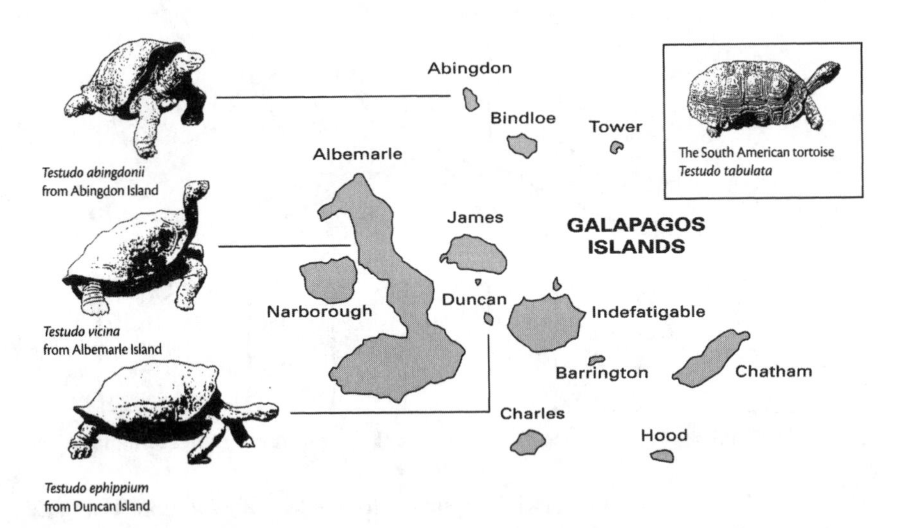

Figure 1.2. Distribution of Galapagos tortoises. Copyright © 2007 Michael Ruse.

tulated, then (as you will see shortly) evolution is a theory. If you mean by "theory," as we sometimes do, "a bit of an iffy hypothesis"—"I've got a theory about Kennedy's assassination"—then evolution is not a theory. Darwin showed it to be a fact. Do I mean that logically, eternally, absolutely without possibility, biologists could never be wrong on this point? No, one never means this in science. But, as in a court of law, the fact of evolution has been established beyond a reasonable doubt. That is good enough for rational people.

Darwin said comparatively little about the path of evolution. Obviously, our knowledge of it does depend, very crucially, on the fossil record, and since this record is very incomplete (not to mention the difficulties in finding what there is of it), we shall always be ignorant about much of life's history. Within limits, however, these difficulties can be circumvented. You can compare similarities (homologies) between organisms, living and fossilized, and work out what would be the most reasonable links. Recently, our much-improved understanding of the molecules of the body has made these techniques of comparison very powerful indeed. Ten years ago, on the basis of a spotty fossil record, it was thought that the human line had broken from the ape line at least fifteen million years ago. Now, because of evidence based on the molecules, it seems that we broke from the chimpanzees a mere six million years ago, and the gorillas broke from both of us about nine million years ago. (This remarkable finding means that we

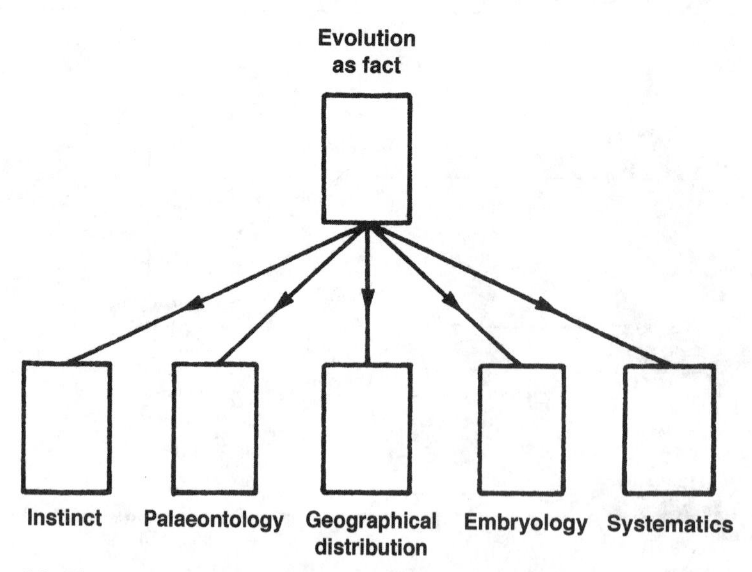

Figure 1.3. The structure of Darwin's argument for the fact of evolution. The fact explains and unifies claims made in the subdisciplines (only some of which are shown), which later in turn yield the "circumstantial evidence" for the fact itself.

are more closely related to chimpanzees than they are to gorillas.)

Third, we come to evolution as mechanism. Darwin thought that he found this mechanism in what he called "natural selection." More organisms are born than can survive and reproduce, survival is the result of special features (not possessed by the losers), and given enough time this all adds up to evolution. But, evolution of what kind? Answers make proper sense only if you know what questions they are addressing. This is particularly true for Darwin and natural selection. As will be seen from the beginning of our selection from the *Origin of Species*, Darwin stresses that he is trying to explain how organisms are *adapted.* For him, and for today's biologists, living beings are not just thrown together randomly. They work, they function, and they are adapted. They have "adaptations," such as the hand and the eye. Natural selection was supposed to address this.

Why did Darwin take adaptation so seriously? Many people think he was writing against a background of Creationism, with everybody believing in the absolutely unchanged word of the Bible, particularly Genesis. This is not true. By the nineteenth century, educated people realized the early chapters of Genesis have to be understood metaphorically. Nor did they consider this a strain on their faith. (More on this in the final section.) It was, rather, the Argument from Design that Darwin was addressing. He was nurtured on Paley and accepted the theologian's key premise, namely that the organic world is *as if* designed. What Darwin wanted to provide was an alternative explanation in terms of natural causes. This he thought he could do through natural selection.

Did Darwin succeed? Many people have questioned whether natural selection really explains adaptation, and the final readings of the section address the most important question of all. Darwin was always adamant that the variations on which selection works, the "raw stuff" of evolution, have no inherent direction or purpose; they just occur without any concern for the needs of their possessors. And the followers of Darwin today agree that the changes in the heritable links of life ("mutations") occur randomly with respect to needs. But, can a process of blind law, working on random differences between organisms, really lead to so much intricacy? Many people think not. Of course, in theory, a monkey randomly striking the keys of a typewriter could produce *Hamlet.* In practice, not. In his discussion, the Australian molecular biologist Michael Denton shows just why he thinks randomness of the kind presupposed by selection could never lead to fully functioning organisms.

The English biologist Richard Dawkins, one of today's most imaginative thinkers, believes that we read the monkey analogy incorrectly. He argues in the concluding reading of this section that, properly understood, random striking can indeed lead to *Hamlet*—or to organisms. Dawkins may or may not make his case.

What impresses me is how he subtly uses one of the familiar symbols of our time, the home computer, to try to make common sense of what many people find to be very uncommon sense. In science, as in literature, the right analogy or metaphor is often the key to a successful argument.

Finally we have a remarkable piece by the Canadian-born, English-residing morphologist Brian Goodwin. Again we return to Aristotle, for it was he who first stressed that there are two ways in which we can regard organisms—either in terms of function or final cause, or in terms of their form or (an aspect of) what Mayr calls proximate cause. A selectionist like Richard Dawkins sees a complicated piece of organic mechanism, like the eye or the hand, and tries to understand it in terms of purpose brought about by natural selection. But what if, when faced with such complexity, the whole search for purpose is mistaken? What if it is all a matter of factors already existing that bring on the shape? We have just seen that forelimbs are influenced in their present shape by the past, namely the ancestral form. What if that is all there is to things? Suppose that any more evolution (nobody is denying the fact of evolution) is simply a matter of various constraints and physical laws and basically has little or nothing to do with purpose or selection? Goodwin suggests that this might often be so. Many plants show patterns that are governed by the so-called Fibonacci series (made famous by the *Da Vinci Code*). Darwinians at once assume that this must be connected to selection. They say that such plants are adapted for survival and reproduction, and the patterns are part of that plan—perhaps they enable the plant to maximize exposure to sunlight or some such thing. Goodwin argues that this might not be so, and that the laws of physics and chemistry rule supreme. The world is nothing like as selection-governed as Darwinians think. We must take more note of form, the shapes of things, and less note of function, the purposes of things. This is not just a matter of science, but of general attitudes to the world around us.

DARWINISM

Having a mechanism is only the first step in building a complete theory. In the *Origin*, Darwin tried to use natural selection as the foundation for a comprehensive explanation of the phenomenon of evolution. But, as critics noted from the start, Darwin was hampered by his lack of understanding of the true nature of heredity. Why do pigs have piglets and not calves, and why do pink pigs tend to have pink piglets and not black ones, but not always? Without such understanding, selection falters. No matter how successful an organism may be in life's struggles, without an adequate method of transmitting its features its efforts are for naught. Moreover, we need some way of introducing new variations in each generation. Without a constant supply of building blocks; selection will exhaust the potential, and evolution will grind to a halt.

Today, because of studies in heredity, we know much about the origination and transmission of organic features. Most importantly, we know that nature precisely follows principles that make selection effective. These principles are collectively known as "Mendelian genetics," named after the Moravian monk Gregor Mendel (a contemporary of Darwin) who first described them. In recent years, as we have seen in Leslie Orgel's piece, genetics has increasingly been placed on a molecular basis, relating heredity to information carried on the macromolecules of nucleic acid.

However, at the beginning of this century, it was thought that genetics and Darwinian selection were rival causes of evolutionary change, and it was not until around 1930 that a number of thinkers (including J. B. S. Haldane) saw that they complemented each other, providing different parts of the overall picture. Rapidly then, because of the work of Theodosius Dobzhansky (in the United States) and Julian Huxley (in England), "neo-Darwinism," or the "synthetic theory of evolution," was born. It is in the light of these developments in Darwin's theory that one should approach the analysis of Darwinism given by the philosopher A. G. N. Flew in the first reading of this section.

Naturally, the mathematics of evolutionary theory is much more sophisticated than Flew demonstrates. One important finding regarding the way the units of heredity ("genes") function in groups is that selection can promote diversity as well as eliminate it! A simple example follows from the supposition that there is a biological advantage to being rare (say, thus escaping a predator's eye). Selection will make the rare less rare and the common less common, creating a balance of forms. Obviously this conclusion is crucially important for the Darwinian. Even if new variations do not appear in order, that such variation is always being held in *groups* of organisms (balanced by selection) means there will usually be an adequate supply of usable material as times change and new evolutionary demands need to be met. Selection does not wait for the appropriate new variation, but generally has a veritable library on which to draw.

But, Darwinism is not yet out of the woods. As indicated, some critics fear that any claim about natural selection cannot be a genuine empirical proposition. It is rather a truism or a tautology, with this latter term being used in the logician's sense of a truth by logical necessity. Just as the claim that "all bachelors are male" holds universally because if something is a bachelor it is necessarily male, Darwin's theory of natural selection is really a tautology, true by definition.

How can this be? Critics seize on natural selection's *alter ego*, the "survival of the fittest." Initially, this was the term of Darwin's contemporary, Herbert Spencer, although Darwin introduced it into later editions of the *Origin.* Both by tradition and in fact it is, literally, natural selection by another name. But now one asks—who are the fittest? And the critics answer that the only way of determining "fitness" is in terms of survival. In other words, the fittest are those who survive. Hence, natural selection collapses into "those who survive are those who survive."

It is precisely this charge that Tom Bethell makes in "Darwin's Mistake." He maintains that Darwinism—that is, any evolutionary theory that depends crucially on the causal mechanism of natural selection—has to be fatally flawed. However, Bethell is called to account by the brilliant biological essayist Stephen Jay Gould (who, sadly, died of cancer in 2002). As we shall learn, Gould was no supporter of ubiquitous adaptationism. He deplored seeking of design-like effects in each and every corner of the living world. For this reason, we shall see Gould draw back from a blanket Darwinism. He nevertheless regarded Darwin's theory as genuinely empirical and basically sound. Gould thought that Bethell, and all similar critics, miss the point and force of natural selection. It is true that the only way we can identify fitness is through survival and reproduction, but this in no way indicates a straight tautological identification. (Note incidentally how Bethell surely misreads Flew. The latter asserts explicitly that the propositions of Darwinism are empirical and nontautological. He sees logically necessary connections *between* the statements of the theory.)

MACROEVOLUTION

I think it is fair to say that neo-Darwinism is still the theory that lies behind evolutionary thought today, although in certain respects it is being challenged, extended, revised, and rejected. This, of course, is the mark of living science. One of the orthodoxy's most persistent critics was Gould, whom we just met in the last section. There he was defending natural selection, so he was certainly no outright opponent of neo-Darwinism. However, as he explained in the first reading of this section, he thought that the Darwinian sees altogether too much adaptation in the living world, and he suggested that we must free ourselves from this. This opened the way for Gould to introduce his own major contribution to evolutionary thought, the theory of "punctuated equilibrium," which he formulated with his fellow paleontologist Niles Eldredge. Unlike the gradual change seen by the Darwinian, Gould and Eldredge saw change on the large scale ("macroevolution") as going in fits and starts. The staccato fossil record is less a function of inadequacy and more one of the way evolution actually occurs.

Of course, if the pattern of evolution is one long period of nonchange ("stasis"), followed by rapid switches from one form to another, the question of what fuels these changes arises. Gould and Eldredge denied vehemently that they want to go beyond the Mendelian pale, arguing rather that in small isolated ("peripatric") populations of organisms change can be much more sudden than is usually the case. (This is known as the "founder principle.") The important consequence of this is that one cannot hope to understand the overall processes of evolution, "macroevolution," simply from the small-scale causes, "micro-evolution." Other phenomena such as external events must be taken into account, thus preventing a simplistic "reduction."

In response, we have an article by Francisco J. Ayala, one of today's most prominent neo-Darwinians. In his analysis of Gould's thesis, he tries to differentiate between those parts he thinks no one would want to deny and those parts he thinks everyone should want to deny. Ayala's objection to Gould is primarily concerned with the notion of "reduction," which he sees as conflating three separate issues. First we have the question of the underlying processes and of whether they operate at all times and levels. Second, there is the question of the adequacy of these processes to explain all of the phenomena of evolution, particularly phylogenies. Third, and most stringently, we have the question of whether everything that happens at the macro level can be shown to be a straight deductive consequence of happenings at the micro level. With this trichotomy made explicit, Ayala is ready to answer Gould.

Details apart, it is clear that Gould and Ayala have a fundamentally different attitude toward the significance and extent of organic adaptation. In some respects, the clash between Darwinians and critics such as Gould resembles conflicting approaches at the beginning of the nineteenth century, when there were two major schools of natural theology: the Paleyites, with their stress on adaptation, and continental thinkers such as the German poet Goethe, forerunner of the school known as *Naturphilosophie*, which always stressed the structure of organisms as much as it stressed their workings. Why, to ask a classic question, do we have homologies, even though they seem to serve no purpose? As we know, Darwinians have been aware of such questions. Indeed, we have seen how Darwin was able to explain homologies through common descent. But, nonadaptive features have never been the primary focus of Darwinians as they were to Goethe and to contemporary biologists such as Gould.

You will surely have sensed that between Gould and Ayala we have an ongoing play of the difference between the ardent Darwinian adaptationists and Brian Goodwin. They want always to stress function. He wanted to stress form. Gould was certainly an adaptationist, we know that. But it is clear that he was a lot less enthused by adaptation than is Ayala. For Gould, formal issues were also of paramount importance, and adaptation must pick up as it may and can. Looking at some strange organic feature, Ayala thinks about purpose. Gould was not so sure. Part of the problem here, obviously, is that it is so hard to test claims about the distant past. If you take a living organism, like a cat for instance, you can study it in the wild and see if (to pick a feature) the claws seem to have any function. But it is a lot less easy with organisms in the past. The fearsome *Tyrannosaurus rex* had puny little arms—like those of a philosopher rather than those you would expect of a brutal predator. Did these arms have any function at all, or were they just useless appendages? Were they used in mating or eating or whatever?

The young philosopher Derek Turner takes up some of these questions in his concluding piece. He thinks that testing is possible, but he doubts that we ever will (or only rarely) find what he calls a "smoking gun." One of the kinds of claims Turner considers are those around supposed trends in the fossil record, and one of the most famous of these is "Cope's law or rule" that claims organisms in evolving lines get bigger over time. I have included a recent discussion article about this rule, both to fill out Turner's discussion and to give you a flavor of how paleontologists reason today.

(A "clade" is a group of organisms like a species or a genus, the members of which stem from a common ancestor. The difference between organisms that have been r-selected and those that have been K-selected is that the former have many offspring but do not much look after them, whereas the latter have few offspring but look after them well. Herrings are r-selected, and humans are K-selected. It is thought that the differences stem from living conditions. If the environment fluctuates wildly from famine to feast, r-selection is the right strategy. If the environment is stable, K-selection is the right strategy. Some have argued that human culture shows the differences. Protestants tend to live in stable environments and have few offspring, whereas Catholics live in fluctuating environments and promote large families. I offer this obviously speculative suggestion—one that much appealed to Charles Darwin—not to get you to accept it, but to prepare you for the controversies we will encounter when we get to evolution and human nature.)

CLASSIFICATION

Whatever the nature of the fossil record, the world around us is not a smooth continuum, with one form grading imperceptibly into another. Rather, we see fairly distinct forms with breaks between them—fish, birds, mammals, dandelions, dogs, humans. In fact, not all of these types or kinds seem to be of quite the same order. I am a human. I am also a mammal and an animal. My dog Toby is likewise a mammal and an animal. He is not a human. In the first reading of this section Ernst Mayr explains that classifiers ("taxonomists") follow the eighteenth-century thinker Carolus Linnaeus in classifying all organisms hierarchically, with each individual being assigned to a sequence of groups (known as "taxa") at ever-higher, more inclusive levels (known as "categories"). The lowest standard level is that of "species." There, humans belong to the taxon *Homo sapiens.* The next category up is "genus." There, we belong to *Homo.* We are now the only living representatives of *Homo.* It is believed that in the past there were other taxa, members of which also were included in *Homo*, namely *Homo habilis* and *Homo erectus.*

Biologists have a sense that there is something special about species, more so than about taxa of any other level. After all, Darwin's great work was not called *On the Origin of Genera.* There is a feeling that somehow species are uniquely "real" or "natural." The breaks between species are objective. Placing an organism in the taxon *Homo sapiens* tells you something about the organism that other assignments—for example, to the group "born on June 21, 1940" (my birthday!)—do not. Even telling you that the organism belongs to the genus *Homo* does not have quite the same impact. Apart from anything else, we know that our ancestors were australopithecines, and some of the earliest members of *Homo* were very similar to some of the latest members of the coexisting *Australopithecus.*

Why are species different? What gives them their special status? To tackle the "species problem," we must first have some clear idea of what is meant by species. Mayr runs crisply through various proposals for characterizing species showing how, in the light of our commitment to evolutionism, some of the more obvious suggestions, resting on physical similarity and the like, just will not do. His own proposal, involving the reproductive barriers between members of different species, has been very influential. Yet, although a good definition is crucial, it is only part of the solution of the species problem. There still remains the question of "naturalness."

Much ink has been spilled in search of an answer to this question. Our second reading presents a discussion by the philosopher Richard Richards that looks at an argument by the philosopher David L. Hull, who (with the biologist Michael Ghiselin) has proposed the most daring solution in recent years. He argues that no solution to the species problem can be forthcoming as long as we misidentify the true nature of species. Commonly, we think they are groups or classes, such as baseball teams. Just as Babe Ruth was a member of the Yankees, so Michael Ruse is a member of *Homo sapiens*. However, according to Hull, species are not classes at all. They are literally *individuals*, such as particular organisms. As my hand is part of me, so I am part of *Homo sapiens.* Once species are taken to be individuals, the original problem is solved; they exist in their own right no less than particular organisms whose reality no one questions. Richards compares this view with that of philosopher Philip Kitcher, who argues for a "pluralist" approach to species, namely that species should be considered special because they represent many different coinciding biological properties. (As I point out in a paper to which Richards refers, this approach—one which I, too, favor—is in the same way of thinking as Whewell's consilience of inductions. If things coincide, you start to think that there must be a real-world reason.)

The dispute between the champions and the opponents of the notion that species are literally individuals is by no means resolved. To this date, no one seems to have won over the other side. What I will draw your attention to is the fact that both sides seem to agree this is not an issue that can be resolved by a simple appeal to common sense, whatever that might be. The status of species must be judged against the background of accepted biological thought. This point leads us back to the discussion of the last section. If you are an orthodox Darwinian, you may be quite happy with the species-as-a-class view. If, to the contrary, you are drawn toward the punctuated equilibria theory, you may well favor viewing species as individuals. In this latter case, you see species coming and going fairly distinctly, just as individuals, and when species are in existence, organisms within are integrated by being subject to shared constraints. Certainly, Gould embraces the Ghiselin/Hull thesis in his article.

The species problem may be the most troublesome question about the foundations of taxonomy. However, in a way it is only the beginning of the problem, for you must then go on to ask questions about classification taken as a whole. In particular, what underlying theme or philosophy should guide you as you attempt to slot organisms into their various levels of the Linnean hierarchy? Why, for example, put whales with mammals rather than fish, or bats with mammals rather than birds, or humans with apes rather than a very social group such as ants? As the biologist Mark Ridley points out in the final reading of this section, essentially taxonomists face two options. On the one hand, they can try to classify entirely on raw physical similarity ("phenetically"). Or, they can bring their knowledge of evolution into play, insisting from the beginning that all classification must in some way reflect evolutionary principles. Ridley himself argues for this second option ("phylogenetic" classification), even though it may not seem the most intuitively attractive. Much of his discussion is given to defending this choice.

ULTIMATE CAUSES

Let's start to dig right into things now. Mayr says there are two kinds of causal understanding, in terms of proximate causes and in terms of ultimate causes. In dealing with the origin of life, we looked mainly at proximate causes. It is understood that DNA codes for RNA, and RNA codes for proteins. What then are the proximate causes of DNA? Could it be that RNA in some sense can function as a proximate cause of itself? But then, with life up and running, we started to look more at the end-related, purposeful side to things. We started to look at ultimate or final causes. How do we explain the design-like nature of the living world? How do we explain the function of things? What is the point of the eyes and the teeth? In the last section, we took things one stage further even, because we moved on from the ultimate ends of individuals to ask if life itself has an ultimate end. Could it be that the whole of evolution is pointed to the appearance of humankind?

Paley's God is pretty helpful here. He designed the world—he made it teleological—so that we have eyes for seeing and teeth for biting. Since He wanted humans around to love and care for, He was going to make sure that we appeared, whether He created us miraculously on the Sixth Day or did it all more slowly through evolution. (More on these sorts of issues in the final section of this collection.) Unfortunately for this kind of approach, none of today's biologists, whether or not they believe in God, want to bring Him into the discussion. This is just not the way you do science these days. So the trick has been to see if you can get an ultimate-cause-type explanation merely using regular laws. That one could, using natural selection, was the claim of Darwin. Suppose, however, that one is successful. Does this now mean the end of science as we know it? Famously, the English philosopher Francis Bacon, at the beginning of the seventeenth century, likened final causes to Vestal Virgins—decorative but sterile. In physics and chemistry there is no talk of ultimate causes. The old joke is that the moon shines at night to light the way home for drunken philosophers. The reason why this is funny—OK, not very funny—is precisely that this is not a real scientific explanation. In physics and chemistry you don't ask what things are for. Things work. They demand proximate causal explanations. Ultimate causal explanations have no place in these sciences.

Yet apparently these sorts of explanations are right there—right there, proudly and always—in evolutionary biological explanation. Even the molecular biologists are into the game. What is the function of the ordering of the units along the DNA molecule? Answer: "It is a code that when broken gives you the information for detecting and using the different amino acids needed to make proteins." Codes are objects of design. Codes have functions. Codes call for final or ultimate cause explanations. So ask again. Does this mean the end of science as we know it? Or does this rather mean that biology—biological understanding—is different? Does this mean that the idea that some day biology could be made part of the physical sciences is simply a pipe dream? Is biological understanding in some way autonomous?

These are the issues of this section.

We start with the greatest of the modern philosophers, the eighteenth-century German, Immanuel Kant. Although he is famous mainly for his thinking about mathematics and physics, he thought deeply about biology. Most particularly, agreeing with Aristotle that we can look at organisms with respect to form (proximate cause) and function (ultimate cause), he also agreed with Aristotle that function is something that is here to stay. It truly does differentiate biology from the physical sciences. It is not so much that biology is second-rate (frankly, I think that Kant could never quite free himself of this prejudice), but that because of the nature of organisms, it calls for a different kind of understanding. This is not a luxury or a shorthand way of thinking, but truly basic.

Kant was no evolutionist. Surely after Darwin this kind of thinking would vanish? Not at all! The language of final or ultimate cause, of a designer, reigns supreme. The extract from George C. Williams' book, *Adaptation and Natural Selection*—one of the most influential works on evolution in the past half century—makes clear that Darwinians are as strong on final cause as were the Christians of yore. But, say what you will, surely this is mere laziness? Granted that evolutionists use design language, is this a good thing? Some biologists feel uncomfortable and urge us to drop all language with anthropomorphic connotations, whether it be talk of goals, strategies, or designs. One who takes this view is the botanist Paul J. Kramer, who believes that such language in this post-Darwinian age is inappropriate. He would therefore eliminate it all. People (and, presumably, Gods) have strategies and goals and ends; plants and animals and their parts do not. For Kramer, it is time biology grew up and took its place as a mature science, alongside physics and chemistry.

Yet, doubts remain. Apart from anything else, it may be much more difficult to eliminate the general anthropomorphism of biology than most reformers allow. Without comment, although having italicized certain interesting terms, I note that Kramer himself says that "plant structure and *function* are *compromises* to *conflicting pressures*." And, in any case, perhaps there is a point to the language and implied understanding of biology that is worth acknowledging and preserving. Biologists face problems that are different from physicists—even the most hardened atheist must agree that plants and animals are design-like in appearance in a way that rocks and planets are not. Could it be that the very nature of biological phenomena forces a distinctive form of understanding on biologists?

This is certainly the feeling of many biologists and philosophers. Like Kant, they argue that in biology, as elsewhere in life, particularly when dealing with human actions and intentions, we can appropriately seek a special kind of understanding. Instead of the usual mode, where we try to explain things in terms of what has gone before—"Why was there a bang?" "Because the door slammed."—We can try to explain things in terms of what will happen—"Why is there this belt?" "To restrain you in an accident." These forward-looking teleological explanations include examples of the kind we have been considering—"Why is an arctic fox white?" "In order to conceal it against the snow." Thus laid out, we can readily see that it is not really the anthropomorphism of design that is a problem. After all, physicists use anthropomorphic terms such as "work" and "force" all the time, and no one objects.

Rather, it is the forward-looking aspect that troubles. In a normal explanation, that which explains is already history. We know it occurred (the slamming door). In a teleological explanation, it (the car accident, the concealed fox) is yet to come. But, what if it never comes? What if the accident never occurs? What if all the snow melts? Then you seem to be left with an explanation without a base. Unless, that is, you can cover yourself against the "missing goal object," as the philosopher Ernest Nagel called the problem.

These are some of the issues I tackle in my contribution. I am not so much arguing about the rightness of teleological understanding—although this is a theme of my essay—but trying to ferret out what is going on and why biologists think in ultimate-cause terms. I argue that the key point is that a metaphor—the organic world as an object of design—is at stake here. One of the chief points I try to make, and I am incidentally following Kant here, is that metaphors have great heuristic power. They help you to make discoveries. The Permian reptile, Dimetrodon, had a large sail on its back. By thinking of the sail as an object of design—by asking "What's it for?"—we move forward. Our answers—it is to scare off predators or control the body temperature—may or may not be right. But at least we have some answers. We have some hypotheses to test. Many people argue that metaphor use is a sign of immaturity. A good theory tells about the world in literal terms. Kramer's failure to do so shows how difficult this is going to be. But my position—and that of many philosophers and scientists—is that we should welcome metaphors. A good scientific theory is not just one that tells us about things, but that stimulates us to go out and find more things. Critics of science fail to realize that in the best work, you start the day with one question and end the day with that question answered and two new questions for the morrow.

HUMAN NATURE

Darwin was always convinced that his theory extended to our own species. Indeed, in his private notebooks the first clear statement showing that he had grasped the significance of natural selection is a passage applying it to human intelligence. However, for a number of fairly obvious reasons, the full application of Darwinism to humankind took much time and effort and indeed in some respects is still in its infancy. On the one hand, to make a convincing case, it was necessary to work out in some detail the actual path (as well as the processes) of human evolution. It was also necessary to extend and develop those areas of evolutionary theory that are particularly relevant to human nature. I refer here in particular to our understanding of social behavior, which is so very much part of what it is to be a human being.

In recent years, great progress has been made on both of these fronts. As mentioned earlier, we now have some good ideas about the break of the human line from that of other apes, and many of the details from the point of the break have been filled in. At the same time, students of social behavior have made impressive strides in understanding such phenomena, showing how natural selection is as much a factor in that realm as it is in other parts of biology. Indeed, this new subbranch of the Darwinian picture, "sociobiology," is one of the most intensively discussed topics in evolutionary studies today.

Yet, for all the advances in understanding and for all the realization that humans are part of the story of evolution, the development of the human branch of sociobiology has been clouded by controversy. In part, this is an accident of history. With the decline of the traditional Christian picture of mankind, various other claimants rushed to fill its place. Particular manifestations of this development have been Marxism and Freudianism, carrying their associated world pictures. Whether these pictures are right or wrong, and whether they are truly rivals to any biological picture, it has meant that the growing Darwinian account of humankind has had to fight for breathing space. And this has led to much discord.

Nevertheless, complaints about rivals aside, even its strongest supporters must agree that human sociobiology has been, in part, the cause of its own misfortunes. Its enthusiasts have rushed in with scant theory and scantier evidence to make bold pronouncements. As critics have rightly pointed out, too frequently human sociobiologists have assumed that contingent aspects of late twentieth-century western culture are manifestations of a deeper biological reality applicable to all people at all times. One thinks in particular of some of the sexual and social inequalities of our society that have been presented as ingrained in human nature.

Having said this, one may still feel that there is a basis for optimism about the biological approach to humankind. This is the position of Edward O. Wilson, chief spokesman for human sociobiology and author of *On Human Nature*, an extract from which opens our section. He does not argue that human beings are blind puppets, determined to strut through life at the mercy of their genes. He does not want to argue that culture and learning are powerless in affecting human lives. However, Wilson does maintain that biology sets certain broad constraints, that these are founded in adaptive strategies in our evolutionary past, and that ignoring them (perhaps for the sake of some ideology) cannot serve any useful purpose.

There are various ways in which human sociobiology might be attacked. It might simply be dismissed on moral or political grounds, with the argument that it is only racist ideology brought fifty years forward. Even if sociobiology contained an element of truth, it is too dangerous to be explored. One may sympathize with this reaction—it is undeniable that genetic pictures of mankind have been used to justify reactionary and extremely vicious ends. However, to refuse absolutely to consider the questions of the sociobiologists is to allow Hitler power from beyond the grave. Alternatively, one might dismiss human sociobiology as bad science. It depends on "ultraadaptationism," suggesting, for instance, that differences between males and females may have a (biological) adaptive origin. As we know, Stephen Jay Gould—who is one of the foremost critics of human sociobiology—denies ultraadaptationism even in the animal world. To deny it in the human world is simply an extension of his general thesis, and it is precisely this that he does in the second selection of this section. He accuses people such as Wilson of following Rudyard Kipling in making up "just so" stories. Kipling suggested that the elephant's trunk grew because its nose was pulled. In the same way, sociobiologists provide pseudo-historical accounts for various kinds of human behavior.

The final piece in this section is by the Australian philosopher Kim Sterelny. He thinks there is something deeply flawed about the original human sociobiological program—something that today often goes under different names like "human behavioral ecology" or "evolutionary psychology." (In major part, these names reflect the departmental affiliations of those who have taken up the program.) One idea that has been taken up with enthusiasm by present-day human sociobiologists is that the mind is made up of separate modules, each of which do different things, and that these modules are programmed as it were by selective pressures of the past. Sterelny is not very keen on this idea either. He thinks the mind is much more flexible and has evolved rapidly to meet today's challenges in their own right, rather than forever trying to handle present things with the tools of the past. Nevertheless, Sterelny sees humans as having made their own environments but within this context believes that our biology is still important. (Note that he relies on something called "group selection." This is the idea that natural selection can act for the benefit of the group rather than just for the benefit of the individual.)

A question that you might want to ask yourself at this point—a question that will hover over the next section's discussion of morality and is of importance to all of the subsequent sections—is about the ultimate nature of the kinds of differences we are considering here. Of course, in a way they are scientific—is the mind modular, for example. Of course, in a way they are philosophical—what kind of evidence would one consider definitive to conclude that the human mind is still stuck in the Pleistocene? But in another, perhaps deeper level, one might almost say we have religious issues at stake here. What is it to be human? Are we basically part of the animal world, or are we somehow different (with perhaps the implicit assumption that we are better)? Even if you do not think that we uniquely have souls or that we are made in the image of God—and I doubt that Gould thought that and do not know whether or not Sterelny thinks that—there is a basic question about human nature that is raised by the great religions like Christianity, Judaism, and Islam. It is still a question that has to be asked even if you have given up religion. Are we different? And if so (or if not), what does this mean—for ourselves, for our fellow humans, for the environment, and for our relationship with the Unknown?

EVOLUTION AND ETHICS

From the time of the *Origin* on, before it in fact, there were people trying to make more of evolution than straight science. Following on the point made in the last paragraph, we find that many thought Christianity is an exhausted paradigm—or at least something in dire need of refurbishment—and they turned to evolution for help. Morality has always been an important part of religion—the Ten Commandments, the Love Commandment—and so those who turned to evolution usually sought some kind of ethical direction from the science. Notorious after the *Origin* was so-called Social Darwinism, although in fact it owed more to Darwin's contemporary, the English man of letters Herbert Spencer. The underlying assumption was that the course of evolution is progressive, from the monad to the man, as they were fond of saying. Hence, since we have now organisms of the highest worth—*Homo sapiens*—it follows that morally we should work to keep up this achievement, perhaps improving it but certainly not letting it decline. This means, since natural selection is the chief means by which evolution has progressed, in the human realm we should let natural selection reign freely, however much it may seem to hurt or go against our natural instincts.

For many, this means that we should adopt some kind of laissez-faire socioeconomic system. Let the market forces prevail, even though some—the widows and children—will go to the wall. More significantly, the talented and hardworking will succeed, and the stupider and lazy will be wiped out. But remember that Christians often differ on the interpretations of their moral dictates. The Love Commandment tells us to love our neighbor as ourselves. But who is our neighbor? The person across the street or the person across the world? And what is love? Faced by the Nazi terror, was Churchill breaking the Love Commandment in urging his fellow citizens to fight and fight again? Likewise with Social Darwinism. If you are a group selectionist, then you might think you ought to help your fellow humans rather than hurt them. This was the position of Alfred Russel Wallace, a Victorian naturalist and socialist who discovered natural selection independently of Charles Darwin. Even those in favor of laissez faire took different stances. The Scottish-American industrialist Andrew Carnegie—he made his fortune in the steel industry—gave millions to found public libraries. These would be places where the poor but talented child could go and learn and thus rise up in society. This would be the survival of the fittest (as Herbert Spencer called natural selection) rather than the non-survival of the non-fittest.

Social Darwinism, if not by that name—we have seen how human sociobiology now hides itself under different labels—persisted right down through the twentieth century. Expectedly, one of its greatest boosters has been Edward O. Wilson. He is, as we will learn, a great supporter of environmentalism, believing that humans need nature to survive and reproduce. Hence, in the name of evolution, he promotes biodiversity. Without it, progress will wither and die. It is this kind of thinking that Philip Kitcher locates in the jointly authored (by me and Wilson) pro-evolution-and-ethics paper reproduced here. Kitcher criticizes us severely, arguing that we have provided no firm foundation for ethics. In an argument that goes back to the great eighteenth-century Scottish philosopher David Hume, Kitcher argues that you simply cannot derive morality—matters of obligation—from the way that the world is—matters of fact. The one simply does not imply the other. Kitcher is not completely negative about the significance of biology for moral understanding, but he thinks it has a more limited role. It can help us in making moral decisions—think of this point when we come to the stem cell and abortion issues—but it cannot provide the basic premises.

Kitcher's criticism is well taken, although I doubt it will much move Wilson. He will think that in the case of evolution and ethics we must make an exception to the is-ought barrier. After all, he will argue, the mere fact that we go from talking about the way the world is to the way the world should be is no big deal. In science, we are always going from talk of one kind of thing—molecules for instance—to talk of another kind of thing—gas temperatures and pressures, for instance. But there is another kind of argument (one which I favor) struggling to get out of the Ruse-Wilson piece. It is simply to conclude that morality is an adaptation to help humans survive and reproduce. (Not such a simple assumption really, given the criticisms of the last section, but ignore these now for the sake of argument.) You can then say that morality exists and has its nature because it is of biological worth. Even if you think that selection always works for the individual rather than the group, morality can be derived because of what evolutionists call "reciprocal altruism." You scratch my back and I'll scratch yours. (This was Darwin's favored solution.)

But what about the justification of morality? The fact that it evolved hardly tells us that it is true. My response is: Could it be that basically there is no justification for morality? It just is, like hands and teeth? If this is so, then Kitcher's criticism is moot. There is no doubt that we humans think morality is justified. We think that killing really and truly is wrong. But perhaps this is all part of the biology, too. If we did not think morality is justified, we would start ignoring it. In other words, not only is morality an adaptation, but so also is the belief that morality is more than a mere adaptation! In other words, selection is deceiving us for our own good, because we function better if we are moral than otherwise.

GENETICALLY MODIFIED FOODS

I have some aged, very respectable English relatives. One day, surfing the net, I was amazed to see them in a group of naked people, lying in a field, spelling out: "NO GM FOODS." Beaming up at us, as one might say. Obviously passions run deeply on these matters. In fact, very interestingly, passions run in different ways on different continents. Genetically Modified foods—GM foods, for short—are crops and animals that have been changed through the insertion of new units into their genotypes. So, for instance, there is a new form of rice ("yellow rice") that has been genetically modified to produce beta-carotene, a substance we humans use to make Vitamin A. There are various reasons why government and industry wants to produce GM foods. Often, by so doing one can increase crop yields or reduce the need for fertilizer. Sometimes the benefits are more directly related to health. Without sufficient quantities of Vitamin A—a problem faced by at least 124 million children in the world—one is at risk for blindness and other diseases.

You might think that everyone would jump at the prospects of the genetic modification of plants and animals. And this has certainly been the case in some parts of the world. In North America, GM foods have been produced and are used without any debate at all. After all, since the beginning of agriculture humans have been changing their crops and plants. If the difference between teosinte—the wild grass of Mexico—and the present-day corn (maize), which was produced from it is not a case of extreme genetic modification, it is hard to know what would be. In any case, in some instances, discussion seems almost moot. If you extract oil from rape seed (canola), you do not retain the DNA, so in a sense it matters not how the plant was manipulated at the genomic level. Yet in other parts of the world, Europe particularly, there has been (and still is) great opposition to GM foods. They are thought to be unnatural. They are thought to be dangerous. They are thought to be products of the worse aspects of materialism and industrialism. If a major firm like Monsanto wants to make GM crops, then there must be something wrong. This cry has been taken up by various campaigners in the Third World. Indian scientist and philosopher Vandana Shiva has been eloquent in her claims that GM crops push out traditional crops, and with this the natural and fruitful way of life of many Indian peasants, especially Indian women. (Hence the reason why people like Shiva are known as "eco-feminists.")

As always in this collection, I choose pieces not only for the surface interest and importance—and if the topic of GM foods is not interesting and important, I do not know what is—but to show how there are related themes running through so many of our discussions. What might earlier have seemed of purely theoretical interest has a funny way of coming right up in the middle of major moral disputes. The opening clash between the Prince of Wales—a man well known for his love of organic foods, old-fashioned architecture, and traditional methods of schooling—and Richard Dawkins—passionate evolutionist, hater of fuzzy thinking—makes this point very clearly. The prince's position is predicated on a progressivist view of life's history. Everything moves upward toward the good. What has evolved naturally must therefore have an innate worth. We should not tamper with it.

Although, as a matter of fact, Dawkins is quite sympathetic to some notions of progress, he will have nothing to do with this. What has evolved is not necessarily what is good. Think of some of the worst viruses and plagues. It is just silly not to use our technology for our own benefits. Of course, the prince could (and probably would) defend his position by invoking the deity. He could say that the world is God's creation (and this holds even if—we shall encounter those who would say, especially if—God creates through evolution), and hence what has unfurled has to be good. As it happens, Dawkins is a passionate atheist, so he would not buy this response. But even the Christian (or Jew or Muslim) might worry about this kind of reply. If we are made in God's image, surely we are supposed to use our abilities—remember Jesus' parable of the talents—and hence using molecular techniques to modify the plants and animals is truly to give praise to the Creator of all things.

Often in discussions like these, the word "natural" comes bubbling up. There is a sense that natural is good and unnatural is bad. These days, organic is even more powerful than natural, and people with money spend lots of it in supermarkets buying vegetables and meat that have been raised without artificial fertilizers and the like. Yet, it is often difficult to tell what is natural and what is unnatural. GM foods would seem to be the height of the unnatural, but now evolutionists realize that genes are carried from one organism to another very differently thanks to microorganisms. The gene goes from one organism to the microorganism and then onto the other organism. This is natural, so is the biologist who takes a gene from a pig (say) and puts it in a potato (say) doing what comes naturally or what comes unnaturally? Lee M. Silver in a splendidly polemical piece takes up some of these issues. For Silver there is virtually nothing but good to be said about GM foods. The modified porker, Enviropig, is a wonderful example of an animal modified to reduce its phosphorus output and thus avoid the massive pollution that normal pigs produce. Down the road, Silver even sees pigs growing on trees, and directly using the energy of the sun's rays. Silver thinks that there is altogether too much half-baked nonsense talk about natural and unnatural, organic and non-organic, and the various health and safety factors. If we are going to go that route, then we should realize that nature is just about the most dangerous place known to humans. "[I]ndeed, every plant and microbe carries a variety of mostly uncharacterized, more or less toxic attack chemicals, and the synthetic chemicals are no more likely to be toxic than the natural ones."

Finally, Canadian philosopher David Castle looks at some of the issues raised by GM foods and, referring specifically to Silver's paper, suggests that often we have a nice balance between risk and benefit, and that also we must take into account people's perceptions of these things. It is rarely if ever a simple matter to decide if GM foods are unambiguously a good (or a bad) thing. Castle is on balance in favor of GM technologies and the benefits they yield, but he advises a broader and deeper discussion than often occurs.

BIOLOGICAL METAPHORS

The question of metaphor in science has already been raised. I have argued that it is a key component of evolutionary biological understanding, and I have defended the design metaphor on the grounds that it has tremendous heuristic power. We return to metaphor in this section, and we start to see that its use is a two-edged sword. It can be incredibly powerful and useful, but it can be deeply misleading and time-wasting. It can also be very, very difficult to say whether a particular metaphor is one or the other. Take Darwin's natural selection for instance—a metaphor taken from the ways in which animal and plant breeders and fanciers pick and choose from which of their stock they will breed. Darwin thought it was a good metaphor, and most Darwinians follow him in thinking so, too. But some people took the metaphor altogether too literally. There were those who accepted Darwin's ideas precisely because they did not think he had expelled God from science! They thought that talk of selection of any kind implies an intelligence, and so what Darwin was saying was fine by them. Indeed, some Presbyterians even thought Darwinism is Calvinism writ large. They always knew that God distinguishes between the goats and the sheep, and so natural selection was a very old and comfortable idea! It was in part because of this that Wallace urged on Darwin that he use the alternative "survival of the fittest," and Darwin himself regretted that he had not used "natural preservation."

In this section we encounter one of the most disputed metaphors of recent times, although the metaphor itself is old, perhaps even going back to Plato's dialogue the *Timaeus*. This is the metaphor of the world as an organism. Literally, of course, the world is not an organism, at least not in the usual sense. But many think that we should think of it in this way. Today, such is the claim of the British scientist James Lovelock, who has postulated the Gaia hypothesis, which sees the world as a living being. "The earth is more than just a home, it's a living system and we are part of it." As Lovelock happily admits, a number of things immediately follow from this. For a start, apparently a whole pile of teleology comes rushing right back in. We are back in a pre-Baconian mind phase. It makes perfectly good sense to ask questions about the purpose of land and sea. For a second, we are now really mixing up matters of fact and matters of morality in a pre-Humean fashion. If we humans are part of the whole, then just as it makes good sense to say something like "a person whose mind leads them to cut off their own hand is really sick or bad," so also it makes good sense to say "a person who pollutes the sea is really sick or bad." Except, of course, whereas you would usually only be affecting yourself if you cut off your hand (perhaps others if you were doing it to avoid conscription), if you pollute the sea you are affecting other parts of the whole including other humans.

To be fair to Lovelock, he is sensitive to these sorts of criticisms. I am not sure how moved he is by the objection that he derives values from facts, but he has addressed the issue of teleology. He argues that it is not genuine teleology in the sense of the vitalists, where you really did have spirit forces or some kind of divine planning or even backwards causation. It is more a matter of feedback systems. As you do one thing, you set in motion other things that attempt to counter and smooth over the first action. To illustrate this, Lovelock offers his "Daisyworld" model. Suppose you have two kinds of daisies covering a planet. One kind is black and absorbs heat but does not do very well when things are really hot. One kind is white and reflects heat and does not do really well when things are very cold. The sun shines on the rather cold planet. The black plants initially do rather well, but they absorb heat and before long they not only cover most of the planet, they heat it up. Now the white plants are at a selective advantage, and so they take over. The point is that, however you interfere with the planet (short of killing off all of the plants of one kind), it has a built-in mechanism to right itself and find a natural balance. There is no unwonted teleology here, although there is certainly the appearance of such.

As you can imagine, conventional scientists are not very keen at all on Lovelock's thesis, teleological or not. They do not find the organic metaphor very helpful. One such critic is the Californian scientist James W. Kirchner. He has a number of objections, starting with the fact that Lovelock has not been very consistent in defining and presenting his Gaia hypothesis. Kirchner therefore offers a taxonomy of hypotheses ranging from the weak (and quite possibly true but not desperately interesting) to the strong (and surely false, so that is the end of that). He is not very taken with the Daisyworld model either. He thinks that Lovelock's ideas are too simple. Juggle around with the initial conditions, for instance make the black daisies happier in warmer conditions, and you do not at all get the kinds of feedback situations that Lovelock supposes. The point I really want to draw your attention to, however, is Kirchner's thinking about metaphors. He is really not desperately taken by them. Although he does allow that metaphors are valuable in discovering things (the heuristic side to science), he thinks them rather flabby and too flexible to lead to real testable hypotheses.

This is the trouble with Gaia. "As hypotheses, metaphors are ill-defined because they can be reinterpreted to explain almost any observed behavior; they fail to specify in what sense the metaphor is true."

Is this claim of Kirchner well taken? After all, if you use—if you must use—metaphor, then you had better find some way of making your hypotheses testable.

Is the problem generic and something against all hypotheses, or is it more specific? Is it the Gaia hypothesis in particular that does things badly? In search of an answer to these sorts of questions, the section concludes with another paper on metaphor, by Derek Turner. Often hovering around Gaia-type discussions are other metaphors, including the one that Turner discusses—the metaphor of being at war with nature. In a way, of course, if you adopt Gaia, we could never be at war with nature. We are part of nature and only metaphorically (!) could we be at war with ourselves. (This point in itself raises the interesting question of how different metaphors fit together. I can be a wizard at maths and a lothario in love, but I cannot be a wizard at maths and a fool at maths, even though both wear pointed hats. For that matter, why are some metaphors appropriate—a wizard at maths—and some simply don't make sense—a snake at maths? One thing we can say is that Darwin was a genius in the way in which he bound together so many metaphors—the struggle for existence, natural selection, division of labor, tree of life, not to mention adaptation and design.)

You will soon see that the reason why I have included Turner's piece in this collection is not simply because of its overt content—although I certainly think the question of war and nature is important. Turner digs carefully into the issues that the Lovelock-Kirchner clash raise, namely about the nature of metaphor and its function in science. We are agreed that metaphor has a heuristic function: it helps you to discover new ideas and stimulates you to think up new hypotheses. But there seems to be a general agreement that metaphor cannot help in proof. Turner invokes a venerable distinction—very popular fifty years ago among philosophers of science—between the context of discovery and the context of justification. I can discover something in all sorts of strange ways, perhaps by looking at tea leaves. But when it comes to supporting my beliefs—justifying them—then something else has to move in. Metaphor goes out and evidence comes in. So it seems that Kirchner is right in his tempered approach.

As a matter of fact, Darwin makes you wonder whether this kind of sharp division really works. He discovered the key to evolution through artificial selection and then used it to justify his belief in natural selection. Somehow there seems a blurring of discovery and justification—similarly, when we think about design. Does it really make sense to say, for instance, that the sail on the back of Dimetrodon functioned to keep the animal at a constant temperature, unless in some way the very question is impregnated with the metaphor of design? These are the sorts of questions Turner takes up. I doubt he will have the final word, but he does offer some real insight into the way or ways in which we find an idea, and how these then go on to influence the way or ways in which we establish it as truth. It would be so nice if we could separate off the rather messy psychological process of discovery from the neat, rational process of justification, but as in real life, things are usually not this clean-cut or easy.

ENVIRONMENTAL ETHICS

You do not have to subscribe to James Lovelock's Gaia hypothesis to worry about the environment. I do not know if the hurricane Katrina that destroyed New Orleans was a direct function of global warming, but other than a few right-wing politicians all today agree that, thanks to the carbon dioxide we are producing from the internal combustion engine, we are heating up the earth's atmosphere at a rate that will have ever-greater bad effects on such things as the ice caps at the poles of the globe and the ever-growing deserts that disfigure the continents. And global warming is just one of the threats that face us today. Every year an area of the Brazilian rainforest larger than a small US state is wiped out and built over. Overpopulation in places like Ethiopia means that often there is not a growing tree or bush for hundreds of miles. The sea has been raped. There was a time when one could virtually walk on the backs of the cod around Newfoundland. Now they are few and far between. And so the story goes.

Not long ago, evangelical Christians took little note of environmental issues. More important was preparing one's soul and waiting for the end. Now they, like many others, are rereading the Bible—especially the early chapters of Genesis (given in the next and final section), seeing that God does not simply give the earth to humans, but expects us to be a steward over His creation. This does not mean that we should not use the earth, but that we should care for it and tend it. The question we must ask, however, regardless of our religious beliefs or non-beliefs, is whether this is enough. Humans have done appalling damage to their home. Is our basic way of thinking outmoded? Someone like Lovejoy would say that it is, and that with the metaphor of the world as an organism, we see that we are part of the problem, destroying essentially what includes us, ourselves.

Others take a somewhat different tack, arguing that what we need is a new ethic, one that is not human-centered (anthropocentric) but starts with the needs and rights of other animals and plants, even ecosystems taken as a whole. This is the position of Holmes Rolston III, a philosopher and a Protestant clergyman who has long argued for a new approach to the environment. The key for him comes in the change of the sign in the park, from don't kill the plants (because other humans will want to enjoy them) to let the plants be (because they have as much right to life and liberty as you do). If you start morality here, then the need and obligation to preserve the environment follows at once.

Edward O. Wilson cares desperately about the environment and the need to care for it. He is the world's leading authority on the ants, and this has led him many times into the jungle, something for which he has deep affection and the preservation of which is for him the top priority. As an evolutionist, he is deeply committed to what he calls biophilia—the love of nature—and he goes so far as to argue that we humans have evolved in symbiotic relationship with the rest of living things. We need nature, for its physical and medical and like benefits. More importantly, we need nature spiritually. A world of plastic would kill us, literally as well as metaphorically. As we also know, the ultimate basis of Wilson's morality lies in the upward progress of evolution. Humans have come out on top. We ought therefore to cherish humans. The cherishing of humans demands a regard for nature. Notice this means that although Wilson is clearly deeply sympathetic to the position of someone like Rolston, in the end the two thinkers are miles apart. For Rolston, we must give up our own special status on this world. For Wilson, our special status on this world is the first and most fundamental truth. We need an "anthropocentric ethic apart from the issue of rights—one based on the hereditary needs of our own species."

Elliott Sober is a professional and, in many respects, old-fashioned philosopher. (Not necessarily old-fashioned in the same sense as the prince. There is a difference between class and kitsch.) Like most of us in this field—and I happily include myself here—he is skeptical about sweeping new proposals for solving life's difficult problems. The problems may be new, but they are not always that different. Since the time of the Greeks, two and a half thousand years ago, the world's brightest humans have been wrestling with the issues of morality, and Sober is convinced that the tools will now be in place to turn to issues even those as great as is posed by the environment. He gives careful consideration to some of the topics and issues on which I have but touched in this introduction. There is, for instance, a thorough discussion of what it means for something to be natural. Also he looks at matters thus far unmentioned, for instance so-called slippery slope arguments. An example would be a plea for a certain species on the grounds that if you start with one, there is no firm point at which you can ever stop killing species. Sober respects this kind of argument—one that incidentally often (as we shall see) lies behind claims that life must start with conception, otherwise there is no firm point at which to draw the line. Yet Sober is not entirely convinced. There is a difference between being bald and being hirsute, even though it may be difficult to say how many hairs change you from one state to the other. In concluding his piece, Sober invites us to think more deeply not just about moral values but about aesthetic values—the values we find in a work of art. Perhaps focusing on these will help us to disentangle some of the knotty problems of environmentalism.

Finally, the philosopher and ecologist Gregory Mikkelson takes us to the scientific cutting-edge of ecology today, discussing the new transdisciplinary subject of ecological economics. Here we are face-to-face with some of the knottier problems of environmentalism. The big issue is over the extent to which the environment constrains and limits the world economy, and ecologists accuse economists of ignoring or belittling this fact. It is Mikkelson's claim that there is more at stake here than pure science and that really we are involved in a debate about the nature of science itself. In particular, Mikkelson feels that economics is stuck with an old paradigm, that of logical positivism that excludes value-issues from science. In looking at such issues as biodiversity—bringing us around to the interests and worries of Wilson—Mikkelson argues that by revising our conceptions of what really is the nature of science we might more profitably look to the future about some of our pressing ecological challenges.

WHEN LIFE BEGINS

We deal in this section with two of the most debated issues in modern American society, stem cell research and abortion. I am certainly not going to pretend that we can solve or end these discussions here. My aim is really important but more limited. I want to show in detail what has started to become apparent in earlier sections—a matter of real concern to people like Greg Mikkelson—namely how difficult it can be to separate out matters of fact and matters of morality. They may be different, but they do get mixed up. Not all of the issues for stem cell research and abortion are the same. Abortion is abortion, whereas stem cell research usually does not demand abortion in its own right, but cleans up and benefits after abortions have occurred. But given our interests, the issues are the same. Killing rhinos for the supposedly aphrodisiac effects of powered horns is wrong. It is very wrong. But, by and large, no one thinks that killing a nonhuman is inherently wrong. The case is different from humans. Unless you take the kind of approach of someone like Rolston (who refuses to distinguish between humans and other living things), killing humans is inherently wrong. The question is: what is a human and what is a nonhuman?

The section is built around two bills, one against abortion in South Dakota, and the other in favor of stem cell research in the federal government. The former was passed and signed by the governor (March 6, 2006) and the latter was passed and vetoed by the president (July 19, 2006). First we have the South Dakota case, where it is claimed that human life—full human life—begins the moment that the sperm, the male sex cell, meets and joins with the ovum, the female sex cell. Particularly the "Abortion Position Paper" tries to justify this in terms of medical science.

It is this kind of endeavor that the historian and philosopher of biology Jane Maienschein addresses in her piece. By considering both history and philosophy, Maienschein shows that it is by no means obvious when life begins. There have been many suggestions, some based on current scientific knowledge and others less so. Theologically and philosophically, therefore, there are no easy or ready answers. Maienschein shows that it was only latish in the nineteenth century—for reasons that are obscure and might have been political as much as biological or theological—that the Roman Catholic church stated absolutely that life begins at conception. (Before that, in the Aristotelian tradition, life did not begin until it started to quicken, at least forty days after conception.) Maienschein shows that paradoxically and somewhat ironically the Church's stance put it in league with materialists of the worst stripe!

She also shows that the position is also out of sync with biological thinking today. It is truly hard to think that conception is and must be the point at which life begins—not before and not after. Everything we have learned from embryology shows that the beginning of life is gradual, not instantaneous. For instance, most fertilized eggs do not get implanted. They just get washed away and die. It is odd indeed to insist that these are as human as you or I. Moreover, until at least the third division of the eggs—making eight in all—we have the potential of each and every one going on to make a human itself. That is how we get twins and larger numbers of same-birth offspring. Because we have eight potential humans, does this mean we are all eight people until after this? Maienschein concludes that science makes a complete mess of much thinking about abortion.

Next we have the stem cell issue and again, and although the United States Senate wants work to go forward, we have opponents of stem cell research arguing against it on supposed biological grounds, this time the president of the United States of America (George W. Bush) no less. Philosopher Christopher Pynes tries to throw light on this debate and ease the moral dilemmas by shifting the terms of debate. He argues that too often we look upon stem cell research as linked to the abortion debate. If we think of it more in terms of fertility treatments, then stem cell use becomes less morally problematic. Others no doubt will dispute this. Agree that, all other things being equal, it is a morally good thing that people want to have children. The question is whether this moral good outweighs the moral evil (as some will regard it) that in order to have one or two thriving children it is necessary to create (let us say) ten fertilized eggs and then discard nine or ten. Those fertilized eggs could equally have been implanted and have led to children. In fact, as Pynes notes, this is sometimes done by nonrelatives—religious people doing what they think right—and is praised by those of a similar mind (like the president). It is a tough point to decide. Do those who oppose abortion want also to say that children who are born thanks to artificial means are themselves morally tainted? And, to bring the discussion around to where we came in, do we want to say that the facts of the matter have any importance in this discussion, or is it all religion and philosophy?

GOD AND BIOLOGY

We come to the final section. Everybody knows that for the past three or four decades America has been caught up in a major battle between science and religion. On the one hand there are the scientists and their supporters (many of whom are liberal Christians and Jews), who think that evolution is true and that Darwinism or some modification is the correct causal explanation. On the other hand, there are the evangelical Christians, in early days called fundamentalists, now more commonly creationists, who believe in the early chapters of Genesis taken absolutely literally—six days of creation, about six thousand years ago, miracles all of the way, humans last, and great big flood that wiped out most living things. You should understand that this latter is not traditional Christianity. Saint Augustine, the great theologian who lived around 400 CE, was very firm on the matter of literalism. If modern science shows that the Bible taken at its face word is false, then we must recognize and live with that. It is not that the Bible now becomes false, but that we must interpret it metaphorically. To take an example already touched on, Genesis talks of God having created Adam and the animals. This He certainly did do, but we should not necessarily think that this was done all at once or that (as Genesis says) God literally brought the animals over to be named. The identification and naming of organisms is an important part of human activity. This is what Genesis acknowledges. It does not necessarily tell you exactly how this all happened.

Creationism is an invention of the Protestant church in America in the nineteenth century. As Joseph Cain shows in his article about the famous Scopes Trial of 1925, you miss the whole point if you do not see this as a cultural struggle. No one really lies awake at night worrying about gaps in the fossil record. But they do worry about the lack of religion, morality, respect, and more that science generally, evolution particularly, seems to bring in its wake. (Whether this is true is for you to judge.) Particularly around and after the time of the Civil War, many Americans (especially in the South and mid-West) started to see evolution as epitomizing all of the things that they hated about modern life—not the least of which was the drive to end slavery, something that most thought was justified by the Bible. (Paul does not tell the runaway slave that he is free. Rather he tells him to return to and obey his master. Of course, a more liberal reading of the Bible condemns slavery. It is hardly consistent with the Sermon on the Mount.)

This cultural divide continues right down to the present. In 1981 in Arkansas there was a trial over whether something known as Creation Science—otherwise known as Genesis taken literally—could be taught in the state schools. Creationism was thrown out, but like the phoenix keeps rising from the flames. Today Creationism has morphed into a smoother version, so-called Intelligent Design Theory. Its supporters, like William Dembski, who has a piece in this section, argue that the world is so complex, so intricately designed, that no natural, lawbound process like natural selection could have produced it. There must have been an intervention from without. Note that although someone like Dembski joins with someone like Brian Goodwin in rejecting selection, Goodwin is still a deeply committed naturalistic evolutionist (as was Gould, who also at times had his doubts about selection). In fact, in respects, Dembski and Goodwin (and Gould) are 180 degrees opposed. Both sides obviously see some selection-caused adaptation in the world. But overall, Dembski sees the world as so adaptively complex (exhibiting what he and his fellows call irreducible complexity) that it cannot be explained by natural selection. Overall, Goodwin and Gould see the world as so nonadaptive, they do not see the need to invoke natural selection.

Robert Pennock is a philosopher. In 2005, a school board in Dover, Pennsylvania, decreed that if the children of the area were to be taught evolution, then they must also be taught Intelligent Design Theory. At once, as had happened in Arkansas, the American Civil Liberties Union (ACLU) objected on the grounds that this is an unconstitutional melding of church and state. The case came to trial, and what you have here is some of the testimony that Pennock gave against Intelligent Design Theory. (Strictly speaking, this was his "deposition," or what he told the opposing lawyers he, as an expert witness on the natures of science and religion, was going to say on the stand.) You can see that Pennock argues strongly not only that Intelligent Design Theory is false, but that it does not qualify as science. For Pennock, as for almost all of the contributors to this volume, science can have no truck with the miraculous. It must appeal to and only to unbroken natural law. At a certain level this stance is based on a definition of science, but it is also a heuristic principle. By assuming that the most difficult cases are explicable by unbroken law, scientists of the past have been stimulated to persist until they can find explanations. Does this mean that by definition there can be no miracles? Some atheists like Richard Dawkins probably think just this. Pennock is Christian, a Quaker, and I suspect that he would say that miracles are possible, but you can never prove them by a scientific argument. They must be taken on faith.

Completing this section is one of America's leading theologians working on the science-religion relationship. A practicing Roman Catholic, John Haught offers us a classification, a taxonomy, of possible relationships (or nonrelationships) between science and religion. These go all of the way from outright conflict—war—to much more accommodating approaches, where science and religion can complement each other and even work together. It is clear that Haught himself favors the last kind of position. As a Christian and as an evolutionist, he feels that the two must support and reinforce each other. Some of the points he makes are historical and perhaps can be held by those who have no belief. It is certainly true, as he says, that it was the Judeo-Christian account of origins that put people to asking questions about where we come from—the very questions to which evolution is an answer. The ancient Greeks really had no idea of history in this way. They tended to think of things as cyclical or as going nowhere in particular—now this way, now that way. But even if we accept this story (and I am inclined to), it is still open for someone like Richard Dawkins to say that now we have the right answer and then they did not. Haught wants to go further than this, however. He concludes by arguing that the nature of God's love is that it cannot be a static, a once-and-for-all phenomenon. It must be something forever giving and renewing itself. In other words, it must be something that begs for an evolutionary background.

Clearly this cannot be the last word. Richard Dawkins is always pointing to the fact that the struggle for existence leads to pain. A favorite quotation is from a letter Darwin wrote to his American Christian friend, the Harvard botanist Asa Gray.

> With respect to the theological view of the question; this is always painful to me.—I am bewildered.—I had no intention to write atheistically. But I own that I cannot see, as plainly as others do, & as I shd. wish to do, evidence of design & beneficence on all sides of us. There seems to me too much misery in the world. I cannot persuade myself that a beneficent & omnipotent God would have designedly created the Ichneumonidae with the express intention of their feeding within the living bodies of caterpillars, or that a cat should play with mice. Not believing this, I see no necessity in the belief that the eye was expressly designed. (Letter to Asa Gray, May 22, 1860)

What Dawkins and others usually omit to mention is that Darwin then goes on to say that nevertheless he believes in some kind of deity (which he still did for a few years, until his beliefs faded into agnosticism). Even so, the real point is that someone like Haught has obviously got his work cut out to reconcile science and religion. But that is a matter for another place and another time. By now you will realize that scholars never stop discussion and arguing. My hope is that you now realize that this can be fun, and the results are too important not to make the effort.

SHORT GLOSSARY

Abiotic: not derived from living organisms

Catalyst: a substance that speeds up a process (enzymes are protein catalysts, and ribozymes are RNA catalysts)

Genome: the genetic makeup of an organism (extended to the makeup of the species as in the Human Genome Project)

Metabolism: the chemical processes in the cell

Monomer: small molecule that can be joined with others to make a long, chainlike molecule (polymer), such as a protein made from amino acids

Morphology: the form or structure of organisms

Nucleotide: a subunit of DNA or RNA (including pyrimidine or purine)

Oligomer: a short stretch of a chain molecule, made from monomers (polymers are much longer stretches)

Polymerase: an enzyme that is used to build DNA and RNA molecules

Prebiotic: prebiological (that is, before there are organisms)

Pyrimidine base: one of the units (along with purine) that make DNA and RNA

BIOLOGICAL EXPLANATION

ERNST MAYR

CAUSE AND EFFECT IN BIOLOGY

Being a practicing biologist, I feel that I cannot attempt the kind of analysis of cause and effect in biological phenomena that a logician would undertake. I would instead like to concentrate on the special difficulties presented by the classical concept of causality in biology. From the first attempts to achieve a unitary concept of cause, the student of causality has been bedeviled by these difficulties. Descartes's grossly mechanistic interpretation of life, and the logical extreme to which his ideas were carried by Holbach and de la Mettrie, inevitably provoked a reaction leading to vitalistic theories which have been in vogue, off and on, to the present day. I have only to mention names like Driesch (entelechy), Bergson (élan vital), and Lecomte du Noüy, among the more prominent authors of the recent past. Though these authors may differ in particulars, they all agree in claiming that living beings and life processes cannot be causally explained in terms of physical and chemical phenomena. It is our task to ask whether this assertion is justified, and if we answer this question with "no," to determine the source of the misunderstanding.

Causality, no matter how it is defined in terms of logic, is believed to contain three elements: (1) an explanation of past events ("a posteriori causality"); (2) prediction of future events; and (3) interpretation of teleological—that is, "goal-directed"—phenomena.

The three aspects of causality (explanation, prediction, and teleology) must be the cardinal points in any discussion of causality and were quite rightly singled out as such by Nagel (1961). Biology can make a significant contribution to all three of them. But before I can discuss this contribution in detail, I must say a few words about biology as a science.

TWO FIELDS

The word *biology* suggests a uniform and unified science. Yet recent developments have made it increasingly clear that biology is a most complex area—indeed, that the word *biology* is a label for two largely separate fields which differ greatly in method, *Fragestellung*, and basic concepts. As soon as one goes beyond the level of purely descriptive structural biology, one finds two very different areas, which may be designated functional biology and evolutionary biology. To be sure, the two fields have many points of contact and overlap. Any biologist working in one of these fields must have a knowledge and appreciation of the other field if he wants to avoid the label of a narrow-minded specialist. Yet in his own research he will be occupied with problems of either one or the other field. We cannot discuss cause and effect in biology without first having characterized these two fields.

FUNCTIONAL BIOLOGY

The functional biologist is vitally concerned with the operation and interaction of structural elements, from molecules up to organs and whole individuals. His ever-repeated question is "How?" How does something operate, how does it function? The functional anatomist who studies an articulation shares this method and approach with the molecular biologist who studies the function of a DNA molecule in the transfer of genetic information. The functional biologist attempts to isolate the particular component he studies, and in any given study he usually deals with a single individual, a single organ, a single cell, or a single part of a cell. He attempts to eliminate, or control, all variables, and he repeats his experiments under constant or varying conditions until he believes he has clarified the function of the element he studies.

The chief technique of the functional biologist is the experiment, and his approach is essentially the same as that of the physicist and the chemist. Indeed, by isolating the studied phenomenon sufficiently from the complexities of the organism, he may achieve the ideal of a purely physical or chemical experiment. In spite of certain limitations of this method, one must agree with the functional biologist that such a simplified approach is an absolute necessity for achieving his particular objectives. The spectacular success of biochemical and biophysical research justifies this direct, although distinctly simplistic, approach.

EVOLUTIONARY BIOLOGY

The evolutionary biologist differs in his method and in the problems in which he is interested. His basic question is "Why?" When we say "why" we must always be aware of the ambiguity of this term. It may mean "How come?" but it may also mean the finalistic "What for?" When the evolutionist asks "Why?" he or she always has in mind the historical "How come?" Every organism, as an individual and as a member of a species, is the product of a long history, a history which indeed dates back more than 3,000 million years. As Max Delbrück (1949) has said, "A mature physicist, acquainting himself for the first time with the problems of biology, is puzzled by the circumstance that there are no 'absolute phenomena' in biology. Everything is time-bound and space-bound. The animal or plant or micro-organism he is working with is but a link in an evolutionary chain of changing forms, none of which has any permanent validity." There is hardly any structure or function in an organism that can be fully understood unless it is studied against this historical background. To find the causes for the existing characteristics, and particularly adaptations, of organisms is the main preoccupation of the evolutionary biologist. He is impressed by the enormous diversity of the organic world. He wants to know the reasons for this diversity as well as the pathways by which it has been achieved. He studies the forces that bring about changes in faunas and floras (as in part documented by paleontology), and he studies the steps by which the miraculous adaptations so characteristic of every aspect of the organic world have evolved.

We can use the language of information theory to attempt still another characterization of these two fields of biology. The functional biologist deals with all aspects of the decoding of the programmed information contained in the DNA of the fertilized zygote. The evolutionary biologist, on the other hand, is interested in the history of these programs of information and in the laws that control the changes of these programs from generation to generation. In other words, he is interested in the causes of these changes.

But let us not have an erroneous concept of these programs. It is characteristic of them that the programming is only in part rigid. Such phenomena as learning, memory, nongenetic structural modification, and regeneration show how "open" these programs are. Yet, even here there is great specificity, for instance with respect to what can be "learned," at what stage in the life cycle "learning" takes place, and how long a memory engram is retained. The program, then, may be in part quite unspecific, and yet the range of possible variation is itself included in the specifications of the program. The programs, therefore, are in some respects highly specific; in other respects they merely specify "reaction norms" or general capacities and potentialities.

Let me illustrate this duality of programs by the difference between two kinds of birds with respect to "species recognition." The young cowbird is raised by foster parents—let us say, in the nest of a song sparrow or warbler. As soon as it becomes independent of its foster parents, it seeks the company of other young cowbirds, even though it has never seen a cowbird before! In contrast, after hatching from the egg, a young goose will accept as its parent the first moving (and preferably also calling) object it can follow and become "imprinted" to. What is programmed is, in one case, a definite "gestalt," in the other, merely the capacity to become imprinted to a "gestalt." Similar differences in the specificity of the inherited program are universal throughout the organic world.

THE PROBLEM OF CAUSATION

Let us now get back to our main topic and ask: Is *cause* the same thing in functional and evolutionary biology?

Max Delbrück (1949), again, has reminded us that as recently as 1870 Helmholtz postulated "that the behavior of living cells should be accountable in terms of motions of molecules acting under certain fixed force laws." Now, says Delbrück correctly, we cannot even account for the behavior of a single hydrogen atom. As he also says, "Any living cell carries with it the experiences of a billion years of experimentation by its ancestors."

Let me illustrate the difficulties of the concept of causality in biology by an example. Let us ask: What is the cause of bird migration? Or more specifically: Why did the warbler on my summer place in New Hampshire start his southward migration on the night of the 25th of August?

I can list four equally legitimate causes for this migration:

(1) *An ecological cause.* The warbler, being an insect eater, must migrate, because it would starve to death if it should try to winter in New Hampshire.

(2) *A genetic cause.* The warbler has acquired a genetic constitution in the course of the evolutionary history of its species which induces it to respond appropriately to the proper stimuli from the environment. On the other hand, the screech owl, nesting right next to it, lacks this constitution and does not respond to these stimuli. As a result, it is sedentary.

(3) *An intrinsic physiological cause.* The warbler flew south because its migration is tied in with photoperiodicity. It responds to the decrease in day length and is ready to migrate as soon as the number of hours of daylight has dropped below a certain level.

(4) *An extrinsic physiological cause.* Finally, the warbler migrated on the 25th of August because a cold air mass, with northerly winds, passed over our area on that day. The sudden drop in temperature and the associated weather conditions affected the bird, already in a general physiological readiness for migration, so that it actually took off on that particular day.

Now, if we look over the four causations of the migration of this bird once more, we can readily see that there is an immediate set of causes of the migration, consisting of the physiological condition of the bird interacting with photoperiodicity and drop in temperature. We might call these the *proximate* causes of migration. The other two causes, the lack of food during winter and the genetic disposition of the bird, are the *ultimate* causes. These are causes that have a history and that have been incorporated into the system through many thousands of generations of natural selection. It is evident that the functional biologist would be concerned with analysis of the proximate causes, while the evolutionary biologist would be concerned with analysis of the ultimate causes. This is the case with almost any biological phenomenon we might want to study. There is always a proximate set of causes and an ultimate set of causes; both have to be explained and interpreted for a complete understanding of the given phenomenon.

Still another way to express these differences would be to say that proximate causes govern the responses of the individual (and his organs) to immediate factors of the environment, while ultimate causes are responsible for the evolution of the particular DNA program of information with which every individual of every species is endowed. The logician will, presumably, be little concerned with these distinctions. Yet, the biologist knows that many heated arguments about the "cause" of a certain biological phenomenon could have been avoided if the two opponents had realized that one of them was concerned with proximate and the other with ultimate causes. I might illustrate this by a quotation from Loeb (1916): "The earlier writers explained the growth of the legs in the tadpole of the frog or toad as a case of adaptation to life on land. We know through Gudernatsch that the growth of the legs can be produced at any time even in the youngest tadpole, which is unable to live on land, by feeding the animal with the thyroid gland."

Let us now get back to the definition of "cause" in formal philosophy and see how it fits with the usual explanatory "cause" of functional and evolutionary biology. We might, for instance, define cause as a nonsufficient condition without which an event would not have happened, or as a member of a set of jointly sufficient reasons without which the event would not happen. Definitions such as these describe causal relations quite adequately in certain branches of biology, particularly in those which deal with chemical and physical unit phenomena. In a strictly formal sense they are also applicable to more complex phenomena, and yet they seem to have little operational value in those branches of biology that deal with complex systems. I doubt that there is a scientist who would question the ultimate causality of all biological phenomena—that is, that a causal explanation can be given for past biological events. Yet such an explanation will often have to be so unspecific and so purely formal that its explanatory value can certainly be challenged. In dealing with a complex system, an explanation can hardly be considered very illuminating that states: "Phenomenon *A* is caused by a complex set of interacting factors, one of which is *b*." Yet often this is about all one can say. We will have to come back to this difficulty in connection with the problem of prediction. However, let us first consider the problem of teleology.

THE PROBLEM OF TELEOLOGY

No discussion of causality is complete which does not come to grips with the problem of teleology. This problem had its beginning with Aristotle's classification of causes, one of the categories being the "final" causes. This category is based on the observation of the orderly and purposive development of the individual from the egg to the "final" stage of the adult. Final cause has been defined as "the cause responsible for the orderly reaching of a preconceived ultimate goal." All goal-seeking behavior has been classified as "teleological," but so have many other phenomena that are not necessarily goal-seeking in nature.

Aristotelian scholars have rightly emphasized that Aristotle—by training and interest—was first and foremost a biologist, and that it was his preoccupation with biological phenomena which dominated his ideas on causes and induced him to postulate final causes in addition to the material, formal, and efficient causes. Thinkers from Aristotle to the present have been challenged by the apparent contradiction between a mechanistic interpretation of natural processes and the seemingly purposive sequence of events in organic growth, reproduction, and animal behavior. Such a rational thinker as Bernard (1885) has stated the paradox in these words.

> There is, so to speak, a preestablished design of each being and of each organ of such a kind that each phenomenon by itself depends upon the general forces of nature, but when taken in connection with the others it seems directed by some invisible guide on the road it follows and led to the place it occupies.
>
> We admit that the life phenomena are attached to physicochemical manifestations, but it is true that the essential is not explained thereby; for no fortu-

> itous coming together of physicochemical phenomena constructs each organism after a plan and a fixed design (which are foreseen in advance) and arouses the admirable subordination and harmonious agreement of the acts of life . . . Determinism can never be [anything] but physicochemical determinism. The vital force and life belong to the metaphysical world.

What is the *x*, this seemingly purposive agent, this "vital force," in organic phenomena? It is only in our lifetime that explanations have been advanced which deal adequately with this paradox.

The many dualistic, finalistic, and vitalistic philosophies of the past merely replaced the unknown *x* by a different unknown, *y* or *z*, for calling an unknown factor *entelechia élan vital* is not an explanation. I shall not waste time showing how wrong most of these past attempts were. Even though some of the underlying observations of these conceptual schemes are quite correct, the supernaturalistic conclusions drawn from these observations are altogether misleading.

Where, then, is it legitimate to speak of purpose and purposiveness in nature, and where it is not? To this question we can now give a firm and unambiguous answer. An individual who—to use the language of the computer—has been "programmed" can act purposefully. Historical processes, however, *cannot* act purposefully. A bird that starts its migration, an insect that selects its host plant, an animal that avoids a predator, a male that displays to a female—they all act purposefully because they have been programmed to do so. When I speak of the programmed "individual," I do so in a broad sense. A programmed computer itself is an "individual" in this sense, but so is, during reproduction, a pair of birds whose instinctive and learned actions and interactions obey, so to speak, a single program.

The completely individualistic and yet also species-specific DNA program of every zygote (fertilized egg cell), which controls the development of the central and peripheral nervous systems, of the sense organs, of the hormones, of physiology and morphology, is the *program* for the behavior computer of this individual.

Natural selection does its best to favor the production of programs guaranteeing behavior that increases fitness. A behavior program that guarantees instantaneous correct reaction to a potential food source, to a potential enemy, or to a potential mate will certainly give greater fitness in the Darwinian sense than a program that lacks these properties. Again, a behavior program that allows for appropriate learning and the improvement of behavior reactions by various types of feedback gives greater likelihood of survival than a program that lacks these properties.

The purposive action of an individual, insofar as it is based on the properties of its genetic code, therefore is no more nor less purposive than the actions of a computer that has been programmed to respond appropriately to various inputs.

It is, if I may say so, a purely mechanistic purposiveness.

We biologists have long felt that it is ambiguous to designate such programmed, goal-directed behavior "teleological," because the word *teleological* has also been used in a very different sense, for the final stage in evolutionary adaptive processes.

The development or behavior of an individual is purposive; natural selection is definitely riot. When MacLeod (1957) stated, "What is most challenging about Darwin, however, is his re-introduction of purpose into the natural world," he chose the wrong word. The word *purpose* is singularly inapplicable to evolutionary change, which is, after all, what Darwin was considering. If an organism is well adapted, if it shows superior fitness, this is not due to any purpose of its ancestors or of an outside agency, such as "Nature" or "God," that created a superior design or plan. Darwin "has swept out such finalistic teleology by the front door," as Simpson (1960) has rightly said.

We can summarize this discussion by stating that there is no conflict between causality and teleonomy, but that scientific biology has not found any evidence that would support teleology in the sense of various vitalistic or finalistic theories (Simpson 1960; 1950; Koch 1957). All the so-called teleological systems which Nagel discusses (1961) are actually illustrations of teleonomy.

THE PROBLEM OF PREDICTION

The third great problem of causality in biology is that of prediction. In the classical theory of causality the touchstone of the goodness of a causal explanation was its predictive value. This view is still maintained in Bunge's modern classic (1959): "A theory can predict to the extent to which it can describe and explain." It is evident that Bunge is a physicist; no biologist would have made such a statement. The theory of natural selection can describe and explain phenomena with considerable precision, but it cannot make reliable predictions, except through such trivial and meaningless circular statements as, for instance: "The fitter individuals will on the average leave more offspring." Scriven (1959) has emphasized quite correctly that one of the most important contributions to philosophy made by the evolutionary theory is that it has demonstrated the independence of explanation and prediction.

Although prediction is not an inseparable concomitant of causality, every scientist is nevertheless happy if his causal explanations simultaneously have high predictive value. We can distinguish many categories of prediction in biological explanation. Indeed, it is even doubtful how to define "prediction" in biology. A competent zoogeographer can predict with high accuracy what animals will be found on a previously unexplored mountain range or island. A paleontologist likewise can predict with high probability what kind of fossils can be expected in a newly accessible geological horizon. Is such correct guessing of the results of past events genuine prediction? A similar doubt pertains to taxonomic predictions, as discussed in the next paragraph. The term *prediction* is, however, surely legitimately used for future events. Let me give four examples to illustrate the range of predictability.

(1) *Prediction in classification.* If I have identified a fruit fly as an individual of *Drosophila melanogaster* on the basis of bristle pattern and the proportions of face and eye, I can "predict" numerous structural and behavioral characteristics which I will find if I study other aspects of this individual. If I find new species with the diagnostic key characters of the genus *Drosophila,* I can at once "predict" a whole set of biological properties.

(2) *Prediction of most physicochemical phenomena on the molecular level.* Predictions of very high accuracy can be made with respect to most biochemical unit processes in organisms, such as metabolic pathways, and with respect to biophysical phenomena in simple systems, such as the action of light, heat, and electricity in physiology.

In examples 1 and 2 the predictive value of causal statements is usually very high. Yet there are numerous other generalizations or causal statements in biology that have low predictive values. The following examples are of this kind.

(3) *Prediction of the outcome of complex ecological interactions.* The statement, "An abandoned pasture in southern New England will be replaced by a stand of grey birch (*Betula populifolia*) and white pine (*Pinus strobus*)" is often correct. Even more often, however, the replacement may be an almost solid stand of *P. strobus,* or *P. strobus* may be missing altogether and in its stead will be cherry (*Prunus*), red cedar (*Juniperus virginianus*), maples, sumac, and several other species.

Another example also illustrates this unpredictability. When two species of flour beetles (*Tribolium confusum* and *T. castaneum*) are brought together in a uniform environment (sifted wheat flour), one of the two species will always displace the other. At high temperatures and humidities, *T. castaneum* will win out; at low temperatures and humidities, *T. confusum* will be the victor. Under intermediate conditions the outcome is indeterminate and hence unpredictable (Park 1954).

(4) *Prediction of evolutionary events.* Probably nothing in biology is less predictable than the future course of evolution. Looking at the Permian reptiles, who would have predicted that most of the more flourishing groups would become extinct (many rather rapidly), and that one of the most undistinguished branches would give rise to the mammals? Which student of the Cambrian fauna would have predicted the revolutionary changes in the marine life of the subsequent geological eras? Unpredictability also characterizes small-scale evolution. Breeders and students of natural selection have discovered again and again that independent parallel lines exposed to the same selection pressure will respond at different rates and with different correlated effects, none of them predictable.

As is true in many other branches of science, the validity of predictions for biological phenomena (except for a few chemical or physical unit processes) is nearly always statistical. We can predict with high accuracy that slightly more than 500 of the next 1,000 newborns will be boys. We cannot predict the sex of a particular child prior to conception.

REASONS FOR INDETERMINACY IN BIOLOGY

Without claiming to exhaust all the possible reasons for indeterminacy, I can list four classes. Although they somewhat overlap each other, each deserves to be treated separately.

(1) *Randomness of an event with respect to the significance of the event.* Spontaneous mutation, caused by an "error" in DNA replication, illustrates this cause for indeterminacy very well. The occurrence of a given mutation is in no way related to the evolutionary needs of the particular organism or of the population to which it belongs. The precise results of a given selection pressure are unpredictable because mutation, recombination, and developmental homeostasis are making indeterminate contributions to the response to this pressure. All the steps in the determination of the genetic contents of a zygote contain a large component of this type of randomness. What we have described for mutation is also true for crossing over, chromosomal segregation, gametic selection, mate selection, and early survival of the zygotes. Neither underlying molecular phenomena nor the mechanical motions responsible for this randomness are related to their biological effects.

(2) *Uniqueness of all entities at the higher levels of biological integration.* In the uniqueness of biological entities and phenomena lies one of the major differences between biology and the physical sciences. Physicists and chemists often have genuine difficulty in understanding the biologist's stress on the unique, although such an understanding has been greatly facilitated by developments in modern physics. If a physicist says "ice floats on water," his statement is true for any piece of ice and any body of water. The members of a class usually lack the individuality that is so characteristic of the organic world, where each individual is unique; each stage in the life cycle is unique; each population is unique; each species and higher category is unique; each interindividual contact is unique; each natural association of species is unique; and each evolutionary event is unique. Where these statements are applicable to man, their validity is self-evident. However, they are equally valid for all sexually reproducing animals and plants. Uniqueness, of course, does not entirely preclude prediction. We can make many valid statements about the attributes and behavior of man, and the same is true for other organisms. But most of these statements (except for those pertaining to taxonomy) have purely statistical validity. Uniqueness is particularly characteristic for evolutionary biology. It is quite impossible to have for unique phenomena general laws like those that exist in classical mechanics.

(3) *Extreme complexity.* The physicist Elsässer stated in a recent symposium: "[An] outstanding feature of all organisms is their well-nigh unlimited structural and dynamical complexity." This is true. Every organic system is so rich in feedbacks, homeostatic devices, and potential multiple pathways that a complete description is quite impossible. Furthermore, the analysis of such a system would require its destruction and would thus be futile.

(4) *Emergence of new qualities at higher levels of integration.* It would lead too far to discuss in this context the thorny problem of "emergence." All I can do here is to state its principle dogmatically: "When two entities are combined at a higher level of integration, not all the properties of the new entity are necessarily a logical or predictable consequence of the properties of the components." This difficulty is by no means confined to biology, but it is certainly one of the major sources of indeterminacy in biology. Let us remember that indeterminacy does not mean lack of cause, but merely unpredictability.

All four causes of indeterminacy, individually and combined, reduce the precision of prediction.

One may raise the question at this point whether predictability in classical mechanics and unpredictability in biology are due to a difference of degree or of kind. There is much to suggest that the difference is, in considerable part, merely a matter of degree. Classical mechanics is, so to speak, at one end of a continuous spectrum, and biology is at the other. Let us take the traditional example of the gas laws. Essentially they are only statistically true, but the population of molecules in a gas obeying the gas laws is so enormous that the actions of individual molecules become integrated into a predictable—one might say "absolute"—result. Samples of 5 or 20 molecules would show definite individuality. The difference in the size of the studied "populations" certainly contributes to the difference between the physical sciences and biology.

CONCLUSIONS

Let us now return to our initial question and try to summarize some of our conclusions on the nature of the cause-and-effect relations in biology.

(1) Causality in biology is a far cry from causality in classical mechanics.
(2) Explanations of all but the simplest biological phenomena usually consist of sets of causes. This is particularly true for those biological phenomena that can be understood only if their evolutionary history is also considered. Each set is like a pair of brackets which contains much that is unanalyzed and much that can presumably never be analyzed completely.
(3) In view of the high number of multiple pathways possible for most biological processes (except for the purely physicochemical ones) and in view of the randomness of many of the biological processes, particularly on the molecular level (as well as for other reasons), causality in biological systems is not predictive, or at best is only statistically predictive.
(4) The existence of complex programs of information in the DNA of the germ plasm permits teleonomic purposiveness. On the other hand, evolutionary research has found no evidence whatsoever for a "goal seeking" of evolutionary lines, as postulated in that kind of teleology which sees "plan and design" in nature. The harmony of the living universe, so far as it exists, is an a posteriori product of natural selection.

Finally, causality in biology is not in real conflict with the causality of classical mechanics. As modern physics has also demonstrated, the causality of classical mechanics is only a very simple, special case of causality. Predictability, for instance, is not a necessary component of causality. The complexities of biological causality do not justify embracing nonscientific ideologies, such as vitalism or finalism, but should encourage all those who have been trying to give a broader basis to the concept of causality.

NOTES

1. This essay is reprinted from *Science* 134:1501–06; copyright 1961 by the American Association for the Advancement of Science.

2. Various philosophers have published seeming refutations of Scriven's claims. No doubt my views on prediction at that time were rather simplistic. I have since revised them considerably (Mayr 1982:57–59). The philosophical problems of prediction have been well stated by Grünbaum (1963:114–49).

REFERENCES

Bernard, C. *Lecons sur les phenomènes de la vie*. Vol. 1. 1885.

Bunge, M. *Causality*. Cambridge, MA: Harvard University Press, 1959.

Delbrück, M. "A Physicist Looks at Biology." *Transactions of the Connecticut Academy of Arts and Sciences* 33 (1949):173–90.

Grünbaum, A. *Induction: Some Current Issues*. Middletown, CT: Wesleyan University Press, 1963.

Huxley, J. "The Openbill's Open Bill: A Teleonomic Enquiry." *Zoologische Jahrbücher. Abteilung für Anatomie und Ontogenie der Tiere* 88 (1960): 9–30.

Koch, L. F. "Vitalistic-mechanistic Controversy." *Scientific Monthly* 85 (1957): 245–55.

Loeb, J. *The Organism as a Whole*. New York: Putnam, 1916.

MacLeod, R. B. "Teleology and Theory of Human Behavior." *Science* 125 (1957): 477–80.

Mayr, E. *The Growth of Biological Thought*. Cambridge, MA: Harvard University Press, 1982.

Nagel, E. *The Structure of Science*. New York: Harcourt, 1961.

Park, T. "Experimental Studies of Interspecies Competition II." *Physiological Zoology* 27 (1954): 177–238.

Pittendrigh, C. S. "Adaptation, Natural Selection, and Behavior." In *Behavior and Evolution*, edited by A. Roe and G.G. Simpson, 390–416. New Haven, CT: Yale University Press, 1958.

Scriven, M. "Explanation and Prediction in Evolutionary Theory." *Science* 130 (1959): 477–82.

Simpson, G. G. "Evolutionary Determinism and the Fossil Record." *Scientific Monthly* 71 (1950): 262–67.
———. "The World into which Darwin Led Us." *Science* 131 (1960): 966–74.

LIFE AND ITS ORIGINS

ARISTOTLE*

THE GENERATION OF ANIMALS

There is a considerable difficulty in understanding how the plant is formed out of the seed or any animal out of the semen. Everything that comes into being or is made must be made out of something, be made by the agency of something, and must become something. Now that out of which it is made is the material; this some animals have in its first form within themselves, taking it from the female parent, as all those which are not only born alive but produced as a grub or an egg; others receive it from the mother for a long time by sucking, as the young of all those which are not only externally but also internally viviparous. Such, then, is the material out of which things come into being, but we now are inquiring not out of what the parts of an animal are made, but by what agency. Either it is something external which makes them, or else something existing in the seminal fluid and the semen; and this must either be soul or a part of soul, or something containing soul.

Now it would appear irrational to suppose that any of either the internal organs or the other parts is made by something external, since one thing cannot set up a motion in another without touching it, nor can a thing be affected in any way by anything that does not set up a motion in it. Something then of the sort we require exists in the embryo itself, being either a part of it or separate from it. To suppose that it should be something else separate from it is irrational. For after the animal has been produced does this something perish or does it remain in it? But nothing of the kind appears to be in it, nothing which is not a part of the whole plant or animal. Yet, on the other hand, it is absurd to say that it perishes after making either all the parts or only some of them. If it makes some of the parts and then perishes, what is to make the rest of them? Suppose this something makes the heart and then perishes, and the heart makes another organ, by the same argument either all the parts must perish or all must remain. Therefore it is preserved. Therefore it is a part of the embryo itself which exists in the semen from the beginning; and if indeed there is no part of the soul which does not exist in some part of the body, it would also be a part containing soul in it from the beginning.

How, then, does it make the other parts? Either all the parts, as heart, lung, liver, eye, and all the rest, come into being together or in succession, as is said in the verse ascribed to Orpheus, for there he says that an animal comes into being in the same way as the knitting of a net. That the former is not the fact is plain even to the senses, for some of the parts are clearly visible as already existing in the embryo while others are not; that it is not because of their being too small that they are not visible is clear, for the lung is of greater size than the heart, and yet appears later than the heart in the original development. Since, then, one is earlier and another later, does the one make the other, and does the later part exist on account of the part which is next to it, or rather does the one come into being only *after* the other? I mean, for instance, that it is not the fact that the heart, having come into being first, then makes the liver, and the liver again another organ, but that the liver only comes into being *after* the heart, and not by the agency of the heart, as a man becomes a man *after* being a boy, not by his agency. An explanation of this is that, in all the productions of nature or of art, what already exists potentially is brought into being only by what exists actually; therefore if one organ formed another the form and the character of the later organ would have to exist in the earlier, e.g. the form of the liver in the heart. And otherwise also the theory is strange and fictitious.

Yet again, if the whole animal or plant is formed from semen or seed, it is impossible that any part of it should exist ready made in the semen or seed, whether that part be able to make the other parts or no. For it is plain that, if it exists in it from the first, it was made by that which made the semen. But semen must be made first, and that is the function of the generating parent. So, then, it is not possible that any part should exist in it, and therefore it has not within itself that which makes the parts.

But neither can this agent be external, and yet it must needs be one or other of the two. We must try, then, to solve this difficulty, for perhaps some one of the statements made cannot be made without qualification, e.g. the statement that the parts cannot be made by what is external to the semen. For if in a certain sense they cannot, yet in another sense they can. (Now it makes no difference whether we say "the semen" or "that from which the semen comes," in so far as the semen has in itself the movement initiated by the other.) It is possible, then, that A should move B, and B move C; that, in fact, the case should be the same as with the automatic puppets. For the parts of such puppets while at rest have a sort of potentiality of motion in them, and when any external force puts the first of them in motion, immediately the next is moved in actuality. As, then, in these automatic puppets the external force moves the parts in a certain sense (not by touching any part at the moment, but by having touched one previously), in like manner also that from which the semen comes, or in other words that which made the semen, sets up the movement in the embryo and makes the parts of it by having first touched something though not continuing to touch it. In a way it is the innate motion that does this, as the act of building builds the house. Plainly, then, while there is something which makes the parts, this does not exist as a definite object, nor does it exist in the semen at the first as a complete part.

But how is each part formed? We must answer this by starting in the first instance from the principle that, in all products of nature or art, a thing is made by something actually existing out of that which is potentially such as the finished product. Now the semen is of such a nature, and has in it such a principle of motion, that when the motion is ceasing each of the parts comes into being, and that as a part having life or soul. For there is no such thing as face or flesh without soul in it; it is only homonymously that they will be called face or flesh if the life has gone out of them, just as if they had been made of stone or wood. And the homogeneous parts and the organic come into being together. And just as we should not say that an axe or other instrument or organ was made by the fire alone, so neither shall we say that foot or hand were made by heat alone. The same applies also to flesh, for this too has a function. While, then, we may allow that hardness and softness, stickiness and brittleness, and whatever other qualities are found in the parts that have life and soul, may be caused by mere heat and cold, yet, when we come to the principle in virtue of which flesh is flesh and bone is bone, that is no longer so; what makes them is the movement set up by the male parent, who is in actuality what that out of which the offspring is made is in potentiality. This is what we find in the products of art; heat and cold may make the iron soft and hard, but what makes a sword is the movement of the tools employed, this movement containing the principle of the art. For the art is the starting-point and form of the product; only it exists in something else, whereas the movement of nature exists in the product itself, issuing from another nature which has the form in actuality.

Has the semen soul, or not? The same argument applies here as in the question concerning the parts. As no part, if it participate not in soul, will be a part except homonymously (as the eye of a dead man is still called an eye), so no soul will exist in anything except that of which it is soul; it is plain therefore that semen both has soul, and is soul, potentially.

But a thing existing potentially may be nearer or further from its realization in actuality, just as a sleeping geometer is further away than one awake and the latter than one actually studying. Accordingly it is not any part that is the cause of the soul's coming into being, but it is the first moving cause from outside. (For nothing generates itself, though when it has come into being it thenceforward increases itself.) Hence it is that only one part comes into being first and not all of them together. But that must first come into being which has a principle of increase (for this nutritive power exists in all alike, whether animals or plants, and this is the same as the power that enables an animal or plant to generate another like itself, that being the function of them all if naturally perfect). And this is necessary for the reason that whenever a living thing is produced it must grow. It is produced, then, by something else of the same name, as e.g. man is produced by man, but it is increased by means of itself. There is, then, something which increases it. If this is a single part, this must come into being first. Therefore if the heart is first made in some animals, and what is analogous to the heart in the others which have no heart, it is from this or its analogue that the first principle of movement would arise.

*J. B. S. HALDANE**

WHAT IS LIFE?

I am not going to answer this question. In fact, I doubt if it will ever be possible to give a full answer because we know what it feels like to be alive, just as we know what redness, or pain, or effort are. So we cannot describe them in terms of anything else. But it is not a foolish question to ask, because we often want to know whether a man is alive or not, and when we are dealing with the microscopic agents of disease, it is clear enough that bacteria are alive, but far from clear whether viruses, such as those which cause measles and smallpox, are so.

So we have to try to describe life in terms of something else, even if the description is quite incomplete. We might try some such expression as "the influence of spirit on matter." But this would be of little use for several reasons. For one thing, even if you are sure that man, and even dogs, have spirits, it needs a lot of faith to find a spirit in an oyster or a potato. For another thing, such a definition would certainly cover great works of art, or books which clearly show their author's mind, and go on influencing readers long after he is dead. Similarly, it is no good trying to define life in terms of a life force. George Bernard Shaw and Professor C. E. M. Joad think there is a life force in living things. But if this has any meaning, which I doubt, you can only detect the life force in an animal or plant by its effects on matter. So we should have to define life in terms of matter. In ordinary life we recognize living things partly by their shape and texture. But these do not change for some hours after death. In the case of mammals and birds we are sure they are dead if they are cold.

This test will not work on a frog or a snail. We take it that they are dead if they will not move when touched. But in the case of a plant the only obvious test is whether it will grow, and this may take months to find out. All these tests agree in using some kind of motion or change as the criterion of life, for heat is only irregular motion of atoms. They also agree in being physical rather than chemical tests. There is no doubt, I think, that we can learn a lot more about life from a chemical than from a physical approach. This does not mean that life has been

From J. B. S. Haldane, *What Is Life?* (London: Acuin Press, 1949), pp. 58–62.

fully explained in terms of chemistry. It does mean that it is a pattern of chemical rather than physical events. Perhaps I can make this clear by an example.

Suppose a blind man and a deaf man both go to performances of *Macbeth* and of *Alexander Nevsky*. The deaf man will understand little of the play. He will not know Duncan was murdered, let alone who did it. The blind man will miss far less. The essential part of Shakespeare's plays are the words. But with the film it will be the other way round. What is common to all life is the chemical events. And these are extraordinarily similar in very different organisms. We may say that life is essentially a pattern of chemical happenings, and that in addition there is some building of a characteristic shape in almost all living things, characteristic motion in most animals, and feeling and purpose in some of them. The chemical make-up of different living things is very different. A tree consists largely of wood, which is not very like any of the constituents of a man, though rather like a stuff called glycogen which is part of most, if not all, of our organs. But the chemical changes which go on in the leaves, bark, and roots of a tree, particularly the roots, are surprisingly like those which go on in human organs. The roots need oxygen just as a man does, and you can see whether a root is alive, just as you can see whether a dog is alive, by measuring the amount of oxygen which it consumes per minute. And the oxygen is used in the same kinds of chemical processes, which may roughly be described as controlled burning of foodstuffs at a low temperature. Under ordinary circumstances oxygen does not combine with sugar unless both are heated. It does so in almost all living things through the agency of what are called enzymes. Most of the oxygen which we use has first to unite with an enzyme consisting mainly of protein, but containing a little iron. Warburg discovered this in yeast in 1924. In 1926 I did some rather rough experiments which showed the same, or very nearly the same, enzyme in green plants, moths, and rats. Since then it has been found in a great variety of living things.

Just the same is true for other processes. A potato makes sugar into starch and your liver makes it into glycogen by substantially the same process. Most of the steps by which sugar is broken down in alcoholic fermentation and muscular contraction are the same. And so on. The end results of these processes are, of course, very different. A factory may switch over from making bren guns to making sewing machines or bicycles without very great changes. Similarly the chemical processes by which an insect makes its skin and a snail its slime are very similar, though the products differ greatly.

In fact, all life is characterised by a fundamentally similar set of chemical processes arranged in very different patterns. Thus, animals use up foodstuffs, while most plants make them. But in both plants and animals the building up and breaking down are both going on all the time. The balance is different.

Engels said that life was the mode of existence of proteins (the word which he used is often translated as "albuminous substances"). This is true in so far as all enzymes seem to be proteins. And it is true in so far as the fundamental similarity of all living things is a chemical one. But enzymes and other proteins can be purified and will carry on their characteristic activities in glass bottles. And no biochemist would say they were alive.

In the same way Shakespeare's plays consist of words, whereas words are a very small part of Eisenstein's films. It is important to know this, as it is important to know that life consists of chemical processes. But the arrangement of the words is even more important than the words themselves. And in the same way life is a pattern of chemical processes.

This pattern has special properties. It begets a similar pattern, as a flame does, but it regulates itself as a flame does not except to a slight extent. And, of course, it has many other peculiarities. So when we have said that life is a pattern of chemical processes, we have said something true and important. It is practically important because we are at last learning how to control some of them, and the first fruits of this knowledge are practical inventions like the use of sulphonamides, penicillin, and streptomycin.

But to suppose that one can describe life fully on these lines is to attempt to reduce it to mechanism, which I believe to be impossible. On the other hand, to say that life does not consist of chemical processes is to my mind as futile and untrue as to say that poetry does not consist of words.

LESLIE E. ORGEL

THE ORIGIN OF LIFE: A REVIEW OF FACTS AND SPECULATIONS

Three popular hypotheses attempt to explain the origin of prebiotic molecules: synthesis in a reducing atmosphere, input in meteorites and synthesis on metal sulfides in deep-sea vents. It is not possible to decide which is correct. It is also unclear whether the RNA world was the first biological world or whether some simpler world preceded it.

The problem of the origin of life on the earth has much in common with a well-constructed detective story. There is no shortage of clues pointing to the way in which the crime, the contamination of the pristine environment of the early earth, was committed. On the contrary, there are far too many clues and far too many suspects. It would be hard to find two investigators who agree on even the broad outline of the events that occurred so long ago and made possible the subsequent evolution of life in all its variety. Here, I outline two of the main questions and some of the conflicting evidence that has been used in attempts to answer them. First, however, I summarize the few areas where there is fairly general agreement.

The earth is slightly more than 4.5 billion years old. For the first half billion years or so after its formation, it was impacted by objects large enough to evaporate the oceans and sterilize the surface.[1,2] Well-preserved microfossils of organisms that have morphologies similar to those of modern blue-green algae, and date back about 3.5 billion years, have been found,[3] and indirect but persuasive evidence supports the proposal that life was present 3.8 billion years ago.[4] Life, therefore, originated on or was transported to the earth at some point within a window of a few hundred million years that opened about four billion years ago. The majority of workers in the field reject the hypothesis that life was transported to the earth from somewhere else in the galaxy and take it for granted that life began de novo on the early earth.

The uniformity of biochemistry in all living organisms argues strongly that all modern organisms descend from a last-common ancestor (LCA). Detailed analysis of protein sequences suggests that the LCA had a complexity comparable to that of a simple modern bacterium and lived 3.2–3.8 billion years ago.[5] If we knew the stages by which the LCA evolved from abiotic components present on the primitive earth, we would have a complete account of the origin of life. In practice, the most ambitious studies of the origins of life address much simpler questions. Here, I discuss two of them: What were the sources of the small organic molecules that made up the first self-replicating systems? How did biological organization evolve from an abiotic supply of small organic molecules?

ABIOTIC SYNTHESIS OF SMALL ORGANIC MOLECULES

Miller, a graduate student who was working with Harold Urey, began the modern era in the study of the origin of life at a time when most people believed that the atmosphere of the early earth was strongly reducing. Miller subjected a mixture of methane, ammonia and hydrogen to an electric discharge and led the products into liquid water.[6] He showed that a substantial percentage of the carbon in the gas mixture was incorporated into a relatively small group of simple organic molecules and that several of the naturally occurring amino acids were prominent among these products. This was a surprising result; organic chemists would have expected a much-less-tractable product mixture. The Urey-Miller experiments were widely accepted as a model of prebiotic synthesis of amino acids by the action of lightning.

Miller and his co-workers went on to study electric-discharge synthesis of amino acids in greater detail.[7,8] Using more powerful analytical techniques, they identified many more amino acids—some, but by no means all of which occur in living organisms. They also showed that a major synthetic route to the amino acids is through the Strecker reaction—that is, from aldehydes, hydrogen cyanide and ammonia. Glycine, for example, is formed from formaldehyde, cyanide and ammonia—all of which can be detected among the products formed in the electric-discharge reaction.

In the years following the Urey-Miller experiments, the synthesis of biologically interesting molecules from products that could be obtained from a reducing gas mixture became the principle aim of prebiotic chemistry (fig. 2.1.). Remarkably, Oro and Kimble were able to synthesize adenine from hydrogen cyanide and ammonia.[9] Somewhat later, Sanchez, Ferris and I showed that cyanoacetylene is a major product of the action of an electric discharge on a mixture of methane and

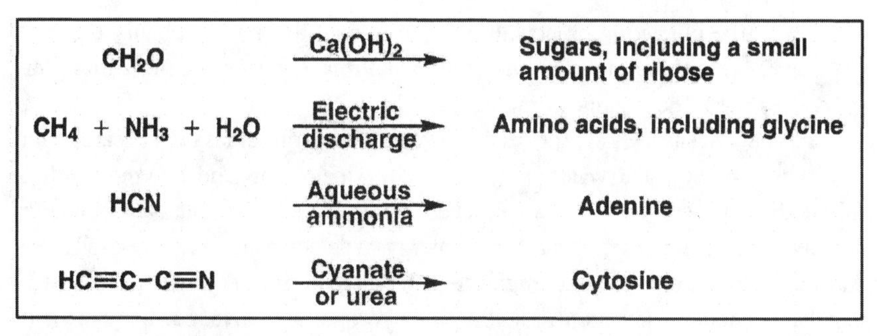

Figure 2.1. Early prebiotic synthesis of biomonors.

nitrogen and that cyanoacetylene is a plausible source of the pyrimidine bases uracil and cytosine.[10] This new information, together with previous studies that showed that sugars are formed readily from formaldehyde,[11,12] convinced many students of the origins of life that they understood the first stage in the appearance of life on the earth: the formation of a prebiotic soup of biomonomers.

As in any good detective story, however, the principle suspect, the reducing atmosphere, has an alibi. Recent studies have convinced most workers concerned with the atmosphere of the early earth that it could never have been strongly reducing.[13] If this is true, Miller's experiments, and most other early studies of prebiotic chemistry, are irrelevant. I believe that the dismissal of the reducing atmosphere is premature, because we do not completely understand the early history of the earth's atmosphere. It is hard to believe that the ease with which sugars, amino acids, purines and pyrimidines are formed under reducing-atmosphere conditions is either a coincidence or a false clue planted by a malicious creator.

Many of those who dismiss the possibility of a reducing atmosphere believe that the crime was an outside job. A substantial proportion of the meteorites that fall on the earth belong to a class known as carbonaceous chondrites.[14] These are particularly interesting because they contain a significant amount of organic carbon and because some of the standard amino acids and nucleic-acid bases are present. Could the prebiotic soup have originated in preformed organic material brought to the earth by meteorites and comets?

Supporters of the impact theory have argued convincingly that sufficient organic carbon must have been present in the meteorites and comets that reached the surface of the early earth to have stocked an abundant soup. However, would this material have survived the intense heating that accompanies the entry of large bodies into the atmosphere and their subsequent collisions with the surface of the earth? The results of theoretical calculations depend strongly on assumptions made about the composition and density of the atmosphere, the distribution of sizes of the impacting objects, etc.[15] The impact theory is probably the most popular at present, but nobody has proved that impacts were the most important sources of prebiotic organic compounds.

The newest suspects are the deep-sea vents, submarine cracks in the earth's surface where superheated water rich in transition-metal ions and hydrogen sulfide mixes abruptly with cold sea water. These vents are sites of abundant biological activity, much of it independent of solar energy. Wächtershäuser has proposed a scenario for the origin of life that might fit such an environment.[16,17] He hypothesizes that the reaction between iron(II) sulfide and hydrogen sulfide [a reaction that yields pyrites (FeS_2) and hydrogen] could provide the free energy necessary for reduction of carbon dioxide to molecules capable of supporting the origin of life. He asserts that life originated on the surface of iron sulfides as a result of such chemistry. The assumptions that complex metabolic cycles self-organize on the surface and that the significant products never escape from the surface are essential parts of this theory; in Wächtershäuser's opinion, there never was a prebiotic soup!

Stetter and colleagues have confirmed the novel suggestion that hydrogen sulfide, in the presence of iron(II) sulfide, acts as a reducing agent.[18] They have reduced, for example, acetylene to ethane, and mercaptoacetic acid to acetic acid, but they have not reported reduction of CO_2. However, in a new study, Wächtershäuser and co-workers[19] have shown that FeS spiked with NiS reduces carbon monoxide. Given that carbon monoxide might well have been present in large amounts in the gases escaping from the vents, Wächtershäuser's findings could well prove important. If metal sulfides can be shown to catalyze the synthesis of a sufficient variety of organic molecules from carbon monoxide, the vent theory of the origins of biomonomers will become very attractive.

In summary, there are three main contending theories of the prebiotic origin of biomonomers (not to mention several other less-popular options). No theory is compelling, and none can be rejected out of hand. Perhaps it is time for a conspiracy theory; more than one of the sources of organic molecules discussed above may have collaborated to make possible the origin of life.

SELF-ORGANIZATION

There is no general agreement about the source of prebiotic organic molecules on the early earth, but there are several plausible theories, each backed by some experimental data. The situation with regard to the evolution of a self-replicating system is less satisfactory; there are at least as many suspects, but there are virtually no experimental data.

The fairly general acceptance of the hypothesis that there was once an RNA world (i.e. a self-contained biological world in which RNA molecules functioned both as genetic materials and as enzyme-like catalysts) has changed the direction of research into the origins of life.[20] The central puzzle is now seen to be the origin of the RNA world. Two specific, but intertwined, questions are central to the debate. Was RNA the first genetic material or was it preceded by one or more simpler genetic materials? How much self-organization of reaction sequences is possible in the absence of a genetic material? I shall concentrate on the first question.

The assumption that a polymer that doubled as a genetic material and as a source of enzyme-like catalytic activity once existed profoundly changes the goals of prebiotic synthesis. The central issue becomes the synthesis of the first genetic monomers: nucleotides or whatever preceded them. The synthesis of amino acids, coenzymes, etc. becomes a side issue, because there is no reason to believe that they were ever synthesized abiotically; some or all of them might have been introduced as direct or indirect consequences of the enzyme-like activities of RNA or its precursor(s). Supporters of the hypothesis that RNA was the first genetic material must explain where the nucleotides came from and how they self-organized. Those who believe in a simpler precursor have the difficult task of identifying such a precursor, but they hope that explaining monomer synthesis will then be simpler.

Returning to the idiom of the detective story, accumulating evidence suggests that RNA, a prime suspect, could have completed the difficult task of organizing itself into a self-contained replicating system. It has proved possible to isolate sequences that catalyze a wide variety of organic reactions from pools of random RNA.[21,22] As regards the origin of the RNA world, the most important reactions are those in which a preformed template-RNA strand catalyzes the synthesis of its complement from monomers or short oligomers. Eklund and co-workers[23] have isolated catalysts for the ligation of short oligonucleotides surprisingly easily, and the catalysts carry out ligation with adequate specificity. These molecules are the RNA equivalents of the RNA and DNA ligases. Considerable progress has also been made in selecting RNA equivalents of RNA polymerases.[24]

If the RNA world evolved de novo, it must have depended initially on an abiotic source of activated nucleotides. However, oxidation–reduction, methylation, oligosaccharide synthesis, etc., supported by nucleotide-containing coenzyme, probably became part of the chemistry of the RNA world before the invention of protein synthesis. Unfortunately, we cannot say just how complex the RNA world could have been until we know more about the that they range of reactions that can be catalyzed by ribozymes. It seems likely that RNA could have catalyzed most of the steps involved in the synthesis of nucleotides,[25] and possibly the coupling of redox reactions to the synthesis of phosphodiesters and peptides, but this remains to be demonstrated experimentally.

The experiments on the selection of ribozymes that catalyze nucleic acid replication (discussed above) use as inputs pools of RNA molecules synthesized by enzymes. Recently, Ferris and coworkers have made considerable progress in the assembly of RNA oligomers from monomers, using an abundant clay mineral, montmorillonite, as a catalyst.[26,27] The substrates that they use, nucleoside 5'-phosphorimidazolides, were probably not prebiotic molecules, but the experiments do indicate that the use of minerals as adsorbents and catalysts could allow the accumulation of long oligonucleotides once suitable activated monomers are available. We have shown that, using activated monomers, non-enzymatic copying of a wide range of oligonucleotide sequences is possible[28] and have obtained similar, but less extensive, results for ligation of short oligomers.

An optimist could propose the following scenario. First, activated mononucleotides oligomerize on montmorillonite or an equivalent mineral. Next, copying of longer templates, using monomers or short oligomers as substrates, leads to the accumulation of a library of dsRNA molecules. Finally, an RNA double helix, one of whose strands has generalized RNA-polymerase activity, dissociates; the polymerase strand copies its complement to produce a second polymerase molecule, which copies the first to produce a second complement—and so on. The RNA world could therefore have arisen from a pool of activated nucleotides.[29] All that would have been needed is a pool of activated nucleotides!

Nucleotides are complicated molecules. The synthesis of sugars from formaldehyde gives a complex mixture, in which ribose is always a minor component. The formation of a nucleoside from a base, and a sugar is not an easy reaction and, at least for pyrimidine nucleosides, has not been achieved under prebiotic conditions; the phosphorylation of nucleosides tends to give a complex mixture of products.[30] The inhibition of the template-directed reactions on D-templates by L-substrates is a further difficulty.[31] It is almost inconceivable that nucleic acid replication could have got started, unless there is a much simpler mechanism for the prebiotic synthesis of nucleotides. Eschenmoser and his colleagues have had considerable success in generating ribose 2,4-diphosphate in a potentially prebiotic reaction from glycolaldehyde mono-phosphate and formaldehyde.[32] Direct prebiotic synthesis of nucleotides by novel chemistry is therefore not hopeless. Nonetheless, it is more likely that some organized form of chemistry preceded the RNA world. This leads us to a discussion of genetic takeover.

Cairns-Smith, long before the argument became popular, emphasized how improbable it is that a molecule as high tech as RNA could have appeared de novo on the primitive earth.[33] He proposed that the first form of life was a self-replicating clay. He suggested that the synthesis of organic molecules became part of the competitive strategy of the clay world and that the inorganic genome was taken over by one of its organic creations. Cairns-Smith's postulate of an inorganic life form has failed to gather any experimental support. The idea lives on in the limbo of uninvestigated hypotheses. However, Cairns-Smith also contemplated the possibility that RNA was preceded by one or more linear organic genomes. This idea has taken root, but its implications have not always been appreciated. . . . The problem of achieving sufficient specificity, whether in aqueous solution or on the surface of a mineral, is so severe that the chance of closing a cycle of reactions as complex as the reverse citric acid cycle, for example, is negligible. The same, I believe, is true for simpler cycles involving small molecules that might be relevant to the origins of life and also for peptide-based cycles.

If RNA was not the first genetic material, biochemistry might provide no clues to the origins of life. Presumably, the biological world that immediately preceded the RNA world already had the capacity to synthesize nucleotides. This should help us to formulate hypotheses about its chemical characteristics. However, if there were two or more worlds before the RNA world, the original chemistry might have left no trace in contemporary biochemistry. In that case, the chemistry of the origins of life is unlikely to be discovered without investigating in detail all the chemistry that might have occurred on the primitive earth—whether or not that chemistry has any relation to biochemistry. This gloomy prospect has not prevented discussion of alternative genetic systems.

The only potentially informational systems, other than nucleic acids, that have been discovered are closely related to nucleic acids. Eschenmoser and his colleagues have undertaken a systematic study of the properties of nucleic acid analogs in which ribose is replaced by another sugar or in which the furanose form of ribose is replaced by the pyranose form (fig. 2.2b).[34] Strikingly, polynucleotides based on the pyranosyl isomer of ribose (p-RNA) form Watson-Crick-paired double helices that are more stable than RNA, and p-RNAs are less likely than the corresponding RNAs to form multiple-strand competing structures. Furthermore, the helices twist much more gradually than those in the standard nucleic acids, which should make it easier to separate strands during replication. Pyranosyl RNA seems to be an excellent choice as a genetic system; in some ways, it might be an improvement on the standard nucleic acids. However, prebiotic synthesis of pyranosyl nucleotides is not likely to prove much easier than synthesis of the standard isomers, although a route through ribose 2,4-diphosphate is being explored by Eschenmoser and his colleagues.

Peptide nucleic acid (PNA) is another nucleic acid analog that has been studied extensively (fig. 2.2c). It was synthesized by Nielsen and colleagues during

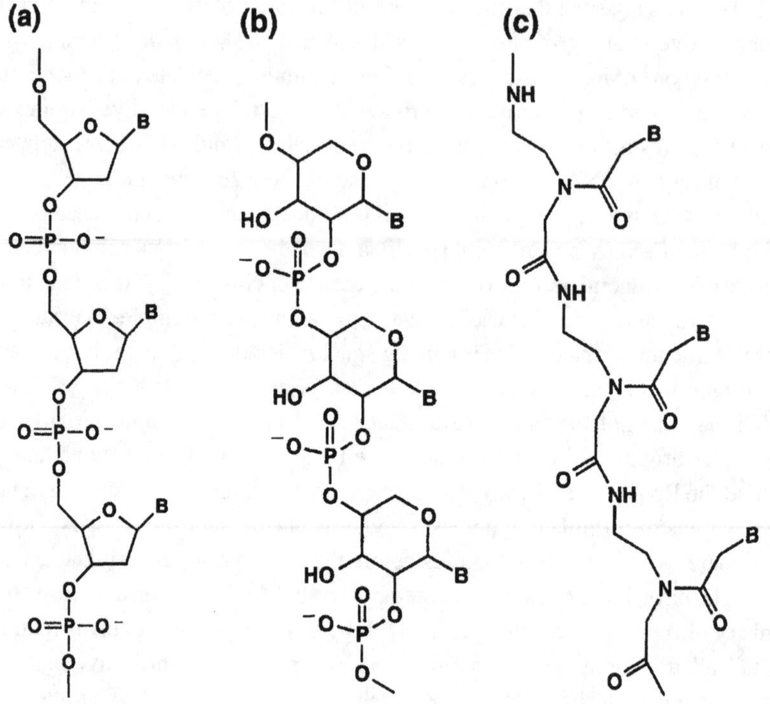

Figures 2.2a, 2.2b, and 2.2c.

work on antisense RNA.[35] PNA is an uncharged, achiral analog of RNA or DNA; the ribose-phosphate backbone of the nucleic acid is replaced by a backbone held together by amide bonds. PNA forms very stable double helices with complementary RNA or DNA.[36,37] We have shown that information can be transferred from PNA to RNA, and vice versa, in template-directed reactions[38,39] and that PNA to DNA chimeras form readily on either DNA or PNA templates.[40] Thus, a transition from a PNA world to an RNA world is possible. Nonetheless, I think it unlikely that PNA was ever important on the early earth, because PNA monomers cyclize when they are activated; this would make oligomer formation very difficult under prebiotic conditions.

The studies described above suggest that there are many ways of linking together nucleotide bases into chains that can form Watson-Crick double helices. Perhaps a structure of this kind will be discovered that can be synthesized easily under prebiotic conditions. If so, it would be a strong candidate for the very first genetic material. However, another possibility remains to be explored: the first genetic material might not have involved nucleoside bases. Two or more very simple molecules could have the pairing properties needed to form a genetic polymer—a positively charged and a negatively charged amino acid, for example. However, it is not clear that stable structures of this kind exist. RNA is clearly adapted to double-helix formation: its constrained backbone permits simultaneous base pairing and stacking. It is unlikely that much simpler molecules could substitute for the nucleotides. Perhaps some other interaction between the chains can stabilize a double helix in the absence of base stacking; binding to a mineral surface might supply the necessary constraints, but this remains to be demonstrated. In the absence of experimental evidence, little useful can be said.

The above discussion reveals a very large gap between the complexity of molecules that are readily synthesized in simulations of the chemistry of the early earth and the molecules that are known to form potentially replicating informational structures. Several authors have therefore proposed that metabolism came before genetics.[41-43] They have suggested that substantial organization of reaction sequences can occur in the absence of a genetic polymer and, hence, that the first genetic polymer probably appeared in an already-specialized biochemical environment. Because it is hard to envisage a chemical cycle that produces D-nucleotides, this theory would fit best if a simpler genetic system preceded RNA.

There is no agreement on the extent to which metabolism could develop independently of a genetic material. In my opinion, there is no basis in known chemistry for the belief that long sequences of reactions can organize spontaneously—and every reason to believe

CONCLUSION/OUTLOOK

In summary, there are several tenable theories about the origin of organic material on the primitive earth, but in no case is the supporting evidence compelling. Similarly, several alternative scenarios might account for the self-organization of a self-replicating entity from prebiotic organic material, but all of those that are well formulated are based on hypothetical chemical syntheses that are problematic. Returning to our detective story, we must conclude that we have identified some important suspects and, in each case, we have some ideas about the method they might have used. However, we are very far from knowing whodunit. The only certainty is that there will be a rational solution.

This review has necessarily been highly selective. I have neglected important aspects of prebiotic chemistry (e.g. the origin of chirality, the organic chemistry of solar bodies other than the earth, and the formation of membranes). The best source for such material is the journal *Origins of Life and Evolution of the*

Biosphere, particularly those issues that contain the papers presented at meetings of the International Society for the Study of the Origin of Life.

ACKNOWLEDGMENTS

This work was supported by NASA (grant number NAG5-4118) and NASA NSCORT/EXOBIOLOGY (grant number NAG5-4546). I thank Aubrey R. Hill, Jr. for technical assistance and Bernice Walker for manuscript preparation.

REFERENCES

1. N. H. Sleep, K. J. Zahnle, J. F. Kasting, and H. J. Morowitz, *Nature* 342 (1989): 139–42.

2. C. F. Chyba, "Geochim. Cosmochim," *Acta* 57 (1993): 3351–58.

3. J. W. Schopf, *The Earth's Earliest Biosphere: Its Origin and Evolution* (Princeton, NJ: Princeton University Press, 1993).

4. S. J. Mojzsis, et al., *Nature* 384 (1996): 55–59.

5. A. F. Doolittle, *Proceedings of the National Academy of Science, United States of America* 94 (1997): 13028–33.

6. S. L. Miller, *Science* 117 (1953): 528–29.

7. D. Ring, Y. Wolman, N. Friedmann, and S. L. Miller, *Proceedings of the National Academy of Science, United States of America* 69 (1972): 765–68.

8. Y. Wolman, H. Haverland, and S. L. Miller, *Proceedings of the National Academy of Science, United States of America* 69 (1972): 809–11.

9. J. Ore and A. P. Kimball, *Biochemical and Biophysical Research Communication* 2 (1960): 407–12.

10. J. P. Ferris, A. Sanchez, and L. E. Orgel, *Journal of Molecular Biology* 33 (1968): 693–704.

11. A. Butlerow, *Comptes rendus hebdomadaires des séances de l'Académie des sciences* 53 (1861): 145–47.

12. Ibid., *Liebigs Annalen der Chemie* 120 (1861): 295.

13. J. F. Kasting, *Science* 259 (1993): 920–26.

14. J. R. Cronin, et al., in *Meteorites and the Early Solar System*, ed. J. F. Kerridge and M. S. Matthew (Tuscon: University of Arizona Press, 1988), pp. 8191–1857.

15. C. Chyba and C. Sagan, *Nature* 355 (1992): 125–32.

16. G. Wächtershäuser, *Microbiology Reviews* 52 (1988): 452–84.

17. Ibid., *Progress in Biophysics and Molecular Biology* 58 (1992): 85–201.

18. E. Blochl, M. Keller, G. Wächtershkuser, and K. Stetter, *Proceedings of the National Academy of Sciences of the United States of America* 89 (1992): 8117–20.

19. C. Huber and G. Wächtershäuser, *Science* 276 (1997): 245–47.

20. R. F. Gesteland and J. F. Atkins, *The RNA World: The Nature of Modern RNA Suggests a Prebiotic World* (Woodbury, NY: Cold Spring Harbor Laboratory Press, 1993).

21. T. Pan, *Current Opinion in Chemical Biology* 1 (1997): 17–25.

22. R. R. Breaker, *Current Opinion in Chemical Biology* 1 (1997): 26–31.

23. E. H. Eklund, J. W. Szostak, and D. P. Bartel, *Science* 269 (1995): 364–70.

24. E. H. Eklund and D. P. Bartel, *Nature* 382 (1996): 373–76.

25. P. J. Unrai and D. P. Bartel, *Nature* 395 (1998): 260–63.

26. J. P. Ferris and G. Ertem, *Journal of the American Chemical Society* 115 (1993): 12270–75.

27. K. Kawamura and J. P. Ferris, *Journal of the American Chemical Society* 116 (1994): 7564–72.

28. A. R. Hill Jr., T. Wu, and L. E. Orgel, *Origins of Life and Evolution of the Biosphere: The Journal of the International Society for the Study of the Origin of Life* 23 (1993): 285–90.

29. L. E. Orgel, *Scientific American* 271 (1994): 52–61.

30. Ferris, *Cold Spring Harbor Symposia on Quantitative Biology* LII (1987): 29–39.

31. G. F. Joyce, et al., *Nature* 310 (1984): 602–04.

32. D. Muller, et al., *Helvetica Chimica Acta* 73 (1990): 1410–68.

33. A. G. Cairns-Smith, *Genetic Takeover and the Mineral Origins of Life* (Cambridge, MA: Cambridge University Press, 1982).

34. A. G. Cairns-Smith and C. J. Davis, in *Encyclopaedia of Ignorance*, ed. A. Duncan and M. Weston-Smith (UK: Pergamon Press, 1977), pp. 397–403.

35. A. Eschenmoser, *Origins of Life and Evolution of the Biosphere: The Journal of the International Society for the Study of the Origin of Life* 27 (1997): 535–53.

36. M. Egholm, O. Buchardt, P. E. Nielsen, and R. H. Berg, *Journal of the American Chemical Society* 114 (1992): 1895–97.

37. M. Egholm, et al., *Nature* 365 (1993): 566–68.

38. J. G. Schmidt, L. Christensen, P. E. Nielsen, and L. E. Orgel, *Nucleic Acids Research* 25 (1997): 4792–96.

39. J. G. Schmidt, P. E. Nielsen, and L. E. Orgel, *Nucleic Acids Research* 25 (1997): 4797–4802.

40. M. Koppitz, P. E. Nielsen, and L. E. Orgel, *Journal of the American Chemical Society* 120 (1998): 4563–69.

41. S. A. Kauffman, *Journal of Theoretical Biology* 119 (1986): 1–24.

42. Wächtershäuser, *Microbiological Reviews* 52 (1988): 452–84.

43. C. De Duve, *Blueprint for a Cell: The Nature and Origin of Life* (N. Patterson, 1991).

EXPLAINING DESIGN

*WILLIAM PALEY**

NATURAL THEOLOGY

STATE OF THE ARGUMENT

In crossing a heath, suppose I pitched my foot against a stone, and were asked how the stone came to be there, I might possibly answer, that for anything I knew to the contrary it had lain there for ever; nor would it, perhaps, be very easy to show the absurdity of this answer. But suppose I had found a watch upon the ground, and it should be inquired how the watch happened to be in that place, I should hardly think of the answer which I had before given, that for anything I knew the watch might have always been there. Yet why should not this answer serve for the watch as well as for the stone; why is it not as admissible in the second case as in the first? For this reason, and for no other, namely, that when we come to inspect the watch, we perceive what we could not discover in the stone—that its several parts are framed and put together for a purpose, e.g. that they are so formed and adjusted as to produce motion, and that motion so regulated as to point out the hour of the day; that if the different parts had been differently shaped from what they are, or placed after any other manner or in any other order than that in which they are placed, either no motion at all would have been carried on in the machine, or none which would have answered the use that is now served by it. To reckon up a few of the plainest of these parts and of their offices, all tending to one result: We see a cylindrical box containing a coiled elastic spring, which, by its endeavor to relax itself, turns round the box. We next observe a flexible chain—artificially wrought for the sake of flexure—communicating the action of the spring from the box to the fusee. We then find a series of wheels, the teeth of which catch in and apply to each other, conducting the motion from the fusee to the balance and from the balance to the pointer, and at the same time, by the size and shape of those wheels, so regulating that motion as to terminate in causing an index, by an equable and measured progression, to pass over a given

*William Paley, *Natural Theology* (London: Rivington, 1802).

space in a given time. We take notice that the wheels are made of brass, in order to keep them from rust; the springs of steel, no other metal being so elastic; that over the face of the watch there is placed a glass, a material employed in no other part of the work, but in the room of which, if there had been any other than a transparent substance, the hour could not be seen without opening the case. This mechanism being observed—it requires indeed an examination of the instrument, and perhaps some previous knowledge of the subject, to perceive and understand it; but being once, as we have said, observed and understood, the inference we think is inevitable, that the watch must have had a maker—that there must have existed, at some time and at some place or other, an artificer or artificers who formed it for the purpose which we find it actually to answer, who comprehended its construction and designed its use.

Suppose, in the next place, that the person who found the watch should after some time discover, that in addition to all the properties which he had hitherto observed in it, it possessed the unexpected property of producing in the course of its movement another watch like itself—the thing is conceivable; that it contained within it a mechanism, a system of parts—a mold, for instance, or a complex adjustment of lathes, files, and other tools—evidently and separately calculated for this purpose; let us inquire what effect ought such a discovery to have upon his former conclusion.

I. The first effect would be to increase his admiration of the contrivance, and his conviction of the consummate skill of the contriver. Whether he regarded the object of the contrivance, the distinct apparatus, the intricate, yet in many parts intelligible mechanism by which it was carried on, he would perceive in this new observation nothing but an additional reason for doing what he had already done—for referring the construction of the watch to design and to supreme art. If that construction *without* this property, or which is the same thing, before this property had been noticed, proved intention and art to have been employed about it, still more strong would the proof appear when he came to the knowledge of this further property, the crown and perfection of all the rest.

II. He would reflect, that though the watch before him were *in some sense* the maker of the watch which was fabricated in the course of its movements, yet it was in a very different sense from that in which a carpenter, for instance, is the maker of a chair—the author of its contrivance, the cause of the relation of its parts to their use. With respect to these, the first watch was no cause at all to the second; in no such sense as this was it the author of the constitution and order, either of the parts which the new watch contained, or of the parts by the aid and instrumentality of which it was produced. We might possibly say, but with great latitude of expression, that a stream of water ground corn; but no latitude of expression would allow us to say, no stretch of conjecture could lead us to think, that the stream of water built the mill, though it were too ancient for us to know who the builder was. What the stream of water does in the affair is neither more nor less than this: by the application of an unintelligent impulse to a mechanism previously arranged, arranged independently of it and arranged by intelligence, an effect is produced, namely, the corn is ground. But the effect results from the arrangement. The force of the stream cannot be said to be the cause or the author of the effect, still less of the arrangement. Understanding and plan in the formation of the mill were not the less necessary for any share which the water has in grinding the corn; yet is this share the same as that which the watch would have contributed to the production of the new watch, upon the supposition assumed in the last section. Therefore,

III. Though it be now no longer probable that the individual watch which our observer had found was made immediately by the hand of an artificer, yet doth not this alteration in anywise affect the inference, that an artificer had been originally employed and concerned in the production. The argument from design remains as it was. Marks of design and contrivance are no more accounted for now than they were before. In the same thing, we may ask for the cause of different properties. We may ask for the cause of the color of a body, of its hardness, of its heat; and these causes may be all different. We are now asking for the cause of that subserviency to a use, that relation to an end, which we have remarked in the watch before us. No answer is given to this question, by telling us that a preceding watch produced it. There cannot be design without a designer; contrivance, without a contriver; order, without choice; arrangement, without any thing capable of arranging; subserviency and relation to a purpose, without that which could intend a purpose; means suitable to an end, and executing their office in accomplishing that end, without the end ever having been contemplated, or the means accommodated to it. Arrangement, disposition of parts, subserviency of means to an end, relation of instruments to a use, imply the presence of intelligence and mind. No one, therefore, can rationally believe that the insensible, inanimate watch, from which the watch before us issued, was the proper cause of the mechanism we so much admire in it—could be truly said to have constructed the instrument, disposed its parts, assigned their office, determined their order, action, and mutual dependency, combined their several motions into one result, and that also a result connected with the utilities of other beings. All these properties, therefore, are as much unaccounted for as they were before.

The conclusion which the first examination of the watch, of its works, construction, and movement, suggested, was, that it must have had, for cause and author of that construction, an artificer who understood its mechanism and designed its use. This conclusion is invincible. A second examination presents us with a new discovery. The watch is found, in the course of its movement, to produce another watch similar to itself and not only so, but we perceive in it a system or organization separately calculated for that purpose. What effect would this discovery have, or ought it to have, upon our former inference? What, as hath already been said, but to increase beyond measure our admiration of the skill which had been employed in the formation of such a machine? Or shall it, instead of this, all at once turn us round to an opposite conclusion, namely, that no art or skill whatever has been concerned in the business, although all other evidences of art and skill remain as they were, and this last and supreme piece of art be now added to the rest? Can this be maintained without absurdity? Yet this is atheism.

APPLICATION OF THE ARGUMENT

This is atheism; for every indication of contrivance, every manifestation of design which existed in the watch, exists in the works of nature, with the difference on the side of nature of being greater and more, and that in a degree which exceeds all computation. I mean, that the contrivances of nature surpass the contrivances of art, in the complexity, subtilty, and curiosity of the mechanism; and still more, if possible, do they go beyond them in number and variety; yet, in a multitude of cases, are not less evidently mechanical, not less evidently contrivances, not less evidently accommodated to their end or suited to their office, than are the most perfect productions of human ingenuity.

I know no better method of introducing so large a subject, than that of comparing a single thing with a single thing: an eye, for example, with a telescope. As far as the examination of the instrument goes, there is precisely the same proof that the eye was made for vision, as there is that the telescope was made for assisting it. They are made upon the same principles; both being adjusted to the laws by which the transmission and refraction of rays of light are regulated. I speak not of the origin of the laws themselves; but such laws being fixed, the construction in both cases is adapted to them. For instance, these laws require, in order to produce the same effect, that the rays of light, in passing from water into the eye, should be refracted by a more convex surface than when it passes out of air into the eye. Accordingly we find that the eye of a fish, in that part of it called the crystalline lens, is much rounder than the eye of terrestrial animals. What plainer manifestation of design can there be than this difference? What could a mathematical instrument maker have done more to show his knowledge of his principle, his application of that knowledge, his suiting of his means to his end—I will not say to display the compass or excellence of his skill and art, for in these all comparison is indecorous, but to testify counsel, choice, consideration, purpose?

*CHARLES DARWIN**

ORIGIN OF SPECIES

STRUGGLE FOR EXISTENCE

Before entering on the subject of this chapter, I must make a few preliminary remarks, to show how the struggle for existence bears on Natural Selection. It has been seen in the last chapter that among organic beings in a state of nature there is some individual variability; indeed I am not aware that this has ever been disputed. It is immaterial for us whether a multitude of doubtful forms be called species or sub-species or varieties; what rank, for instance, the two or three hundred doubtful forms of British plants are entitled to hold, if the existence of any well-marked varieties be admitted. But the mere existence of individual variability and of some few well-marked varieties, though necessary as the foundation for the work, helps us but little in understanding how species arise in nature. How have all those exquisite adaptations of one part of the organization to another part, and to the conditions of life, and of one distinct organic being to another being, been perfected? We see these beautiful co-adaptations most plainly in the woodpecker and missletoe; and only a little less plainly in the humblest parasite which clings to the hairs of a quadruped or feathers of a bird; in the structure of the beetle which dives through the water; in the plumed seed which is wafted by the gentlest breeze; in short, we see beautiful adaptations everywhere and in every part of the organic world.

Again, it may be asked, how is it that varieties, which I have called incipient species, become ultimately converted into good and distinct species, which in most cases obviously differ from each other far more than do the varieties of the same species? How do those groups of species, which constitute what are called distinct genera, and which differ from each other more than do the species of the same genus, arise? All these results, as we shall more fully see in the next chapter, follow inevitably from the struggle for life. Owing to this struggle for life, any variation, however slight and from whatever cause proceeding, if it be in any

---

*Charles Darwin, *Origin of Species* (London: John Murray, 1859), extracts from chapters 3 and 4, pp. 60–150.

degree profitable to an individual of any species, in its infinitely complex relations to other organic beings and to external nature, will tend to the preservation of that individual, and will generally be inherited by its offspring. The offspring, also, will thus have a better chance of surviving, for, of the many individuals of any species which are periodically born, but a small number can survive. I have called this principle, by which each slight variation, if useful, is preserved, by the term of Natural Selection, in order to mark its relation to man's power of selection. We have seen that man by selection can certainly produce great results, and can adapt organic beings to his own uses, through the accumulation of slight but useful variations, given to him by the hand of Nature. But Natural Selection, as we shall hereafter see, is a power incessantly ready for action, and is as immeasurably superior to man's feeble efforts, as the works of Nature are to those of Art.

We will now discuss in a little more detail the struggle for existence. In my future work this subject shall be treated, as it well deserves, at much greater length. The elder De Candolle and Lyell have largely and philosophically shown that all organic beings are exposed to severe competition. In regard to plants, no one has treated this subject with more spirit and ability than W. Herbert, Dean of Manchester, evidently the result of his great horticultural knowledge. Nothing is easier than to admit in words the truth of the universal struggle for life, or more difficult—at least I have found it so—than constantly to bear this conclusion in mind. Yet unless it be thoroughly engrained in the mind, I am convinced that the whole economy of nature, with every fact on distribution, rarity, abundance, extinction, and variation, will be dimly seen or quite misunderstood. We behold the face of nature bright with gladness, we often see superabundance of food; we do not see, or we forget, that the birds which are idly singing round us mostly live on insects or seeds, and are thus constantly destroying life; or we forget how largely these songsters, or their eggs, or their nestlings, are destroyed by birds and beasts of prey; we do not always bear in mind, that though food may be now superabundant, it is not so at all seasons of each recurring year.

I should premise that I use the term Struggle for Existence in a large and metaphorical sense, including dependence of one being on another, and including (which is more important) not only the life of the individual, but success in leaving progeny. Two canine animals in a time of dearth, may be truly said to struggle with each other which shall get food and live. But a plant on the edge of a desert is said to struggle for life against the drought, though more properly it should be said to be dependent on the moisture. A plant which annually produces a thousand seeds, of which on an average only one comes to maturity, may be more truly said to struggle with the plants of the same and other kinds which already clothe the ground. The missletoe is dependent on the apple and a few other trees, but can only in a far-fetched sense be said to struggle with these trees, for if too many of these parasites grow on the same tree, it will languish and die. But several seedling missletoes, growing close together on the same branch, may more truly be said to struggle with each other. As the missletoe is disseminated by birds, its existence depends on birds; and it may metaphorically be said to struggle with other fruit-bearing plants, in order to tempt birds to devour and thus disseminate its seeds rather than those of other plants. In these several senses, which pass into each other, I use for convenience sake the general term of struggle for existence.

A struggle for existence inevitably follows from the high rate at which all organic beings tend to increase. Every being, which during its natural lifetime produces several eggs or seeds, must suffer destruction during some period of its life, and during some season or occasional year, otherwise, on the principle of geometrical increase, its numbers would quickly become so inordinately great that no country could support the product. Hence, as more individuals are produced than can possibly survive, there must in every case be a struggle for existence, either one individual with another of the same species, or with the individuals of distinct species, or with the physical conditions of life. It is the doctrine of Malthus applied with manifold force to the whole animal and vegetable kingdoms; for in this case there can be no artificial increase of food, and no prudential restraint from marriage. Although some species may be now increasing, more or less rapidly, in numbers, all cannot do so, for the world would not hold them.

Summary of Chapter. If during the long course of ages and under varying conditions of life, organic beings vary at all in the several parts of their organisation, and I think this cannot be disputed; if there be, owing to the high geometrical powers of increase of each species, at some age, season, or year, a severe struggle for life, and this certainly cannot be disputed; then, considering the infinite complexity of the relations of all organic beings to each other and to their conditions of existence, causing an infinite diversity in structure, constitution, and habits, to be advantageous to them, I think it would be a most extraordinary fact if no variation ever had occurred useful to each being's own welfare, in the same way as so many variations have occurred useful to man. But if variations useful to any organic being do occur, assuredly individuals thus characterized will have the best chance of being preserved in the struggle for life; and from the strong principle of inheritance they will tend to produce offspring similarly characterized. This principle of preservation, I have called, for the sake of brevity, natural selection. Natural selection, on the principle of qualities being inherited at corresponding ages, can modify the egg, seed, or young, as easily as the adult. Among many animals, sexual selection will give its aid to ordinary selection, by assuring to the most vigorous and best adapted males the greatest number of offspring. Sexual selection will also give characters useful to the males alone, in their struggles with other males.

Whether natural selection has really thus acted in nature, in modifying and adapting the various forms of life to their several conditions and stations, must be judged of by the general tenor and balance of evidence given in the following chapters. But we already see how it entails extinction; and how largely extinction has acted in the world's history, geology plainly declares. Natural selection, also, leads to divergence of character; for more living beings can be supported on the same area the more they diverge in structure, habits, and constitution, of which we see proof by looking at the inhabitants of any small spot or at naturalized productions. Therefore during the modification of the descendants of any one species, and during the incessant struggle of all species to increase in numbers, the more diversified these descendants become, the better will be their chance of succeeding in the battle of life. Thus the small differences distinguishing varieties of the same species, will steadily tend to increase till they come to equal the greater differences between species of the same genus, or even of distinct genera.

We have seen that it is the common, the widely diffused, and widely ranging species, belonging to the larger genera, which vary most; and these will tend to transmit to their modified offspring that superiority which now makes them dominant in their own countries. Natural selection, as has just been remarked, leads to divergence of character and to much extinction of the less improved and intermediate forms of life. On these principles, I believe, the nature of the affinities of all organic beings may be explained. It is a truly wonderful fact—the wonder of which we are apt to overlook from familiarity—that all animals and all plants throughout all time and space should be related to each other in group subordinate to group, in the manner which we everywhere behold—namely, varieties of the same species most closely related together, species of the same genus less closely and unequally related together, forming sections and sub-genera, species of distinct genera much less closely related, and genera related in different degrees, forming sub-families, families, orders, sub-classes, and classes. The several subordinate groups in any class cannot be ranked in a single file, but seem rather to be clustered round points, and these round other points, and so on in almost endless cycles. On the view that each species has been independently created, I can see no explanation of this great fact in the classification of all organic beings, but, to the best of my judgment, it is explained through inheritance and the complex action of natural selection, entailing extinction and divergence of character, as we have seen illustrated in the diagram (see fig. 3.1).

The affinities of all the beings of the same class have sometimes been represented by a great tree. I believe this simile largely speaks the truth. The green and budding twigs may represent existing species; and those produced during each former year may represent the long succession of extinct species. At each period of growth all the growing twigs have tried to branch out on all sides, and to overtop and kill the surrounding twigs and branches, in the same manner as species and groups of species have tried to overmaster other species in the great battle for life. The limbs divided into great branches, and these into lesser and lesser branches, were themselves once, when the tree was small, budding twigs; and this connection of the former and present buds by ramifying branches may well represent the classification of all extinct and living species in groups subordinate to groups. Of the many twigs which flourished when the tree was a mere bush, only two or three, now grown into great branches, yet survive and bear all the other branches; so with the species which lived during long-past geological periods, very few now have living and modified descendants. From the first growth of the tree, many a limb and branch has decayed and dropped off; and these lost branches of various sizes may represent those whole orders, families, and genera which have now no living representatives, and which are known to us only from having been found in a fossil state. As we here and there see a thin straggling branch springing from a fork low down in a tree, and which by some chance has been favoured and is still alive on its summit, so we occasionally see an animal like the Ornithorhynchus or Lepidosiren, which in some small degree connects by its affinities two large branches of life, and which has apparently been saved from fatal competition by having inhabited a protected station. As buds give rise by growth to fresh buds, and these, if vigorous, branch out and overtop on all sides many a feebler branch, so by generation I believe it has been with the great Tree of Life, which fills with its dead and broken branches the crust of the earth, and covers the surface with its ever branching and beautiful ramifications.

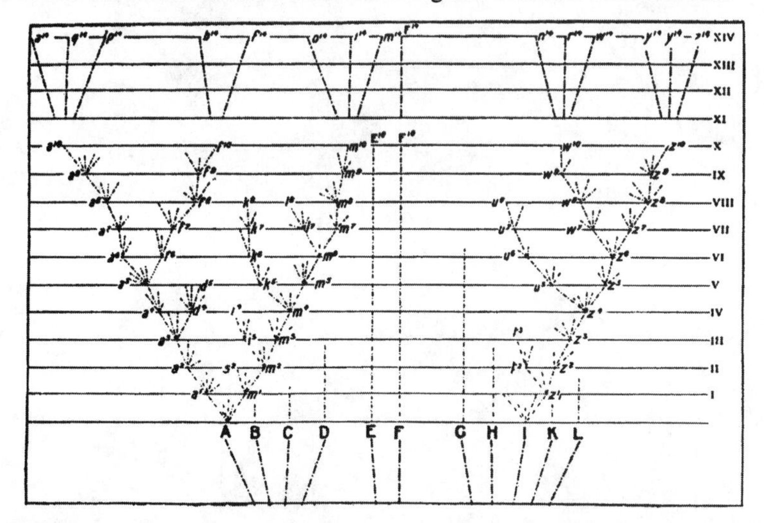

Figure 3.1. In the *Origin*, Darwin gave this figure showing how he thought of evolution as a (branching) tree of life.

MICHAEL DENTON*

BEYOND THE REACH OF CHANCE

He who believes that some ancient form was transformed suddenly . . . will further be compelled to believe that many structures beautifully adapted to all the other parts of the same creature and to the surrounding conditions, have been suddenly produced. . . . To admit all this is, as it seems to me, to enter into the realms of miracle, and to leave those of Science.

According to the central axiom of Darwinian theory, the initial elementary mutational changes upon which natural selection acts are entirely random, completely blind to whatever effect they may have on the function or structure of the organism in which they occur, "drawn," in Monod's words,[1] from "the realm of pure chance." It is only after an innovation has been disclosed by chance that it can be seen by natural selection and conserved.

Thus it follows that every adaptive advance, big or small, discovered during the course of evolution along every phylogenetic line must have been found as a result of what is in effect a purely random search strategy. The essential problem with this "gigantic lottery" conception of evolution is that all experience teaches that searching for solutions by purely random search procedures is hopelessly inefficient.

Consider first the difficulty of finding by chance English words within the infinite space of all possible combinations of letters. A section of this space would resemble the following block letters:

FLNWCYTQONMCDFEUTYLDWPQXCZNMIPQZXHCOT
IRJSALXMZVTNCTDHEKBUZRLHAJCFPTQOZPNOTJXD
WHYGCBZUDKGTWIBMZGPGLAOTDJZKXUEMWBCNX
YTKGHSBQJVUCPDLWKSMYJVGXUZIEMTJBYGLMPSJS
KFURYEBWNQPCLXKZUFMTYBUDISTABWNCPDOMS
MXKALQJAUWNSPDYSHXMCKFLQHAVCPDYRTSIZSJR
YFMAHZLVPRITMGYGBFMDLEPE

*From Michael Denton, *Evolution: A Theory in Crisis* (London: Burnett Books, 1985), chapter 13.

Within the total letter space would occur every single English word and every single English sentence and indeed every single English book that has been or will ever be written. But most of the space would consist of an infinity of pure gibberish.

Simple three-letter English words would be relatively common. There are 26^3 = 17,000 combinations three letters long, and, as there are about five hundred three-letter words in English, then about one in thirty combinations will be a three-letter word. All other three-letter combinations are nonsense. To find by chance three-letter words, eg "not," "bud," "hut," would be a relatively simple task necessitating a search through a string of only about thirty or so letters.

Because three letter words are so probable it is very easy to go from one three letter word to another by making random changes to the letter string. In the case of the word "hat" for example, by randomly substituting letters in the position occupied by h in the word we soon hit on a new three letter word:

hat
aat
bat
cat
dat
eat
fat

Thus not only is it possible to find three-letter English words by chance but because the probability gaps between them are small, it is easy to transform any word we find into a quite different word through a sequence of probable intermediates:

$$\text{hat} \rightarrow \text{cat} \rightarrow \text{can} \rightarrow \text{tan} \rightarrow \text{tin}$$

However, to find by chance longer words, say seven letters long such as "English" or "require," would necessitate a vastly increased search. There are 26^7 or 10^9, that is, one thousand million combinations of letters seven letters long. As there are certainly less than ten thousand English words seven letters long, then to find one by chance we would have to search through letter strings in the order of one hundred thousand units long. Twelve-letter words such as "construction" or "unreasonable" are so rare that they occur only once by chance in strings of letters 10^{14} units long; as there are about 10^{14} minutes in one thousand million years one can imagine how long a monkey at a typewriter would take to type out by chance one English word twelve letters long. Intuitively it seems unbelievable that such apparently simple entities as twelve-letter English words could be so rare, so inaccessible to a random search.

The problem of finding words by chance arises essentially because the space of all possible letter combinations is immense and the overwhelming majority are complete nonsense; consequently meaningful sequences are very rare and the probability of hitting one by chance is exceedingly small.

Moreover, even if by some lucky fluke we were to find, say, one twelve-letter word by chance, because each word is so utterly isolated in a vast ocean of nonsense strings it is very difficult to get another meaningful letter string by randomly substituting letters and testing each new string to see if it forms an English word. Take the word "unreasonable." There are a few closely related words such as "reasonable," "reason," "season," "treason," or "able," which can be reached by making changes to the letter sequence but the necessary letter changes are unfortunately highly specific and finding them by chance involves a far longer and more difficult search than was the case with three-letter words.

Sentences, of course, even short ones, are even rarer and long sentences rare almost beyond imagination. Linguists have estimated a total of 10^{25} possible English sentences one hundred letters long, but as there are a total of 26^{100} or 10^{130} possible sequences one hundred letters long, then less than one in about 10^{100} will be an English sentence. The figure of 10^{100} is beyond comprehension—some idea of the immensity it represents can be grasped by recalling that there are only 10^{70} atoms in the entire observable universe.

Each English sentence is a complex system of letters which are integrated together in highly specific ways: firstly into words, then into word phrases, and finally into sentences. If the subsystems are all to be combined in such a way that they will form a grammatical English sentence then their integration must follow rigorously the a priori rules of English grammar. For example, one of the rules is that the letters in the sentence must be combined in such a way that they form words belonging to the lexicon of the English language.

However, random strings of English words, eg "horse," "cog," "blue," "fly," "extraordinary," do not form sentences because there exists a further set of rules—the rules of syntax which dictate, among other things, that a sentence must possess a subjective and a verbal clause.

On top of this there exists a further set of rules which governs the semantic relationship of the components of a sentence. Obviously, not all strings of English words which are arranged correctly according to the rules of English syntax are meaningful. For example: "The raid (subject) ate (verb) the sky (object)." Each word is from the English lexicon and their arrangement satisfies the rules of syntax. However, the sequence disobeys the rules of semantics and is as nonsensical as a completely random string of letters.

The rules of English grammar are so stringent that only highly specialized letter combinations can form grammatical sentences and consequently, because of the immensity of the space of all possible letter combinations, such highly specialized strings are utterly lost within it, infinitely rare and isolated, absolutely beyond the reach of any sort of random search that could be conceivably carried out in a finite time even with the most advanced computers on earth. Moreover, because sentences are so rare and isolated, even if one was discovered by chance the probability gap between it and the nearest related sentence is so immeasurable that no conceivable sort of random change to the letter or word sequence will ever carry us across the gap.

Consider the sentence: "Because of the complexity of the rules of English grammar most English sentences are completely isolated." If we set out to reach another sentence by randomly substituting a new word in place of an existing word and then testing the newly created sequence of words to see if it made a grammatical sentence, we would find very few substitutions were grammatically acceptable and even to find *one* grammatical substitution would take an unbelievably long time if we searched by pure chance. Some of the few grammatical substitutions are shown below:

because	English
of	grammar
the	most → all → some
complexity → nature	english
of	sentences
the	are
rules → algorithms	completely → totally → invariably → always
of	isolated → immutable

There are about 10^5 words in the lexicon of the English language and, as there are sixteen words in the above sentence, we would have 1.6×10^6 possibilities to test. If there are, say, two hundred individual words out of the 1.6×100^6 which can be substituted grammatically, we would have to test about eight thousand words on average before we found a grammatical substitution.

Testing one new word per minute, it would take us five days working day and night to find by chance our first grammatical substitution, and to test all the possible words in every position in the sentence would take about three years, and after three years of searching all we would have achieved would be a handful of sentences closely related to the one with which we started.

Sentences are not the only complex systems which are beyond the reach of chance. The same principles apply, for example, to watches, which are also highly improbable, and where consequently each different functional watch is intensely isolated by immense probability gaps from its nearest neighbors.

To see why, we must begin by trying to envisage a universe of mechanical objects containing all possible combinations of watch components: springs, gears, levers, cogwheels, each of every conceivable size and shape. Such a universe would contain every functional watch that has ever existed on earth and every functional watch that could possibly exist at any time in the future. Although we cannot in this case calculate the rarity of functional combinations (watches that work) as we could in the case of words and sentences, common sense tells us that they would be exceedingly rare. Our imaginary universe would mostly consist of combinations of gears and cogwheels which would be entirely useless; each functional watch would, like a meaningful sentence, be an isolated island separated from all other islands of function by a surrounding infinity of junk composed only of incoherent and functionless combinations.

Again, as with sentences, because the total number of incoherent nonsense combinations of components vastly exceeds, by an almost inconceivable amount, the tiny fraction which can form coherent combinations, function is exceedingly rare. If we were to look by chance for a functional watch we would have to search for an eternity amid an infinity of combinations until we hit upon a functional watch.

The basic reason why functional watches are so exceedingly improbable is because, to be functional, a combination of watch components must satisfy a number of very stringent criteria (equivalent to the rules of grammar), and these can only be satisfied by highly specialized unique combinations of components which are coadapted to function together. One rule might be that all cogwheels must possess perfect regularly shaped cogs; another rule might be that all the cogs must fit together to allow rotation of one wheel to be transmitted throughout the system.

It is obviously impossible to contemplate using a random search to find combinations which will satisfy the stringent criteria which govern functionality in watches. Yet, just as a speaker of a language cognizant with the rules of grammar can generate a functional sentence with great ease, so too a watchmaker has little trouble in assembling a watch by following the rules which govern functionality in combinations of watch components.

What is true of sentences and watches is also true of computer programs, airplane engines, and in fact of all known complex systems. Almost invariably, function is restricted to unique and fantastically improbable combinations of subsystems, tiny islands of meaning lost in an infinite sea of incoherence. Because the number of nonsense combinations of component subsystems vastly exceeds by unimaginable orders of magnitude the infinitesimal fraction of combinations in which the components are capable of undergoing coherent or meaningful interactions. Whether we are searching for a functional sentence or a functional watch or the best move in a game of chess, the goals of our search are in each case so far lost in an infinite space of possibilities that, unless we guide our search by the use of algorithms which direct us to very specific regions of the space, there is no realistic possibility of success.

Discussing a well-known checker-playing program, Professor Marvin Minsky of the Massachusetts Institute of Technology comments:[2]

> This game exemplifies the fact that many problems can in principle be solved by trying all possibilities—in this case exploring all possible moves, all the opponent's replies all the player's possible replies to the opponent's replies and so on. If this could be done, the player could see which move has the best chance of winning. In practice, however, this approach is out of the question, even for a computer; the tracking down of every possible line of play would involve some 10^{40} different board positions. A similar analysis for the game of chess would call for some 10^{120} positions. Most interesting problems present far too many possibilities for complete trial and error analysis.

Nevertheless, as he continues, a computer can play checkers if it is capable of making intelligent limited searches:

> Instead of tracking down every possible line of play the program uses a partial analysis (a "static evaluation") of a relatively small number of carefully selected features of a board position—how many men there are on each side, how advanced they are and certain other simple relations. This incomplete analysis is not in itself adequate for choosing the best move for a player in a current position. By combining the partial analysis with a limited search for some of the consequences of the possible moves from the current position, however, the program selects its move as if on the basis of a much deeper analysis The program contains a collection of rules for deciding when to continue the search and when to stop. When it stops, it assesses the merits of the "terminal position" in terms of the static evaluation. If the computer finds by this search that a given move leads to an advantage for the player in all the likely positions that may occur a few moves later, whatever the opponent does, it can select this move with confidence.

The inability of unguided trial and error to reach anything but the most trivial of ends in almost every field of interest obviously raises doubts as to its validity in the biological realm. Such doubts were recently raised by a number of mathematicians and engineers at an international symposium entitled "Mathematical Challenges to the Neo-Darwinian Interpretation of Evolution,"[3] a meeting which also included many leading evolutionary biologists. The major argument presented was that Darwinian evolution by natural selection is merely a special case of the general procedure of problem solving by trial and error. Unfortunately, as the mathematicians present at the symposium such as Schutzenberger and Professor Eden from MIT pointed out, trial and error is totally inadequate as a problem solving technique without the guidance of specific algorithms, which has led to the consequent failure to simulate Darwinian evolution by computer analogues. For similar reasons, the biophysicist Pattee has voiced skepticism over natural selection at many leading symposia over the past two decades. At one meeting entitled "Natural Automata and Useful Simulations," he made the point:[4]

> Even some of the simplest artificial adaptive problems and learning games appear practically insolvable even by multistage evolutionary strategies.

Living organisms are complex systems, analogous in many ways to non-living complex systems. Their design is stored and specified in a linear sequence of symbols, analogous to coded information in a computer program. Like any other system, organisms consist of a number of subsystems which are all coadapted to interact together in a coherent manner: molecules are assembled into multimolecular systems, multimolecular assemblies are combined into cells, cells into organs and organ systems finally into the complete organism. It is hard to believe that the fraction of meaningless combinations of molecules, of cells, of organ systems, would not vastly exceed the tiny fraction that can be combined to form assemblages capable of exhibiting coherent interactions. Is it really possible that the criteria for function which must be satisfied in the design of living systems are at every level far less stringent than those which must be satisfied in the design of functional watches, sentences or computer programs? Is it possible to design an automaton to construct an object like the human brain, laying down billions of specific connections, without having to satisfy criteria every bit as exacting and restricting as those which must be met in other areas of engineering?

Given the close analogy between living systems and machines, particularly at a molecular level, there cannot be any objective basis to the assumption that functional organic systems are likely to be less isolated or any easier to find by chance. Surely it is far more likely that functional combinations in the space of all organic possibilities are just as isolated, just as rare and improbable, just as inaccessible to a random search and just as functionally immutable by any sort of random process. The only warrant for believing that functional living systems are probable, capable of undergoing functional transformation by random mechanisms, is belief in evolution by the natural selection of purely random changes in the structure of living things. But this is precisely the question at issue.

If complex computer programs cannot be changed by random mechanisms, then surely the same must apply to the genetic programs of living organisms. The fact that systems in every way analogous to living organisms cannot undergo evolution by pure trial and error and that their functional distribution invariably conforms to an improbable discontinuum comes, in my opinion, very close to a formal disproof of the whole Darwinian paradigm of nature. By what strange capacity do living organisms defy the laws of chance which are apparently obeyed by all analogous complex systems?

We now have machines which exhibit many properties of living systems. Work on artificial intelligence has advanced and the possibility of constructing a self-reproducing machine was discussed by the mathematician von Neumann in his now famous *Theory of Self-Reproducing Automata*.[5] Although some advanced machines can solve simple problems, none of them can undergo evolution by the selection of random changes in their structure without the guidance of already existing programs. The only sort of machine that might, at some future date, undergo some sort of evolution would be one exhibiting artificial intelligence. Such a machine would be capable of altering its own organization in an intelligent way. However evolution of this sort would be more akin to Lamarckian, but by no stretch of the imagination could it be considered Darwinian. The construction of a self-evolving intelligent machine would only serve to underline the insufficiency of unguided trial and error as a causal mechanism of evolution.

It was the close analogy between living systems and complex machines and the impossibility of envisaging how objects could have been assembled by chance that led the natural theologians of the eighteenth and early nineteenth centuries to reject as inconceivable the possibility that chance would have played any role in the origin of the complex adaptations of living things. William Paley, in his classic analogy between an organism and a watch, makes precisely this point:[6]

> Nor would any man in his senses think the existence of the watch, with its various machinery, accounted for, by being told that it was one out of possible combinations of material forms; that whatever he had found in the place where he found the watch, must have contained some internal configuration or other; and that this configuration might be the structure now exhibited, viz, of the works of a watch, as well as a different structure.

It is true that some authorities have seen an analogy to evolution by natural selection in gradual technological advances. Jukes, for instance, in a recent letter to *Nature* drew an analogy between the evolution of the Boeing 747 from Bleriots' 1909 monoplane through the Boeing Clippers in the 1930s to the first Boeing jet airliner, the 707, which started in service in 1959 and which was the immediate predecessor of the 747s, and biological evolution. In his words:[7]

> The brief history of aircraft technology is filled with branching processes, phylogeny and extinctions that are a striking counterpart of three billion years of biological evolution.

Unfortunately, the analogy is false. At no stage during the history of the aviation industry was the design of any flying machine achieved by chance, but only by the most rigorous applications of all the rules which govern function in the field of aerodynamics. It is true, as Jukes states, that "wide-bodied jets evolved from small contraptions made in bicycle shops, or in junkyards," but they did not evolve by chance.

There is no way that a purely random search could ever have discovered the design of an aerodynamically feasible flying machine from a random assortment of mechanical components—again, the space of all possibilities is inconceivably large. All such analogies are false because in all such cases the search for function is intelligently guided. It cannot be stressed enough that evolution by natural selection is analogous to problem solving without any intelligent guidance, without any intelligent input whatsoever. No activity which involves an intelligent input can possibly be analogous to evolution by natural selection.

The above discussion highlights one of the fundamental flaws in many of the arguments put forward by defenders of the role of chance in evolution. Most of the classic arguments put forward by leading Darwinists, such as the geneticist H. J. Muller and many other authorities including C. G. Simpson, in defence of natural selection make the implicit assumption that islands of function are common, easily found by chance in the first place, and that it is easy to go from island to island through functional intermediates.

This is how Simpson, for example, envisages evolution by natural selection:[8]

> How natural selection works as a creative process can perhaps best be explained by a very much oversimplified analogy. Suppose that from a pool of all the letters of the alphabet in large, equal abundance you tried to draw simultaneously the letters *c*, *a*, and *t*, in order to achieve a purposeful combination of these into the word "cat." Drawing out three letters at a time and then discarding them if they did not form this useful combination, you obviously would have very little chance of achieving your purpose. You might spend days, weeks, or even years at your task before you finally succeeded. The possible number of combinations of three let-

> ters is very large and only one of these is suitable for your purpose. Indeed, you might well never succeed, because you might have drawn all the *c*'s, *a*'s, or *t*'s in wrong combinations and have discarded them before you succeeded in drawing all three together. But now suppose that every time you draw a *c*, an *a*, or a *t* in a wrong combination, you are allowed to put these desirable letters back in the pool and to discard the undesirable letters. Now you are sure of obtaining your result, and your chances of obtaining it quickly are much improved. In time there will be only *c*'s, *a*'s, and *t*'s in the pool, but you probably will have succeeded long before that. Now suppose that in addition to returning *c*'s, *a*'s, and *t*'s to the pool and discarding all other letters, you are allowed to clip together any two of the desirable letters when you happen to draw them at the same time. You will shortly have in the pool a large number of clipped *ca*, *ct*, and *at* combinations plus an also large number of the *t*'s, *a*'s, and *c*'s needed to complete one of these if it is drawn again. Your chances of quickly obtaining the desired result are improved still more, and by these processes you have "generated a high degree of improbability"—you have made it probable that you will quickly achieve the combination *cat*, which was so improbable at the outset. Moreover, you have created something. You did not create the letters *c*, *a*, and *t*, but you have created the word "cat," which did not exist when you started.

The obvious difficulty with the whole scheme is that Simpson assumes that finding islands of function in the first place (the individual letters *c*, *a*, and *t*) is highly probable and that the functional island "cat" is connected to the individual letter islands by intermediate functional islands *ca*, *ac*, *ct*, *at*, *ta*, *fc*, so that we can cross from letters to islands by natural selection in unit mutational steps. In other words, Simpson has assumed that islands of function are very probable, but this is the very assumption which must be proved to show that natural selection would work.

Obviously, if islands of function in the space of all organic possibilities are common, like three or four letter words, then of course functional biological systems will be within the reach of chance; and because the probability gaps will be small, random mutations will easily find a way across. However, as is evident from the above discussion, Simpson's scheme, and indeed the whole Darwinian framework, collapse completely if islands of function are like twelve-letter words or English sentences.

These considerations of the likely rarity and isolation of functional systems within their respective total combinatorial spaces also reveal the fallacy of the current fashion of turning to saltational models of evolution to escape the impasse of gradualism. For as we have seen, in the case of every kind of complex functional system the total space of all combinatorial possibilities is so nearly infinite and the isolation of meaningful systems so intense, that it would truly be a miracle to find one by chance Darwin s rejection of chance saltations as a route to new adaptive innovations is surely right. For the combinatorial space of all organic possibilities is bound to be so great that the probability of a sudden macromutational event transforming some existing structure or converting de novo some redundant feature into a novel adaptation exhibiting that perfection of coadaptation in all its component parts so obvious in systems like the feather, the eye or the genetic code and which is necessarily ubiquitous in the design of all complex functional systems biological or artifactual, is bound to be vanishingly small. Ironically, in any combinatorial space, it is the very same restrictive criteria of function which prevent gradual functional change which also isolate all functional systems to vastly improbable and inaccessible regions of the space.

To determine, finally, whether the distribution of islands of function in organic nature conforms to a probable continuum or an improbable discontinuum and to assess definitively the relevance of chance in evolution would be a colossal task. Just as in the case of the sentences and watches we would have to begin by constructing a multi-dimensional universe filled with all possible combinations of organic chemicals. Within this space of all possibilities there would exist every conceivable functional biological system, including not only those which exist on earth, but all other functional biological systems which could possibly work elsewhere in the universe. The functional systems would range from simple protein molecules capable of particular catalytic functions right up to immensely complex systems such as the human brain. Within this universe of all possibilities we would find many strange biological systems, such as enzymes capable of transforming unique substrates not found on earth, and perhaps nervous systems resembling those found among vertebrates on earth, but far more advanced. We would also find many sorts of complex aggregates, the function of which may not be clear but which we could dimly conceive as being of some value on some alien planet.

Such a space would of course consist mainly of combinations which would have no conceivable biological function—merely junk aggregates ranging from functionless proteins to entirely disordered nervous systems reminiscent of Cuvier's incompatibilities. From the space we would be able to calculate exactly how probable functional biological systems are and how easy it is to go from one functional system to another, Darwinian fashion, in a series of unit mutational steps through functional intermediates. Of course if analogy is any guide then the space would in all probability conjure up a vision of nature more in harmony with the thinking of Cuvier and the early nineteenth-century typology than modern evolutionary thought in which each island of meaning is intensely isolated unlinked by transitional forms and quite beyond the reach of chance.

At present we are very far from being able to construct such a space of all organic possibilities and to calculate the probability of functional living systems. Nevertheless, for some of the lower order functional systems, such as individual proteins, their rarity in the space can be at least tentatively assessed.

A protein . . . is fundamentally a long chain-like molecule built up out of twenty different kinds of amino acids. After its assembly the long amino acid chain automatically folds into a specific stable 3D configuration. Particular protein functions depend on highly specific 3D shapes and, in the case of proteins which possess catalytic functions, depend on the protein possessing a particular active site, again of highly specific 3D configuration.

Although the exact degree of isolation and rarity of functional proteins is controversial it is now generally conceded by protein chemists that most functional proteins would be difficult to reach or to interconvert through a series of successive individual amino acid mutations. Zuckerkandl comments:[9]

> Although, abstractly speaking, any polypeptide chain can be transformed into any other by successive amino acid substitutions and other mutational events, in concrete situations the pathways between a poorly and a highly adapted molecule will be mostly impracticable. Any such pathway, whether the theoretically shortest, or whether a longer one, will perforce include stages of favorable change as well as hurdles. Of the latter some will be surmountable and some not. Some of the latter will presumably be present along the pathways of adaptive change in a very large majority of ill adapted de novo polypeptide chains.
>
> Consequently, when a protein molecule is selected for its weak enzymatic activity and in spite of limited substrate specificity, it will most often represent a dead end road.

The impossibility of gradual functional transformation is virtually self-evident in the case of proteins: mere casual observation reveals that a protein is an interacting whole, the function of every amino acid being more or less (like letters in a sentence or cogwheels in a watch) essential to the function of the entire system. To change, for example, the shape and function of the active site (like changing the verb in a sentence or an important cogwheel in a watch) in isolation would be bound to disrupt all the complex intramolecular bonds throughout the molecule, destabilizing the whole system and rendering it useless. Recent experimental studies of enzyme evolution largely support this view, revealing that proteins are indeed like sentences, and are only capable of undergoing limited degrees of functional change through a succession of individual amino acid replacements. The general consensus of opinion in this field is that significant functional modification of a protein would require several simultaneous amino acid replacements of a relatively improbable nature. The likely impossibility of major functional transformation through individual amino acid steps was raised by Brian Hartley, a specialist in this area, in an article in the journal *Nature* in 1974. From consideration of the atomic structure of a family of closely related proteins which, however, have different amino acid arrangements in the central region of the molecule, he concluded that their functional interconversion would be impossible:[10]

> It is hard to see how these alternative arrangements could have evolved without going through an intermediate that could not fold correctly (i.e. would be non functional).

Here then, is at least one functional subset of the space of all organic possibilities which almost certainly conforms to the general discontinuous pattern observed in the case of other complex systems. But how discontinuous is the pattern of the distribution of proteins within the space of all organic possibilities? Might functional proteins be beyond the reach of chance?

In attempting to answer the question—how rare are functional proteins?—we must first decide what general restrictions must be imposed on a sequence of amino acids before it can form a biologically functional protein. In other words, what are the rules or criteria which govern functionality in an amino acid sequence?

First, a protein must be a stable structure so that it can hold a particular 3D shape for a sufficiently long period to allow it to undergo a specific interaction with some other entity in the cell. Second, a protein must be able to fold into its proper shape. Third, if a protein is to possess catalytic properties it must have an active site which necessitates a highly specific arrangement of atoms in some region of its surface to form this site.

From the tremendous advances that have been made over the past two decades in our knowledge of protein structure and function, there are compelling reasons for believing that these criteria for function would inevitably impose severe limitations on the choice of amino acids. It is very difficult to believe that the criteria for stability and for a folding algorithm would not require a relatively severe restriction of choice in at least twenty percent of the amino acid chain. To get the precise atomic 3D shape or active sites may well require an absolute restriction in between one and five percent of the amino acid sequence.

There is a considerable amount of empirical evidence for believing that the criteria for function must be relatively stringent. One line of evidence, for example, is the very strict conservation of overall shape and the exact preservation of the configuration of active sites in homologous proteins such as the cytochromes in very diverse species. Further evidence suggests that most mutations which cause changes in the amino acid sequence of proteins tend to damage function to a greater or lesser degree. The effects of such mutations have been carefully documented in the case of hemoglobin, and some of them were described in an excellent article in *Nature* by Max Perutz,[11] who himself pioneered the X-ray crystallographic work which first revealed the detailed 3D structure of proteins. As Perutz shows, although many of the amino acids occupying positions on the surface of the molecule can be changed with little effect on function, most of the amino acids in the centre of the protein cannot be changed without having drastic deleterious effects on the stability and function of the molecule.

There are, in fact, both theoretical and empirical grounds for believing that the a priori rules which govern function in an amino acid sequence are relatively stringent. If this is the case, and all the evidence points to this direction, it would mean that functional proteins could well be exceedingly rare. The space of all possible amino acid sequences (as with letter sequences) is unimaginably large and consequently sequences which must obey particular restrictions which can be defined, like the rules of grammar, are bound to be fantastically rare. Even short unique sequences just ten amino acids long only occur once by chance in about 10^{13} average-sized proteins; unique sequences twenty amino acids long once in about 10^{26} proteins; and unique sequences thirty amino acids long once in about 10^{39} proteins!

As it can easily be shown that no more than 10^{40} possible proteins could have ever existed on earth since its formation, this means that, if proteins, functions reside in sequences any less probable than 10^{-40}, it becomes increasingly unlikely that any functional proteins could ever have been discovered by chance on earth.

We have seen . . . that envisaging how a living cell could have gradually evolved through a sequence of simple protocells seems to pose almost insuperable problems. If the estimates above are anywhere near the truth then this would undoubtedly mean, that the alternative scenario—the possibility of life arising suddenly on earth by chance—is infinitely small. To get a cell by chance would require at least one hundred functional proteins to appear simultaneously in one place. That is one hundred simultaneous events each of an independent probability which could hardly be more than 10^{-20} giving a maximum combined probability of 10^{-2000}. Recently, Hoyle and Wickramasinghe in *Evolution from Space* provided a similar estimate of the chance of life originating, assuming functional proteins to have a probability of 10^{-20}:[12]

> By itself, this small probability could be faced, because one must contemplate not just a single shot at obtaining the enzyme, but a very large number of trials

> such as are supposed to have occurred in an organic soup early in the history of the Earth. The trouble is that there are about two thousand enzymes, and the chance of obtaining them all in a random trial is only one part in $(10^{20})^{2000} = 10^{40,000}$ an outrageously small probability that could not be faced even if the whole universe consisted of organic soup.

Although at present we still have insufficient knowledge of the rules which govern function in amino acid sequences to calculate with any degree of certainty the actual rarity of functional proteins, it may be that before long quite rigorous estimates may be possible. Over the next few decades advances in molecular biology are inevitably going to reveal in great detail many more of the principles and rules which govern the function and structure of protein molecules. In fact, by the end of the century, molecular engineers may be capable of specifying quite new types of functional proteins. From the first tentative steps in this direction it already seems that, in the design of new functional proteins, chance will play as peripheral a role as it does in any other area of engineering.[13]

The Darwinian claim that all the adaptive design of nature has resulted from a random search, a mechanism unable to find the best solution in a game of checkers, is one of the most daring claims in the history of science. But it is also one of the least substantiated. No evolutionary biologist has ever produced any quantitive proof that the designs of nature are in fact within the reach of chance. There is not the slightest justification for claiming, as did Richard Dawkins recently:[14]

> . . . Charles Darwin showed how it is possible for blind physical forces to mimic the effects of conscious design, and, by operating as a cumulative filter of chance variations, to lead eventually to organised and adaptive complexity, to mosquitoes and mammoths, to humans and therefore, indirectly, to books and computers.

Neither Darwin, Dawkins nor any other biologist has ever calculated the probability of a random search finding in the finite time available the sorts of complex systems which are so ubiquitous in nature. Even today we have no way of rigorously estimating the probability or degree of isolation of even one functional protein. It is surely a little premature to claim that random processes could have assembled mosquitoes and elephants when we still have to determine the actual probability of the discovery by chance of one single functional protein molecule!

NOTES

1. J. Monod, *Chance and Necessity* (London: Collins, 1972), p. 114.

2. M. Minsky, "Artificial Intelligence," *Scientific American* 215 (September 1966): 246–60. See ibid., pp. 247–48.

3. P. S. Moorhead and M. M. Kaplan, eds., "Mathematical Challenges to the Darwinian Interpretation of Evolution," *Wistar Institute Symposium Monograph* (1967).

4. H. H. Pattee, "Introduction to Session One," in *Natural Automata and Useful Simulations*, ed. H. H. Pattee et. al (Washington: Spartan Books, 1966), pp. 1–2.

5. J. Von Neumann, *Theory of Self-Reproducing Automata* (Champaign: University of Illinois Press, 1966).

6. W. Paley, *Natural Theology on Evidence and Attributes of Deity*, 18th ed. (Edinburgh, UK: Lackington, Allen, and Co. and James Sawers, 1818), p. 13.

7. T. H. Jukes, "Aircraft Evolution," *Nature* 295 (1982): 548.

8. C. C. Simpson, "The Problem of Plan and Purpose in Nature," *Scientific Monthly* 64 (1947): 481–95. See ibid., p. 493.

9. E. Zuckerkandl, "The Appearance of New Structures in Proteins During Evolution," *Journal of Molecular Evolution* 7 (1975): 1–57. See ibid., p. 21.

10. P. W. J. Rigby, B. D. Burleigh Jr., and B. S. Hartley, "Gene Duplication in Experimental Enzyme Evolution," *Nature* 251 (1974): 200–204. See ibid., p. 200.

11. M. F. Perutz and H. Lehmann, "Molecular Pathology of Human Haemoglobin," *Nature* 219 (1968): 902–909.

12. F. Hoyle and C. Wickramasinghe, *Evolution from Space* (London: J. M. Dent and Sons, 1981), p. 24.

13. C. Paba, "Designing Proteins and Peptides," *Nature* 301 (1983): 200.

14. R. Dawkins, "The Necessity of Darwinism," *New Scientist* 94 (April 15, 1982): 130–32. See ibid., p. 130.

Editor's note: The reference on p. 103 is to the early nineteenth century French biologist Georges Curvier, who denied the very possibility of evolution on the grounds that transitional organisms would simply be functionally inadequate. A reptile-bird intermediate, for instance, would supposedly fail both as a reptile and as a bird.

RICHARD DAWKINS*

ACCUMULATING SMALL CHANGE

We have seen that living things are too improbable and too beautifully "designed" to have come into existence by chance. How, then, did they come into existence? The answer, Darwin's answer, is by gradual, step-by-step transformations from simple beginnings, from primordial entities sufficiently simple to have come into existence by chance. Each successive change in the gradual evolutionary process was simple enough, *relative to its predecessor*, to have arisen by chance. But the whole sequence of cumulative steps constitutes anything but a chance process, when you consider the complexity of the final end-product relative to the original starting point. The cumulative process is directed by nonrandom survival. The purpose of this chapter is to demonstrate the power of this *cumulative selection* as a fundamentally nonrandom process.

If you walk up and down a pebbly beach, you will notice that the pebbles are not arranged at random. The smaller pebbles typically tend to be found in segregated zones running along the length of the beach, the larger ones in different zones or stripes. The pebbles have been sorted, arranged, selected. A tribe living near the shore might wonder at this evidence of sorting or arrangement in the world, and might develop a myth to account for it, perhaps attributing it to a Great Spirit in the sky with a tidy mind and a sense of order. We might give a superior smile at such a superstitious notion, and explain that the arranging was really done by the blind forces of physics, in this case the action of waves. The waves have no purposes and no intentions, no tidy mind, no mind at all. They just energetically throw the pebbles around, and big pebbles and small pebbles respond differently to this treatment so they end up at different levels of the beach. A small amount of order has come out of disorder, and no mind planned it.

---

*From *The Blind Watchmaker: Why the Evidence of Evolution Reveals a Universe Without Design* by Richard Dawkins.

The waves and the pebbles together constitute a simple example of a system that automatically generates non-randomness. The world is full of such systems. The simplest example I can think of is a hole. Only objects smaller than the hole can pass through it. This means that if you start with a random collection of objects above the hole, and some force shakes and jostles them about at random, after a while the objects above and below the hole will come to be nonrandomly sorted. The space below the hole will tend to contain objects smaller than the hole, and the space above will tend to contain objects larger than the hole. Mankind has, of course, long exploited this simple principle for generating non-randomness, in the useful device known as the sieve.

The Solar System is a stable arrangement of planets, comets, and debris orbiting the sun, and it is presumably one of many such orbiting systems in the universe. The nearer a satellite is to its sun, the faster it has to travel if it is to counter the sun's gravity and remain in stable orbit. For any given orbit, there is only one speed at which a satellite can travel and remain in that orbit. If it were travelling at any other velocity, it would either move out into deep space, or crash into the Sun, or move into another orbit. And if we look at the planets of our solar system, lo and behold, every single one of them is traveling at exactly the right velocity to keep it in its stable orbit around the sun. A blessed miracle of provident design? No, just another natural "sieve." Obviously all the planets that we see orbiting the sun must be traveling at exactly the right speed to keep them in their orbits, or we wouldn't see them there because they wouldn't be there! But equally obviously this is not evidence for conscious design. It is just another kind of sieve.

Sieving of this order of simplicity is not, on its own, enough to account for the massive amounts of nonrandom order that we see in living things. Nowhere near enough. [Think of] the analogy of the combination lock. The kind of non-randomness that can be generated by simple sieving is roughly equivalent to opening a combination lock with one dial: it is easy to open it by sheer luck. The kind of non-randomness that we see in living systems, on the other hand, is equivalent to a gigantic combination lock with an almost uncountable number of dials. To generate a biological molecule like hemoglobin, the red pigment in blood, by simple sieving would be equivalent to taking all the amino-acid building blocks of hemoglobin, jumbling them up at random, and hoping that the hemoglobin molecule would reconstitute itself by sheer luck. The amount of luck that would be required for this feat is unthinkable, and has been used as a telling mind-boggler by Isaac Asimov and others.

A hemoglobin molecule consists of four chains of amino acids twisted together. Let us think about just one of these four chains. It consists of 146 amino acids. There are 20 different kinds of amino acids commonly found in living things. The number of possible ways of arranging 20 kinds of things in chains 146 links long is an inconceivably large number, which Asimov calls the "hemoglobin number." It is easy to calculate, but impossible to visualize the answer. The first link in the 146-long chain could be any one of the 20 possible amino acids. The second link could also be any one of the 20, so the number of possible 2-link chains is 20 × 20, or 400. The number of possible 3-link chains is 20 × 20 × 20, or 8,000. The number of possible 146-link chains is 20 times itself 146 times. This is a staggeringly large number. A million is a 1 with 6 noughts after it. A billion (1,000 million) is a 1 with 9 noughts after it. The number we seek, the "hemoglobin number," is (near enough) a 1 with 190 noughts after it! This is the chance against happening to hit upon hemoglobin by luck. And a hemoglobin molecule has only a minute fraction of the complexity of a living body. Simple sieving, on its own, is obviously nowhere near capable of generating the amount of order in a living thing. Sieving is an essential ingredient in the generation of living order, but it is very far from being the whole story. Something else is needed. To explain the point, I shall need to make a distinction between "single-step" selection and "cumulative" selection. The simple sieves we have been considering so far in this chapter are all examples of single-step selection. Living organization is the product of cumulative selection.

The essential difference between single-step selection and cumulative selection is this. In single-step selection the entities selected or sorted, pebbles or whatever they are, are sorted once and for all. In cumulative selection, on the other hand, they "reproduce"; or in some other way the results of one sieving process are fed into a subsequent sieving, which is fed into . . . , and so on. The entities are subjected to selection of sorting over many "generations" in succession. The end-product of one generation of selection is the starting point for the next generation of selection, and so on for many generations. It is natural to borrow such words as "reproduce" and "generation," which have associations with living things, because living things are the main examples we know of things that participate in cumulative selection. They may in practice be the only things that do. But for the moment I don't want to beg that question by saying so outright.

Sometimes clouds, through the random kneading and carving of the winds, come to look like familiar objects. There is a much published photograph, taken by the pilot of a small airplane, of what looks a bit like the face of Jesus, staring out of the sky. We have all seen clouds that reminded us of something—a sea horse, say, or a smiling face. These resemblances come about by single-step selection, that is to say by a single coincidence. They are, consequently, not very impressive. The resemblance of the signs of the zodiac to the animals after which they are named, Scorpio, Leo, and so on, is as unimpressive as the predictions of astrologers. We don't feel overwhelmed by the resemblance, as we are by biological adaptations—the products of cumulative selection. We describe as weird, uncanny, or spectacular, the resemblance of, say, a leaf insect to a leaf or a praying mantis to a cluster of pink flowers. The resemblance of a cloud to a weasel is only mildly diverting, barely worth calling to the attention of our companion. Moreover, we are quite likely to change our mind about exactly what the cloud most resembles.

> *Hamlet.* Do you see yonder cloud that's almost in shape of a camel?
> *Polonius.* By the mass, and 'tis like a camel, indeed.
> *Hamlet.* Methinks it is like a weasel.
> *Polonius.* It is backed like a weasel.
> *Hamlet.* Or like a whale?
> *Polonius.* Very like a whale.

I don't know who it was first pointed out that, given enough time, a monkey bashing away at random on a typewriter could produce all the works of Shakespeare. The operative phrase is, of course, given enough time. Let us limit the task facing our monkey somewhat. Suppose that he has to produce, not the complete works of Shakespeare but just the short sentence "Methinks it is like a weasel," and we shall make it relatively easy by giving him a typewriter with a restricted keyboard, one with just the 26 (capital) letters, and a space bar. How long will he take to write this one little sentence?

The sentence has 28 characters in it, so let us assume that the monkey has a series of discrete "tries," each consisting of 28 bashes at the keyboard. If he types the phrase correctly, that is the end of the experiment. If not, we allow him another "try" of 28 characters. I don't know any monkeys, but fortunately my 11-month old daughter is an experienced randomizing device, and she proved only too eager to step into the role of monkey typist. Here is what she typed on the computer:

> UMMK JK CDZZ F ZD DSDSKSM
> S SS FMCV PU I DDRGLKDXRRDO
> RDTE QDWFDVIOY UDSKZWDCCVYT
> H CHVY NMGNBAYTDFCCVD D
> RCDFYYYRM N DFSKD LD K WDWK
> JJKAUIZMZI UXDKIDISFUMDKUDXI

She has other important calls on her time, so I was obliged to program the computer to simulate a randomly typing baby or monkey:

WDLDMNLT DTJBKWIRZREZLMQCO P
Y YVMQKZPGJXWVHGLAWFVCHQYOPY
MWR SWTNUXMLCDLEUBXTQHNZVJQF
FU OVAODVYKDGXDEKYVMOGGS VT
HZQZDSFZIHIVPHZPETPWVOVPMZGF
GEWRGZRPBCTPGQMCKHFDBGW ZCCF

And so on and on. It isn't difficult to calculate how long we should reasonably expect to wait for the random computer (or baby or monkey) to type METHINKS IT IS LIKE A WEASEL. Think about the total number of *possible* phrases of the right length that the monkey or baby or random computer *could* type. It is the same kind of calculation as we did for hemoglobin, and it produces a similarly large result. There are 27 possible letters (counting "space" as one letter) in the first position. The chance of the monkey happening to get the first letter—M—right is therefore 1 in 27. The chance of it getting the first two letters—ME—right is the chance of it getting the second letter—E—right (1 in 27) *given that* it has also got the first letter—M—right, therefore 1/27 × 1/27, which equals 1/729. The chance of it getting the first word—METHINKS—right is 1/27 for each of the 8 letters, therefore (1/27) × (1/27) × (1/27) . . . , etc. 8 times, or (1/27) to the power 8. The chance of it getting the entire phrase of 28 characters right is (1/27) to the power 28, i.e. (1/27) multiplied by itself 28 times. These are very small odds, about 1 in 10,000 million million million million million million. To put it mildly, the phrase we seek would be a long time coming, to say nothing of the complete works of Shakespeare.

So much for single-step selection of random variation. What about cumulative selection; how much more effective should this be? Very very much more effective, perhaps more so than we at first realize, although it is almost obvious when we reflect further. We again use our computer monkey, but with a crucial difference in its program. It again begins by choosing a random sequence of 28 letters, just as before:

WDLMNLT DTJBKWIRZREZLMQCO P

It now "breeds from" this random phrase. It duplicates it repeatedly, but with a certain chance of random error—"mutation"—in the copying. The computer examines the mutant nonsense phrases, the "progeny" of the original phrase, and chooses the one which, *however slightly*, most resembles the target phrase, METHINKS IT IS LIKE A WEASEL. In this instance the winning phrase of the next "generation" happened to be:

WDLTMNLT DTJBSWIRZREZLMQCO P

Not an obvious improvement! But the procedure is repeated, again mutant "progeny" are "bred from" the phrase, and a new "winner" is chosen. This goes on, generation after generation. After 10 generations, the phrase chosen for "breeding" was:

MDLDMNLS ITJISWHRZREZ MECS P

After 20 generations it was:

MELDINLS IT ISWPRKE Z WECSEL

By now, the eye of faith fancies that it can see a resemblance to the target phrase. By 30 generations there can be no doubt:

MEFHINGS IT ISWLIKE B WECSEL

Generation 40 takes us to within one letter of the target:

METHINKS IT IS LIKE I WEASEL

And the target was finally reached in generation 43. A second run of the computer began with the phrase:

Y YVMQKZPFJXWVHGLAWFVCHQXYOPY,

passed through (again reporting only every tenth generation):

Y YVMQKSPFTXWSHLIKEFV HQYSPY
YETHINKSPITXISHLIKEFA WQYSEY
METHINKS IT ISSLIKE A WEFSEY
METHINKS IT ISBLIKE A WEASES
METHINKS IT ISJLIKE A WEASEO
METHINKS IT IS LIKE A WEASEP

and reached the target phrase in generation 64. In a third run the computer started with:

GEWRCZRPBCTPGQMCKHFDBGW ZCCF

and reached METHINKS IT IS LIKE A WEASEL in 41 generations of selective "breeding."

The exact time taken by the computer to reach the target doesn't matter. If you want to know, it completed the whole exercise for me, the first time, while I was out to lunch. It took about half an hour. (Computer enthusiasts may think this unduly slow. The reason is that the program was written in BASIC, a sort of computer baby-talk. When I rewrote it in Pascal, it took 11 seconds.) Computers are a bit faster at this kind of thing than monkeys, but the difference really isn't significant. What matters is the difference between the time taken by *cumulative* selection, and the time which the same computer, working flat out at the same rate, would take to reach the target phrase if it were forced to use the other procedure of *single-step selection*: about a million million million million million years. This is more than a million million million times as long as the universe has so far existed. Actually it would be fairer just to say that, in comparison with the time it would take either a monkey or a randomly programmed computer to type our target phrase, the total age of the universe so far is a negligibly small quantity, so small as to be well within the margin of error for this sort of back-of-an-envelope calculation. Whereas the time taken for a computer working randomly but with the constraint of *cumulative selection* to perform the same task is of the same order as humans ordinarily can understand, between 11 seconds and the time it takes to have lunch.

There is a big difference, then, between cumulative selection (in which each improvement, however slight, is used as a basis for future building), and single-step selection (in which each new "try" is a fresh one). If evolutionary progress had had to rely on single-step selection, it would never have got anywhere. If, however, there was any way in which the necessary conditions for *cumulative* selection could have been set up by the blind forces of nature, strange and wonderful might have been the consequences. As a matter of fact that is exactly what happened on this planet, and we ourselves are among the most recent, if not the strangest and most wonderful, of those consequences.

It is amazing that you can still read calculations like my hemoglobin calculation, used as though they constituted arguments *against* Darwin's theory. The people who do this, often expert in their own field, astronomy or whatever it may be, seem sincerely to believe that Darwinism explains living organization in terms of chance—"single-step selection"—alone. This belief, that Darwinian evolution is "random," is not merely false. It is the exact opposite of the truth. Chance is a minor ingredient in the Darwinian recipe, but the most important ingredient is cumulative selection which is quintessentially nonrandom.

Clouds are not capable of entering into cumulative selection. There is no mechanism whereby clouds of particular shapes can spawn daughter clouds resembling themselves. If there were such a mechanism, if a cloud resembling a weasel or a camel could give rise to a lineage of other clouds of roughly the same shape, cumulative selection would have the opportunity to get going. Of course, clouds do break up and form "daughter" clouds sometimes, but this isn't enough for cumulative selection. It is also necessary that the "progeny" of any given cloud should resemble its "parent" *more* than it resembles any old "parent" in the "population." This vitally important point is apparently misunderstood by some of the philosophers who have, in recent years, taken an interest in the theory of natural selection. It is further necessary that the chances of a given cloud's surviving and spawning copies should depend upon its shape. Maybe in some distant galaxy these conditions did arise, and the result, if enough millions of years have gone by, is an ethereal, wispy form of life. This might make a good science fiction story—*The White Cloud*, it could be called—but for our purposes a computer model like the monkey/Shakespeare model is easier to grasp.

DARWINISM

A. G. N. FLEW*

THE STRUCTURE OF DARWINISM

Those who do philosophy of science tend to equate science with physics.[1] The same thing also seems often to be true of those who try to present scientific thought to non-scientists. Yet one of the most important of all scientific theories is that developed by Darwin in the *Origin of Species.*[2] Covering the entire range of biological phenomena its scope is enormous. While if any scientific theory is interesting philosophically this one is.[3] It happens also to be exceptionally suitable as an elementary example for teaching. No outlandish or elaborate concepts are involved. The nature of most of the supporting evidence is familiar, and its relevance fairly easy to grasp. The deductive moves made are short and simple. The essential material can be studied in a single original source—conveniently brief, interesting, and written well. Darwin introduces no "theoretical entities" (photons, electro-magnetic waves, the libido, or what have you)[4] and employs no mathematics either within his theory or in presenting his case for it. The lack of these two generally essential features of modern science is at an elementary stage a positive advantage. Especially when, as here, these defects in the example may be made good afterward by considering the later development of genetics; which both provides a hypothetical entity, the gene, and applies mathematics abundantly. The object of the present paper is not to say anything strikingly novel, but simply to suggest some of the general morals which may be drawn from an examination of this particular conceptual structure.

---

*A. G. N. Flew, "The Structure of Darwinism," *New Biology* 28 (London: Penguin Books Ltd., 1959), pp. 25–44. Reprinted with permission of the author.

THE DEDUCTIVE CORE OF DARWINISM

Darwin himself remarked in the last chapter: "this whole volume is one long argument" (p. 413). Again in the *Autobiography*, in a typically modest and engaging passage, he claims: "*The Origin of Species* is one long argument from the beginning to the end."[5] A recent interpreter goes further: "The old arguments for evolution were only based on circumstantial evidence. . . . But the core of Darwin's argument was of a different kind. It did not make it more probable—it made it a certainty. Given his facts his conclusion *must* follow: like a proposition in geometry. You do not show that any two sides of a triangle are very *probably* greater than the third. You show they *must* be so. Darwin's argument was a deductive one—whereas an argument based on circumstantial evidence is inductive."[6]

This is a challenging contention. For surely Darwin was a great empirical naturalist, concerned to discover what as a matter of contingent fact is the case, though it might not have been? What business had he with deductive a priori arguments purporting to demonstrate as in a theorem in geometry that some things *must* be so? We must consider what precisely this deductive core is: how much it does prove and how; and how much it leaves open to be settled by further research.

That Darwin's argument does indeed contain such a deductive core is suggested by a passage in the "Introduction": "As many more individuals of each species are born than can possibly survive; and as, *consequently*, there is a frequently recurring struggle for existence, it *follows that* any being, if it vary however slightly in any manner profitable to itself, under the complex and sometimes varying conditions of life will have a better chance of surviving and thus be naturally selected. From the strong principle of inheritance, any selected variety will tend to propagate its new and modified form" (p. 4: my italics). He promises that in the chapter "Struggle for Existence" he will treat this struggle "amongst all organic beings throughout the world, *which inevitably follows from* the high geometrical ratio of their increase" (p. 4: my italics). In that chapter he develops the argument: "A struggle for existence *inevitably follows* from the high rate at which all organic beings tend to increase . . . as more individuals are produced than can possibly survive, *there must in every case be* a struggle for existence, either one individual with another of the same species, or with the individuals of distinct species, or with the physical conditions of life. It is the doctrine of Malthus applied with manifold force to the whole animal and vegetable kingdoms, for in this case there can be no artificial increase of food, and no prudential restraint from marriage" (p. 59: my italics). Just as the struggle for existence is derived as a consequence of the combination of a geometrical ratio of increase with the finite possibilities of survival: so in the chapter "Natural Selection" this in turn is derived as a consequence of the combination of the struggle for existence with variation. Darwin summarizes his argument here:

> If . . . organic beings vary at all in the several parts of their organization, and I think this cannot be disputed; if there be . . . a severe struggle for life . . . and this certainly cannot be disputed; then . . . I think it would be a most extraordinary fact if no variation had ever occurred useful to each being's own welfare, in the same manner as so many variations have occurred useful to man. But if variations useful to any organic being do occur, assuredly individuals thus characterized will have the best chance of being preserved in the struggle for life: and from the strong principle of inheritance they will tend to produce offspring similarly characterized. This principle of . . . Natural Selection . . . leads to the improvement of each creature in relation to its organic and inorganic conditions of life. (p. 115)

Since Darwin's argument does indeed contain this deductive core the pedagogic device used by Julian Huxley in his latest exposition of evolutionary theory is exceptionally apt. It can, he claims, "be stated in the form of two general evolutionary equations. The first is that reproduction plus mutation produces natural selection; and the second that natural selection plus time produces the various degrees of biological improvement that we find in nature."[7] The idea is excellent, but the execution here is curiously slapdash. For the first equation, as Huxley gives it (R + M→NS), is not valid. Reproduction plus mutation would not necessarily lead to natural selection. It is necessary also to add the struggle for existence: and that in turn has to be derived from the sum of the geometrical ratio of increase plus the limited resources for living—limited *Lebensraum* as Hitler's Germans might have called it. So to represent the core of Darwin's argument we need something more like: GRI + LR, SE; SE + V→ NS; and, not on all fours with the first two, NS + T →B. To avoid anachronism V (for hereditable variation) must be substituted for M (mutation). The third equation represents part of the argument which does not properly belong to what we are calling the deductive core and so from now on we shall neglect it.

Of course to make this core and the equations used to represent it schematically ideally rigorous one would have to construct for all the crucial terms definitions to include explicitly every necessary assumption. There are in fact several many of which when uncovered and noticed may seem too obvious to have been worth stating. Take for instance one to which Darwin himself refers rather obliquely in a passage already quoted: "I think it would be a most extraordinary fact if no variation had ever occurred useful to each being's own welfare" (p. 115). It would indeed. He offers a very powerful reason for believing that this has not in fact been the case, adding after the phrase just quoted; "in the same manner as so many variations have occurred useful to man" (p. 115). Nevertheless, in this particular case bringing out the assumption may have a value other than that of rigorization for its own sake. For it may suggest, what has in fact proved to be the case, that one of the main effects of natural selection is to eliminate unfavorable variations. It not only helps to generate biological improvement. It is essential to prevent biological degeneration.

We shall not attempt here to develop Darwin's argument quite rigorously or to formalize the result. This is an exercise which might be instructive. But it would of course involve transforming, and hence in a way misrepresenting, what Darwin actually said. It is perfectly possible to point the main morals without this.

(i) Though the argument itself proceeds a priori, because the premises are empirical it can yield conclusions which are also empirical. That living organisms all tend to reproduce themselves at a geometrical ratio of increase; that the resources they need to sustain life are limited; and that while each usually reproduces after its kind sometimes there are variations which in their turn usually reproduce after their kind: all these propositions are nonetheless contingent and empirical for being manifestly and incontestably true. That there is a struggle for existence; and that through this struggle for existence natural selection occurs: both these propositions equally are nonetheless contingent and empirical for the fact that it follows, necessarily as a matter of logic a priori, that wherever the first three hold the second two must hold also.

(ii) The premises are matters of obvious fact. The deductive steps are short and simple. The conclusions are enormously important. Yet these conclusions, and that they were implied by these premises, was before Darwin very far indeed from being obvious to able men already sufficiently familiar with the necessary premises. This should give us a greater respect for the power of simple logic working on the obvious. Of course, he did not just have to make some short deductive moves from a few very wide-ranging empirical premises already provided as such. He had first to recognize that these propositions did constitute essential premises; and then, after making the deductions from them, to appreciate that these premises and these conclusions contained and linked together concepts crucial for understanding the problem of the origin of species. The premises, the concepts, the deductions, the conclusions, all are simple. To bring them together and to see the importance of the theoretical scheme so constructed was a simple matter too. But this simplicity is the simplicity of genius.

(iii) The conclusions of the deductive argument are proved beyond dispute:

for though the premises are as empirical generalizations in principle open to revision, in fact, as Darwin urged in a passage quoted already, they cannot reasonably be questioned. It is therefore all the more important to appreciate what it does not prove; and hence what, at least as far as this argument is concerned, is left open to be settled by further inquiry.

It certainly does not prove that all "the various degrees of biological improvement that we find in nature"[8] can be accounted for in these terms. It proves at most only that some "biological improvement" must occur. Darwin needed, and provided, other facts and considerations to support his far wider and more revolutionary conclusion: "that species are not immutable; but that those belonging to what are called the same genera are lineal descendants of some other and generally extinct species, in the same manner as the acknowledged varieties of any one species are the descendants of that species . . . I am convinced that Natural Selection has been the main but not exclusive means of modification" (pp. 5–6). Though he built up a mighty case for this sweeping conclusion the case is one which could not in principle be complete unless the whole science of evolutionary biology were complete. Here the field for inquiry is open and without limit.

Again, this deductive argument has nothing to say about the causes and mechanisms of variation. Indeed the *Origin of Species* as a whole has on these little to say. Darwin begins his summary of the chapter "Laws of Variation": "Our ignorance of the laws of variation is profound" (p. 151). While in "Recapitulation and Conclusion" one of the advantages of accepting his theory urged is: "A grand and almost untrodden field of inquiry will be opened, on the causes and laws of variation, on correlation of growth, on the effects of use and disuse, on the direct action of external conditions, and so forth" (p. 437). Thus Lamarckism is quite compatible with Darwin's theory; and Darwin indeed does himself accept in the chapter on variation the inheritance of acquired characters. There would not, of course, be this elasticity had Darwin, like Julian Huxley, held a theory in which V (variation) would be replaced in the equations by M (mutation); and "mutation" would be so defined as to exclude Lamarckism—the infamous thing! Thus, again, it would be strictly compatible with this deductive core of Darwinism to maintain that some or all favorable variations were the results of special interventions by the Management. Though any such arbitrary and anti-scientific postulations would be entirely out of harmony with Darwin's own thoroughly naturalistic spirit, and his Lyellian insistence on continuity of development: ". . . species are produced and exterminated by slowly acting and still existing causes, and not by miraculous acts of creation and by catastrophes . . ." (p. 439). Notwithstanding the fact that in the *Origin of Species* he always concedes a special creation for the first life: ". . .

life, with its several powers, having been breathed by the Creator into a few forms or into one . . . " (p. 441); yet "I should infer from analogy that probably all the organic beings which have ever lived on this earth have descended from some one primordial form, into which life was first breathed by the Creator" (p. 436).

(iv) Darwinism provides an outstanding example to show how a good theory guides and stimulates inquiry, setting whole new ranges of fruitful questions. This Darwin sees clearly. He rightly claims as a great advantage attending the acceptance of his theory, that: "A grand and almost untrodden field of inquiry will be opened, on the causes and laws of variation. . . . A new variety raised by man will be a more important and interesting subject of study than one more species added to the infinitude of already recorded species. Our classifications will come to be, as far as they can be so made, genealogies . . . we have to discover and trace the many diverging lines of descent in our natural genealogies. . . . Rudimentary organs will speak infallibly with respect to the nature of long lost structures. . . . Embryology will reveal to us the structure, in some degree obscured, of the prototypes of each great class" (pp. 437–38). Ranging further: "In the distant future I see open fields for far more important researches. Psychology will be based on a new foundation, that of the necessary acquirement of each mental power and capacity by gradation" (p. 439).[9]

(v) Again, Darwinism makes an excellent text-book example to show how a theory explains: by showing that the elements to be explained are not after all just a lot of separate brute facts—just one damn thing after or along with another—but rather what, granted the assumptions of the theory, is to be expected.[10]

Whereas before Darwin the general opinion even among biologists was that species constituted natural kinds, created separately:[11] he tried to show that "Although much remains obscure, and will long remain obscure . . . the view which most naturalists entertain, and which formerly I entertained—namely, that each species has been independently created—is erroneous" (p. 5); and that, on the contrary, all species have their places on one single family tree—or at most on four or five—and are thus in the most strict and literal sense related. In what we have called the deductive core of his theory he shows how, on certain assumptions, a struggle for existence, natural selection, and some biological improvement must be expected. In applying this conceptual framework in detail in the attempt to account for the origin of species he shows how not only the existence of an enormous number of species but also many other very general biological facts previously isolated and brute are just what, granted its premises and some other equally plausible assumptions, is to be expected.

In his "Recapitulation" (pp. 422–32) Darwin reviews some of these very general facts which he has considered earlier in more detail. For instance: "As natural selection acts solely by accumulating slight, successive, favourable variations, it can produce no great or sudden modification. . . . Hence the canon of *"Natura non facit saltum"* [Literally: Nature does not take a jump—A. F.], which every fresh addition to our knowledge tends to make truer, is on this theory simply intelligible. We can see why nature is prodigal in variety, though niggard in innovation. But why this should be a law of nature if each species has been independently created, no man can explain" (p. 424: italics mine). Then:

> Looking to geographical distribution, if we admit that there has been during the long course of ages much migration . . . then we can understand, on the theory of descent with modification, most of the great leading facts in Distribution . . . we can understand, by the aid of the Glacial period, the identity of some few plants, and the close alliance of many others, on the most distant mountains, and likewise the close alliance of some of the inhabitants of the sea in the northern and southern temperate zones, though separated by the whole intertropical ocean. (pp. 428–29)

And so on.

If we ask how Darwinism explains, or indeed how any other theory in the natural sciences explains, the answer seems to lie in its powers: to provide connections between elements which without it would be unconnected—just a lot of loose and separate facts; and to show how the phenomena to be explained are, on certain assumptions, exactly what is to be expected.[12]

(vi) Darwin, always conscientious and generous in acknowledging his debts, refers several times to Malthus: "the Struggle for Existence amongst all organic beings throughout the world, which inevitably follows from the high geometrical ratio of their increase, . . . This is the doctrine of Malthus, applied to the whole animal and vegetable kingdoms" (p. 4: cf. passage on p. 9 quoted above).[13] It is therefore perhaps not surprising that the logical skeleton of theory which provided the organizing and supporting framework for all Malthus' inquiries and recommendations about population resembles in almost every respect so far considered the theoretical framework of the *Origin of Species.*

Even for those who are not antecedently inclined, either by their political and religious ideology or by their deep emotional drives, to eschew Malthusian ideas, these similarities may be obscured by the fact that Malthus' theory as he presents it himself contains several logical faults, albeit easily remediable ones. Also, unfortunately and unnecessarily, it has built into it various always controversial and now generally obnoxious value commitments. Nevertheless Malthus, like Darwin, in his theory does proceed a priori from very general, scarcely contestable, empirical premises. From the manifest power of the human animal to reproduce in a geometrical ratio of increase; and from an assumption, too precisely formulated as an arithmetical ratio, which expresses the fact that the possibilities of increasing the resources necessary for living are, though elastic, finite, Malthus would deduce some very important empirical conclusions (cf. [i] above). The deductive moves once the premises are assembled are not elaborate (cf. [ii] above). The limitations of the purely deductive argument are important (cf. [iii] above). For neither in Malthus' original form nor in a slightly amended and corrected form will the premises yield the conclusion that the power of increase must be checked *everywhere and always* absolutely. The same is true of the premises from which Darwin derives the conclusion that there must be a struggle for existence. These alone are not sufficient to prove that no species ever has or ever could have enjoyed *even for a short time* an environment in which its possibilities of increase were not checked by competition either from other species or from other members of the same species. They prove only that such a condition if it did ever occur must necessarily be very short lived. But whereas this qualification is on the evolutionary time scale with which Darwin was dealing insignificant: in the field of human affairs with which Malthus was concerned it may sometimes be very important indeed. For here a generation is a lifetime; and in determining practical policy some weight may properly be given to Keynes' reminder that in the long run we are all dead.

Again, the theory of Malthus, like that of Darwin, has the power to guide and stimulate inquiry, setting ranges of fruitful new questions (cf. [iv] above). Recognition of the enormous animal power of reproductive increase, and of the inescapable fact that this power must always in the fairly short run be checked at the finite though elastic limits of the possibility of increasing the resources necessary for living—if it is not checked earlier by something else—generates the master question: "How in fact is this enormous power checked?" And it was precisely this question which provoked all the empirical inquiries the results of which Malthus embodied in the second version of the *Essay on Population.*[14]

Finally, Malthus' theory too makes an excellent text-book example. It can be used to show how a theory picks out and explains by showing connections between the vital elements in the situation. It can be used to show how granted certain conditions a theory can explain why certain other conditions also hold; by revealing that in these circumstances and on this theory these things are what is to be expected (cf. [v] above). Furthermore, it has the advantage of belonging to the field of social studies, still even more neglected by philosophers of science than biology. It can illustrate excellently the temptation confusingly and implicitly to incorporate controversial value commitments into the very structure of a theory, thereby determining the directions to be taken by any policy founded thereon. At this point the analogy between Malthus and Darwin begins to break down. It would be illuminating to develop a full Plutarchian comparison, to distinguish and catalog the similarities and dissimilarities between the two theories. But here it is sufficient simply to mention a few of the similarities between Darwinism and a system of ideas which, partly because it is uncongenial both to the two most powerful ideological groups in the world[15] and to certain psychologically more elemental sources of prejudice,[16] is still very generally misunderstood and underestimated.[17]

THE CONCEPTUAL CHANGES REQUIRED BY DARWINISM

Darwinism, as again Darwin himself clearly saw, implies that we must abandon assumptions implicit in the previous use of certain categorial terms such as *genus* and *species:* "When the views advanced by me in this volume . . . are generally admitted . . . there will be a considerable revolution in natural history. Systematists will be able to pursue their labours at present; but they will not be incessantly haunted by the shadowy doubt whether this or that form be in essence a species" (p. 436). Which, he remarks, feelingly: "I feel sure, and I speak after experience, will be no slight relief" (p. 436). Hence: "I look at the term *species* as one arbitrarily given for the sake of convenience to a set of individuals closely resembling each other, and that it does not essentially differ from the term *variety,* which is given to less distinct and more fluctuating forms. The term *variety,* again, in comparison with mere individual differences, is also applied arbitrarily, and for mere convenience's sake" (p. 49: italics mine). "Hereafter we shall be compelled to acknowledge that the only distinction between species and well-marked varieties is, that the latter are known, or believed, to be connected at the present day by intermediate gradations, whereas species were formerly thus connected" (p. 436). "In short, we shall have to treat species in the same manner as those naturalists treat genera, who admit that genera are merely artificial combinations made for convenience" (p. 437).

The assumptions which Darwin's theory commits him to challenge are more particular cases of those very general prejudices about language and classification which Locke had begun to uncover and to question in his *Essay Concerning Human Understanding* (1690). These are the assumptions: that all things belong to certain natural kinds, in virtue of their "essential natures"; that there are no marginal cases falling outside and between these sharply delimited collections of individuals; that there must always be straight yes or no answers to the question

"Is this individual a so and so or not?"; and that men have only to uncover and, as it were, write the labels for, the classes to which nature has antecedently allocated every individual thing. To such assumptions in the biological field the pictures appropriate are: that of *Genesis*, of all creatures created after their sharply different kinds; and that of the tradition, illustrated by William Blake, of Adam naming the beasts. They would justify the approach to classification epitomized in the citation of the Swedish Academy in honouring Linnaeus: "he discovered the essential nature of insects." Or that of his own famous but dark saying, quoted by Darwin: ". . . the characters do not make the genus, but . . . the genus gives the characters . . . " (p. 372).

In one aspect Darwin's work can be seen as a continuation and application of that of Locke, particularly of the third book of the *Essay* "Of Words"; though there seems to be no evidence that he had ever read the philosopher. For he is insisting on "that old canon in natural history, '*Natura non facit saltum*'" (p. 185; cf. pp. 174–75 and 424); and arguing that on his theory "we shall at least be freed from the vain search for the undiscovered and undiscoverable essence of the term *species*" (p. 437: italics mine in this and the previous quotation). Sir Arthur Keith in his Plutarchian comparison of Locke with Darwin fails surprisingly to notice this continuity.[18]

As an example of philosophy within scientific theory this analysis and reinterpretation of the concepts of *species*, *variety*, and *genus*, might be compared with Einstein's analysis of *motion* and *simultaneity* in relativity theory. In both cases the analysis is required by the theory. In both cases it had been to a greater or lesser extent anticipated by a philosopher before it was reworked and put to use by a scientist. But whereas Darwin seems never to have read Locke, Einstein certainly did read Mach; and acknowledged indebtedness to him.

THE MODEL CONTAINED AND "DEPLOYED" IN DARWINISM

Recently the useful notion of the "deployment" of a model has been introduced into the philosophy of the physical sciences. "At the stage at which a new model is introduced the data that we have to go on, the phenomena which it is used to explain, do not justify us in prejudging, either way, which of the questions which normally make sense when asked of things which, say, travel will eventually be given a meaning on the new theory also. . . . One might speak of models in physics as more or less 'deployed.'"[19] How far a given model is said to be deployed is a matter of how far the analogy, between that model and the phenomena to which it is applied, is believed to hold. This idea is relevant also elsewhere than in the physical sciences. The acceptance of Darwin's theory made possible a massive deployment of one model which had been curiously boxed up and impotent for an extraordinarily long time. This was the model of the family: with which was associated the method of representation employed to express both familial relationships ("the family tree") and one system of classification ("the tree of Porphyry").[20] Once again Darwin saw what he was doing: "The terms used by naturalists of affinity, relationship, community of type, paternity . . . will cease to be metaphorical and will have a plain signification"; while "Our classifications will come to be, as far as they can be made, genealogies . . ." (p. 437). The remarkable points about this particular model are: first, that it does not seem originally to have been introduced as any part of an attempt at scientific explanation; and, second, that its suggestive force does not seem to have had much influence towards the production of the theory which made possible its deployment. The various terms appropriate to this model were, apparently, introduced because naturalists noticed analogies which made the idea of family relationship seem apt as a metaphor. But the suggestion that the metaphor might be considerably more than a mere metaphor, that the model could be deployed, seems almost always to have been blocked by the strong resistance of the accepted doctrine of the fixity of species.

We can find in Darwin's letters and other papers indications of the strength of this resistance: though, of course, we with the advantage of hindsight see before Darwin the occasional deviations to a doctrine of descent standing out. It is a mistake to interpret this resistance simply as a matter of religion. Certainly belief in the evolution of species would be incompatible with the acceptance of the creation myths of *Genesis*, interpreted literally: (to say nothing of the incompatibility between the two myths themselves). But the evolutionary geology of Lyell is equally irreconcilable with a literal reading of the Pentateuch; and that was already becoming generally accepted among scientific men when Darwin began to work on the origin of species. Quite apart from any considerations of ideology it *is* an inescapable biological fact that living things as we now see them around us do seem, with only comparatively infrequent exceptions, to belong to various natural kinds; which mate with and reproduce, again with only comparatively minor exceptions, after their kinds. It was precisely this massive and stubborn fact which the aetiological myth makers of *Genesis* were trying to account for with their theory of the special creation of fixed species. Even after the evolutionary geologists had opened up the vast time scale needed for a smooth and uniformitarian process of biological evolution, the way of this concept was still blocked by the difficulty of suggesting any mechanisms which could possibly have brought about such an enormous development. To this problem Darwin found the clue in the concept of natural selection, suggested to him by his reading of Malthus.

Another bulwark of the doctrine of fixed species would presumably be the unevolutionary view of language referred to above. For while the general fact of the apparent stability and separateness of seeming "natural kinds" is a main source and stay for certain false assumptions about language: those assumptions, usually hidden, would in turn support the idea of the fixity of species. For to the extent that we assume that, because a question of the "Is this an X?" form can usually be answered by a straight "Yes" or "No," such a question of identification must always be susceptible of a straight yes or no answer: to this extent we are mistaking it that all the things in the world must without exception either definitely fit or definitely not fit into the classificatory pigeonholes provided by our language, labeled with the words presently available in the vocabulary of that language. This is an assumption which has only to be revealed and recognized to be challenged. Nevertheless it is one which has been, and remains, protean, powerful, and pervasive.

THE "PHILOSOPHICAL IMPLICATIONS" OF DARWINISM

Darwin revolutionized biological studies. But in addition, rightly or wrongly, his work has had considerable effects in shaping world outlooks and determining the general climate of opinion. This influence has been felt in two spheres. First, Darwin succeeded in indicating, at least in outline, how the appearance of design in living things might perhaps be accounted for without any appeal to actual interventions by some Supernatural Designer. Second, his ideas have been taken, or mistaken, to justify various moral and political policies.

Now it may well be that David Hume in his masterpiece, the *Dialogues Concerning Natural Religion* (first edition, 1779), provided all the instruments needed to dismantle the Argument to Design. But Hume's subtle speculative arguments, presented as they are in the discreet and curiously difficult form of a philosophical dialogue, have never made much impact directly on popular thinking. Alternatively, it may well be that some version of this hardiest and most perennial of all the arguments of natural theology can be salvaged and refined for reuse, even if you have rejected absolutely the idea that supernatural interventions may be postulated to explain the particular course of nature. Nevertheless, the popular version of this argument, developed perhaps most powerfully in the *Natural Theology: or Evidences of the Existence and Attributes of the Deity Collected from the Appearances of Nature* (first edition, 1802), though not of course originated by Paley, urges that just as from the observation of a watch we may infer the existence of a watchmaker, so by parity of reasoning, from the existence of mechanisms so marvelous as the human eye we must infer the existence of a Designer. Paley explicitly repudiates as an alternative any suggestion "that the eye, the animal to which it belongs, every other animal, every plant . . . are only so many out of the possible varieties and combinations of being which the lapse of infinite ages has brought into existence: that the present world is the relic of that variety; millions of other bodily forms and other species having perished, being by the defect of their constitutions incapable of preservation. . . . Now there is no foundation whatever for this conjecture in anything which we observe in the works of nature . . ." (*Paley's Works*, Vol. I [1838], p. 32). It is this version of the Argument to Design to which the *Origin of Species* is crucially relevant. For in it Darwin offers a demonstration, backed by a mass of empirical material, of how adaptation however remarkable might be the result not of supernatural artifice but of natural selection.

Again, the doctrine of Darwinism has been taken to justify the most extraordinarily diverse, and often mutually incompatible, moral and political policies. As a purely scientific theory by itself it could not entail any normative conclusions (conclusions, that is, about what *ought* to be); because it would not, so long as it remained a purely scientific theory, contain any normative premises. Coming, however, from this important truism in the abstract to our particular case it is perhaps just worth mentioning that one or two of Darwin's scientific ideas are peculiarly open to ethical misinterpretation; though he himself is usually careful to avoid and to discourage such misconstructions.

Thus the concept of the survival of the fittest in the struggle for existence is easily mistaken to imply that Nature favors the survival of the most admirable and best. Whereas of course "fittest" here is to be defined neutrally as "having whatever as a matter of fact it may take to survive." Again, because natural selection in a struggle for existence enables the fittest (in this neutral sense) to win out against their competitors, it has seemed that evolutionary biology provides a ready-made justification for unrestricted competition, and extinction take the hindmost. So it has been found necessary to compile books on mutual aid, to show that co-operation within and between species has also sometimes paid off in the biological rat race. While, on the other side, the descriptive law of the jungle has been accepted as the prescriptive law of a natural order of Nature: ". . . one general law, leading to the advancement of all organic beings, namely, multiply, vary, let the strongest live and the weakest die" (p. 219). In fact Darwinism provides no reason: either for saying that what will survive in unrestricted competition will be the most excellent and worthy; or for denying that some co-operation may pay off within the general struggle for existence. One dramatic turn of phrase does not commit Darwin to the error of regarding the laws of nature which *describe* what does go on, as laws *prescribing* what ought to go on. Again, the not particularly Darwinian notion of higher animals, combined with the Darwinian idea of their evolution by natural selection, may raise a hope that Nature is somehow in favor of progress. Of this Darwin himself was not altogether innocent. While in considering "Geological Succession" he remarks noncommittally: "The inhabitants of each successive period in the world's history have beaten their predecessors in the race for life, and are, insofar, higher in the scale of nature; and this may account for that vague yet ill-defined sentiment, felt by many palaeontologists, that organization on the whole has progressed" (p. 309). In his final peroratory paragraph he goes further into the world of evaluation, claiming: ". . . from the war of nature, from famine and death, the most exalted object we are capable of conceiving, the higher animal, directly follows. There is grandeur in this view of life . . ." (p. 441). Yet in pure evolutionary theory nothing is valuable and nothing is without value. By it nothing is justified, and no values are guaranteed.[21] Things just happen.

ACKNOWLEDGMENT

I should like to express here my thanks to my colleague at Keele, Dr. R. G. Evans for his helpful criticism of an earlier draft of this paper. He is in no way responsible for the biological and other errors which no doubt it still contains.

NOTES

1. See for example Stephen Toulmin's *The Philosophy of Science* (London: Hutchinson, 1953).

2. All references are given to Charles Darwin, *The Origin of Species*, World's Classics ed. (Oxford: Oxford University Press, 1902).

3. Contrast Wittgenstein in the *Tractatus Logico-Philosophicus* (Kegan Paul, 1922): "The Darwinian theory has no more to do with philosophy than has any other hypothesis of natural science" (4.1122).

4. I particularly do not want to suggest that all these, and all the others classed sometimes as "scientific" or "theoretical" entities, enjoy exactly the same ontological status.

5. Charles Darwin, *Autobiography of Charles Darwin* (Watts, 1929), p. 75.

6. C. F. A. Pantin, in *The History of Science* (London: Cohen and West, 1953), p. 137 (italics in original).

7. Julian Huxley, *The Process of Evolution* (Chatto and Windus, 1953), p. 38.

8. Ibid.

9. "About thirty years ago there was much talk that geologists ought only to observe and not to theorize; and I well remember someone saying that at this rate a man might as well go into a gravel pit and count the pebbles and describe the colours. How odd it is that anyone should not see that all observation must be for or against some view if it is to be of any service" (Francis Darwin and A. Sewards, eds., *More Letters of Charles Darwin*, Vol. 1 [Whitefish, MT: Kessinger, 2005], p. 195).

10. The best short study of explanation which I know is by John Hospers in *Essays in Conceptual Analysis*, ed. Flew (Macmillan, 1956). With this may be compared Norman Campbell, *What Is Science?* (New York: Dover, 1952), ch. 5 (the original 1921 UK edition has been out of print for years). Campbell's thinking is oriented toward physics and chemistry: so it may not be obvious immediately how or how far his emphatic distinction between laws (which do not explain) and theories (which do) can be applied to evolutionary biology. It is also unfortunate that he fails to make any distinction between explaining why something occurs and explaining to some particular person the meaning of some notion. He thus commits the howler: "Explanation in general is the expression of an assertion in a more acceptable and satisfactory form" (p. 77). If this were so, no explanation could ever give us any new information.

11. I do not want here or elsewhere in this paper to enter into the difficult question of where exactly Darwin's originality lay. See on this J. Arthur Thomson, "Darwin's Predecessors," in *Darwin and Modern Science*, ed. A. C. Seward (Cambridge: Cambridge University Press, 1909).

12. I insert the qualification "in the natural sciences" simply in order to bypass the recent attempts to show that the second clause does not hold good of explanations in history.

13. The well-known passage in the *Autobiography* relevant here is particularly worth quoting for its timely moral to specialists inclined to neglect their general studies: "In October 1838, that is, fifteen months after I had begun my systematic enquiry, I happened to read for amusement Malthus on *Population*, and being well prepared to appreciate the struggle for existence which everywhere goes on from long-continued observation of the habits of animals and plants, it at once struck me that under these circumstances favourable variations would tend to be preserved, and unfavourable ones to be destroyed. The result of this would be the formation of new species. Here, then, I had at last got a theory by which to work . . ." (p. 57). Compare A. B. Wallace, *My Life; A Record of Events and Opinions*, Vol. 1 (London: Dodd, Mead, 1905), p. 232, for his account of the operation of the same stimulus in his parallel case.

14. The second and all later editions are so substantially different from the first as to constitute different books. It would be good if the practice of distinguishing them as *First Essay* and *Second Essay* became general. Since the Appendices are important, the *Second Essay* is best studied in one of the later editions in which these are included: for example, the sixth edition of 1826.

15. At the UNO World Population Conference in Rome in 1954 it was notable how often Roman Catholic and Communist delegates were found standing side by side in rejecting Malthusian ideas. For a very sober comment on the opposition of this Holy-

Unholy Alliance see *World Population and Resources* (Political and Economic Planning, 1955), pp. 307 ff.

16. See J. C. Flugel, *Population, Psychology, and Peace* (Watts and Co., 1947).

17. For an exposition and an attempt at reconstructive criticism of Malthus' theory I may perhaps be allowed to refer to my "The Structure of Malthus' Population Theory," in *Australasian Journal of Philosophy* (1957).

18. Arthur Keith, "Darwin's Place among Philosophers," in *The Rationalist Annual* (1955).

19. Toulmin, *The Philosophy of Science*, p. 37.

20. On the importance of "methods of representation" we may again refer to Toulmin, *The Philosophy of Science*, throughout. The "tree of Porphyry" was the name given both to a certain method of classification and to a way of representing this method by a sort of family tree.

21. On the subject of this last section see: Julian Huxley, "Progress," in *Essays of a Biologist* (Chatto and Windus, 1923) and *Evolution and Ethics* (Pilot Press, 1947). This last contains a reprint of T. H. Huxley's "Romanes Lecture." For a rather unsympathetic criticism see Toulmin, "World Stuff and Nonsense," *Cambridge Journal* I.

EDITORIAL NOTE

Flew makes particular reference to two men whose ideas much influenced Darwin. The first is the Reverend Thomas Robert Malthus who argued in his *Essay on a Principle of Population* (6th. ed., 1826) that state welfare systems only compound the problem of the poor because (human) population numbers have a tendency to increase geometrically, thus outstripping food supplies which at most can only increase arithmetically. There is therefore an invariable struggle for existence, and no change in the human state is possible. Darwin took Malthus' premises but turned his conclusion on its head, arguing that the struggle is precisely the key to change. Darwin made similar use of the ideas of Charles Lyell, who argued in his *Principles of Geology* (pp. 1830–33) for a uniformitarian perspective namely that of a very old world cycling on indefinitely with the only causes of change being natural, like rain and frost and earthquakes and volcanoes. Darwin's biological uniformitarianism, which likewise eschewed non-natural causes, translated into evolutionism!

Flew refers also to the ideas of the early nineteenth-century evolutionist Jean Baptiste de Lamarck. "Lamarckism" sees evolutionary change as coming through the inheritance of acquired characters, as in the long neck of the giraffe through stretching. This idea, rejected by today's evolutionists, was accepted by Darwin, and he may well have been influenced by Lamarck, for he read his work.

*TOM BETHELL**

DARWIN'S MISTAKE

ALL CHANGE IS NOT PROGRESS

How do we come to have horses and tigers and things? There are at least a million species in existence today, according to the paleontologist George Gaylord Simpson, and for every one extant, perhaps 100 are extinct. Such profusion! Such variety! How did it come about? The old answer was that they are created by God. But with the increasingly scientific temper of the eighteenth and nineteenth centuries, this explanation began to look insufficient. God was invisible, and so could not be part of any scientific explanation.

So an alternative explanation was proposed by a number of savants, among them Jean Baptiste Lamarck and Erasmus Darwin: the various forms of life did not just appear (as at the tip of a magician's wand), but evolved by a process of gradual transformation. Horses came from something slightly less horselike, tigers from something slightly less tigerlike, and so on back, until finally, if you went back far enough in time, you would come to a primitive blob of life which itself got started (perhaps) by lightning striking the primeval soup.

"Either each species of crocodile has been specially created," said Thomas Henry Huxley, "or it has arisen out of some pre-existing form by the operation of natural causes. Choose your hypothesis; I have chosen mine."

That's all very well, replied more conservative thinkers. If all of this life got here by evolution from more primitive life, then how did evolution occur? No answer was immediately forthcoming. Genesis prevailed, then Charles Darwin (grandson of Erasmus) furnished what looked like the solution. He proposed the machinery of evolution, and claimed that it existed in nature. Natural selection, he called it.

His idea was accepted with great rapidity. Once stated it seemed only too obvious. The survival of the fittest—of course! Some types are fitter than others,

---

*T. Bethell, "Darwin's Mistake," *Harper's Magazine*.

and given the competition—the "struggle for existence"—the fitter ones will survive to propagate their kind. And so animals, plants, all life in fact, will tend to get better and better. They would have to, with the fitter ones inevitably replacing those that are less fit. Nature itself, then, had evolving machinery built into it. "How extremely stupid not to have thought of that!" Huxley commented, after reading the *Origin of Species*. Huxley had coined the term *agnostic*, and he remained one. Meanwhile, the Genesis version didn't entirely fade away, but it inevitably took on a slightly superfluous air.

THE EVOLUTION DEBATE

That was a little over 100 years ago. By the time of the Darwin Centennial Celebrations at the University of Chicago in 1959, Darwinism was triumphant. At a panel discussion Sir Julian Huxley (grandson of Thomas Henry) affirmed that "the evolution of life is no longer a theory; it is a fact." He added sternly: "We do not intend to get bogged down in semantics and definitions." At about the same time, Sir Gavin de Beer of the British Museum remarked that if a layman sought to "impugn" Darwin's conclusions, it must be the result of "ignorance or effrontery." Garrett Hardin of the California Institute of Technology asserted that anyone who did not honor Darwin "inevitably attracts the speculative psychiatric eye to himself." Sir Julian Huxley saw the need for "true belief."

So that was it, then. The whole matter was settled—as I assumed, and as I imagined most people must. Darwin had won. No doubt there were backward folk tucked away in the remoter valleys of Appalachia who still clung to their comforting beliefs, but they, of course, lacked education. Not everyone was enlightened—goodness knows the Scopes trial had proved that, if nothing else. And some of them still wouldn't let up, apparently—they were trying to change the textbooks and get the Bible back into biology. Well, there are always diehards.

So it was only casually, about a year ago, that I picked up a copy of *Darwin Retried*, a slim volume by one Norman Macbeth, a Harvard trained lawyer. An odd field for a lawyer, certainly. But an endorsement on the cover by Karl Popper caught my eye. "I regard the book as . . . a really important contribution to the debate," Popper had written.

The debate? What debate? This interested me. I had studied philosophy, and in my undergraduate days Popper was regarded as one of the top philosophers—especially important for having set forth "rules" for discriminating between genuine and pseudo science. And Popper evidently thought there had been a "debate" worth mentioning. In his bibliography Macbeth listed a few articles that had appeared in academic philosophy journals in recent years and evidently were a part of this debate.

That was, as I say, a year ago, and by now I have read these articles and a good many others. In fact, I have spent a good portion of the last year familiarizing myself with this debate. It is surprising that so little word of it has leaked out, because it seems to have been one of the most important academic debates of the 1960s, and as I see it the conclusion is pretty staggering: Darwin's theory, I believe, is on the verge of collapse. In his famous book, *On the Origin of Species by Means of Natural Selection, or The Preservation of Favored Races in the Struggle for Life*, Darwin made a mistake sufficiently serious to undermine his theory. And that mistake has only recently been recognized as such. The machinery of evolution that he supposedly discovered has been challenged, and it is beginning to look as though what he really discovered was nothing more than the Victorian propensity to believe in progress. At one point in his argument, Darwin was misled. I shall try to elucidate here precisely where Darwin went wrong.

What was it, then, that Darwin discovered? What was this mechanism of natural selection? Here it comes as a slight shock to learn that Darwin really didn't "discover" anything at all, certainly not in the same way that Kepler, for example, discovered the laws of planetary motion. The *Origin of Species* was not a demonstration but an argument—"one long argument," Darwin himself said at the end of the book—and natural selection was an idea, not a discovery. It was an idea that occurred to him in London in the late 1830s which he then pondered in the Home Counties over the next twenty years. As we now know, several other thinkers came up with the same or a very similar idea at about the same time. The most famous of these was Alfred Russel Wallace, but there were several others.

The British philosopher Herbert Spencer was one who came within a hair's breadth of the idea of natural selection, in an essay called "The Theory of Population" published in the *Westminster Review* seven years before the *Origin of Species* came out. In this article Spencer used the phrase "the survival of the fittest" for the first time. Darwin then appropriated the phrase in the fifth edition of the *Origin of Species*, considering it an admirable summation of his argument. This argument was in fact an analogy, as follows:

While in his country retreat Darwin spent a good deal of time with pigeon fanciers and animal breeders. He even bred pigeons himself. Of particular relevance to him was that breeders bred for certain characteristics (length of feather, length of wool, coloring), and that the offspring of the selected mates often tended to have the desired characteristic more abundantly, or more noticeably, than its parents. Thus, it could perhaps be said, a small amount of "evolution" had occurred between one generation and the next.

By analogy, then, the same process occurred in nature, Darwin thought. As he wrote in the *Origin of Species:* "How fleeting are the wishes of man! how short his time! and consequently how poor will his productions be, compared with those accumulated by nature during whole geological periods. Can we wonder, then, that nature's productions should be far 'truer' in character than man's productions?"

Just as the breeders selected those individuals best suited to the breeders' needs to be the parents of the next generation, so, Darwin argued, nature selected those organisms that were best fitted to survive the struggle for existence. In that way evolution would inevitably occur. And so there it was: a sort of improving machine inevitably at work in nature, "daily and hourly scrutinizing," Darwin wrote, "silently and insensibly working . . . at the improvement of each organic being." In this way, Darwin thought, one type of organism could be transformed into another—for instance, he suggested, bears into whales. So that was how we came to have horses and tigers and things—by natural selection.

THE GREAT TAUTOLOGY

For quite some time Darwin's mechanism was not seriously examined, until the renowned geneticist T. H. Morgan, winner of the Nobel Prize for his work in mapping the chromosomes of fruit flies, suggested that the whole thing looked suspiciously like a tautology. "For, it may appear little more than a truism," he wrote, "to state that the individuals that are the best adapted to survive have a better chance of surviving than those not so well adapted to survive."

The philosophical debate of the past ten to fifteen years has focused on precisely this point. The survival of the fittest? Any way of identifying the fittest other than by looking at the survivors? The preservation of "favored" races? Any way of identifying them other than by looking at the preserved ones? If not, then Darwin's theory is reduced from the status of scientific theory to that of tautology.

Philosophers have ranged on both sides of this critical question: are there criteria of fitness that are independent of survival? In one corner we have Darwin himself, who assumed that the answer was yes, and his supporters, prominent among them David Hull of the University of Wisconsin. In the other corner are those who say no, among whom may be listed A. G. N. Flew, A. R. Manser, and A. D. Barker. In a nutshell here is how the debate has gone:

Darwin, as I say, just assumed that there really were independent criteria of fitness. For instance, it seemed obvious to him that extra speed would be useful for a wolf in an environment where prey was scarce, and only those wolves first on the scene of a kill would get enough to eat and, therefore, survive. David Hull has supported this line of reasoning, giving the analogous example of a creature that was better able than its mates to withstand desiccation in an arid environment.

The riposte has been as follows: a mutation that enables a wolf to run faster than the pack only enables the wolf to survive better if it does, in fact, survive better. But such a mutation could also result in the wolf outrunning the pack a couple of times and getting first crack at the food, and then abruptly dropping dead of a heart attack, because the extra power in its legs placed an extra strain on its heart. Fitness must be identified with survival, because it is the overall animal that survives, or does not survive, not individual parts of it.

However, we don't have to worry too much about umpiring this dispute, because a look at the biology books shows us that the evolutionary biologists themselves, perhaps in anticipation of this criticism, retreated to a fortified position some time ago, and conceded that "the survival of the fittest" was in truth a tautology. Here is C. H. Waddington, a prominent geneticist, speaking at the aforementioned Darwin Centennial in Chicago:

"Natural selection, which was at first considered as though it were a hypothesis that was in need of experimental or observational confirmation turns out on closer inspection to be a tautology, a statement of an inevitable although previously unrecognized relation. It states that the fittest individuals in a population (defined as those which leave most offspring) will leave most offspring."

The admission that Darwin's theory of natural selection was tautological did not greatly bother the evolutionary theorists, however, because they had already taken the precaution of redefining natural selection to mean something quite different from what Darwin had in mind. Like the philosophical debate of the past decade, this remarkable development went largely unnoticed. In its new form, natural selection meant nothing more than that some organisms have more offspring than others: in the argot, differential reproduction. This indeed was an empirical fact about the world, not just something true by definition, as was the case with the claim that the fittest survive.

The bold act of redefining selection was made by the British statistician and geneticist R. A. Fisher in a widely heralded book called *The Genetical Theory of Natural Selection.* Moreover, by making certain assumptions about birth and death rates, and combining them with Mendelian genetics, Fisher was able to qualify the resulting rates at which population ratios changed. This was called population genetics, and it brought great happiness to the hearts of many biologists, because the mathematical formulae looked so deliciously scientific and seemed to enhance the status of biology, making it more like physics. But here is what Waddington recently said about *this* development: "The theory of neo-Darwinism is a theory of the evolution of the population in respect to leaving offspring and not in respect to anything else. . . . Everybody has it in the back of his mind that the animals that leave the largest number of offspring are going to be those best adapted also for eating peculiar vegetation, or something of this sort, but this is not explicit in the theory. . . . There you do come to what is, in effect, a vacuous statement: Natural selection is that some things leave more offspring than others; and, you ask, which leave more offspring than others; and it is those that leave more offspring, and there is nothing more to it than that. *The whole real guts of evolution—which is how do you come to have horses and tigers and things—is outside the mathematical theory* [my italics]."

Here, then, was the problem. Darwin's theory was supposed to have answered this question about horses and tigers. They had gradually developed, bit by bit, as it were, over the eons, through the good offices of an agency called natural selection. But now, in its new incarnation, natural selection was only able to explain how horses and tigers became more (or less) numerous—that is, by "differential reproduction." This failed to solve the question of how they came into existence in the first place.

This was no good at all. As T. H. Morgan had remarked, with great clarity: "Selection, then, has not produced anything new, but only more of certain kinds of individuals. Evolution, however, means producing new things, not more of what already exists."

One more quotation should be enough to convince most people that Darwin's idea of natural selection was quietly abandoned, even by his most ardent supporters, some years ago. The following comment, by the geneticist H. J. Muller, another Nobel Prize winner, appeared in the Proceedings of the American Philosophical Society in 1949. It represents a direct admission by one of Darwin's greatest admirers that, however we come to have horses and tigers and things, it is not by natural selection. "We have just seen," Muller wrote, "that if selection could be somehow dispensed with, so that all variants survived and multiplied, the higher forms would nevertheless have arisen."

I think it should now be abundantly clear that Darwin made a mistake in proposing his natural-selection theory, and it is fairly easy to detect the mistake. We have seen that what the theory so grievously lacks is a criterion of fitness that is independent of survival. If only there were some way of identifying the fittest beforehand, without always having to wait and see which ones survive, Darwin's theory would be testable rather than tautological.

But as almost everyone now seems to agree, fittest inevitably means "those that survive best." Why, then, did Darwin assume that there were independent criteria? And the answer is, because in the case of artificial selection, from which he worked by analogy, *there really are independent criteria*. Darwin went wrong in thinking that this aspect of his analogy was valid. In our sheep example, remember, long wool was the "desirable" feature—the independent criterion. The lambs of woolly parental sheep may possess this feature even more than their parents, and so be "more evolved"—more in the desired direction.

In nature, on the other hand, the offspring may differ from their parents in any direction whatsoever and be considered "more evolved" than their parents, provided only that they survive and leave offspring themselves. There is, then, no "selection" by nature at all. Nor does nature "act," as it is so often said to do in biology books. One organism may indeed be "fitter" than another from an evolutionary point of view, but the only event that determines this fitness is death (or infertility). This, of course, is not something which helps *create* the organism, but is something that terminates it. It occurs at the end, not the beginning of life.

ONWARD AND UPWARD

Darwin seems to have made the mistake of just assuming that there were independent criteria of fitness because he lived in a society in which change was nearly always perceived as being for the good. B. C. Lewontin, Agassiz Professor of Zoology at Harvard, has written on this point: "The bourgeois revolution not only established change as the characteristic element of the cosmos, but added direction and progress as well. A world in which a man could rise from humble origins must have seemed, to him at least, a good world. Change per se was a moral quality. In this light, Spencer's assertion that change *is* progress is not surprising." One may note also James D. Watson's remark in *The Double Helix* that "cultural traditions play major roles" in the development of science.

Lewontin goes on to point out that "the bourgeois revolution gave way to a period of consolidation, a period in which we find ourselves now." Perhaps that is why only relatively recently has the concept of natural selection come under strong attack.

There is, in a way, a remarkable conclusion to this brief history of natural selection. The idea started out as a way of explaining how one type of animal gradually changed into another, but then it was redefined to be an explanation of how a given type of animal became more numerous. But wasn't natural selection supposed to have a *creative* role? the evolutionary theorists were asked. Darwin had thought so, after all. Now watch how they responded to this:

The geneticist Theodosius Dobzhansky compared natural selection to "a human activity such as performing or composing music." Sir Gavin de Beer described it as a "master of ceremonies." George Gaylord Simpson at one point likened selection to a poet, at another to a builder. Ernst Mayr, Lewontin's predecessor at Harvard, compared selection to a sculptor. Sir Julian Huxley topped them all, however, by comparing natural selection to William Shakespeare.

Life on Earth, initially thought to constitute a sort of prima facie case for a creator, was, as a result of Darwin's idea, envisioned merely as being the outcome of a process and a process that was, according to Dobzhansky, "blind, mechanical, automatic, impersonal," and, according to de Beer, was "wasteful, blind, and blundering." But as soon as these criticisms were leveled at natural selection, the "blind process" itself was compared to a poet, a composer, a sculptor, Shakespeare—to the very notion of creativity that the idea of natural selection had originally replaced. It is clear, I think, that there was something very, very wrong with such an idea.

I have not been surprised to read, therefore, in Lewontin's recent book, *The Genetic Basis of Evolutionary Change* (1974), that in some of the latest evolutionary theories "natural selection plays no role at all." Darwin, I suggest, is in the process of being discarded, but perhaps in deference to the venerable old gentleman, resting comfortably in Westminster Abbey next to Sir Isaac Newton, it is being done as discreetly and gently as possible, with a minimum of publicity.

*STEPHEN JAY GOULD**

DARWIN'S UNTIMELY BURIAL

In one of the numerous movie versions of *A Christmas Carol*, Ebenezer Scrooge encounters a dignified gentleman sitting on a landing, as he mounts the steps to visit his dying partner, Jacob Marley. "Are you the doctor?" Scrooge inquires. "No," replies the man, "I'm the undertaker; ours is a very competitive business." The cutthroat world of intellectuals must rank a close second, and few events attract more notice than a proclamation that popular ideas have died. Darwin's theory of natural selection has been a perennial candidate for burial. Tom Bethell held the most recent wake in a piece called "Darwin's Mistake" (*Harper's*, February 1976): "Darwin's theory, I believe, is on the verge of collapse . . . Natural selection was quietly abandoned, even by his most ardent supporters, some years ago." News to me, and I, although I wear the Darwinian label with some pride, am not among the most ardent defenders of natural selection. I recall Mark Twain's famous response to a premature obituary: "The reports of my death are greatly exaggerated."

Bethell's argument has a curious ring for most practicing scientists. We are always ready to watch a theory fall under the impact of new data, but we do not expect a great and influential theory to collapse from a logical error in its formulation. Virtually every empirical scientist has a touch of the Philistine. Scientists tend to ignore academic philosophy as an empty pursuit. Surely, any intelligent person can think straight by intuition. Yet Bethell cites no data at all in sealing the coffin of natural selection, only an error in Darwin's reasoning: "Darwin made a mistake sufficiently serious to undermine his theory. And that mistake has only recently been recognized as such. . . . At one point in his argument, Darwin was misled."

Although I will try to refute Bethell, I also deplore the unwillingness of scientists to explore seriously the logical structure of arguments. Much of what passes for evolutionary theory is as vacuous as Bethell claims. Many great theories are held together by chains of dubious metaphor and analogy. Bethell has correctly identified the hogwash surrounding evolutionary theory. But we differ

---

*From Stephen Jay Gould, "Darwin's Untimely Burial," *Natural History* 85, no 8 (October 1976): 24–30. Reprinted with permission of Rhonda Roland Shearer.

in one fundamental way: for Bethell, Darwinian theory is rotten to the core; I find a pearl of great price at the center.

Natural selection is the central concept of Darwinian theory—the fittest survive and spread their favored traits through populations. Natural selection is defined by Spencer's phrase "survival of the fittest," but what does this famous bit of jargon really mean? Who are the fittest? And how is "fitness" defined? We often read that fitness involves no more than "differential reproductive success"—the production of more surviving offspring than other competing members of the population. Whoa! cries Bethell, as many others have before him. This formulation defines fitness in terms of survival only. The crucial phrase of natural selection means no more than "the survival of those who survive"—a vacuous tautology. (A tautology is a phrase—like "my father is a man"—containing no information in the predicate ("a man") not inherent in the subject ("my father"). Tautologies are fine as definitions, but not as testable scientific statements—there can be nothing to test in a statement true by definition.)

But how could Darwin have made such a monumental, two-bit mistake? Even his severest critics have never accused him of crass stupidity. Obviously, Darwin must have tried to define fitness differently—to find a criterion for fitness independent of mere survival. Darwin did propose an independent criterion, but Bethell argues quite correctly that he relied upon analogy to establish it, a dangerous and slippery strategy. One might think that the first chapter of such a revolutionary book as *Origin of Species* would deal with cosmic questions and general concerns. It doesn't. It's about pigeons. Darwin devotes most of his first forty pages to "artificial selection" of favored traits by animal breeders. For here an independent criterion surely operates. The pigeon fancier knows what he wants. The fittest are not defined by their survival. They are, rather, allowed to survive because they possess desired traits.

The principle of natural selection depends upon the validity of an analogy with artificial selection. We must be able, like the pigeon fancier, to identify the fittest beforehand, not only by their subsequent survival. But nature is not an animal breeder; no preordained purpose regulates the history of life. In nature, any traits possessed by survivors must be counted as "more evolved"; in artificial selection, "superior" traits are defined before breeding even begins. Later evolutionists, Bethell argues, recognized the failure of Darwin's analogy and redefined "fitness" as mere survival. But they did not realize that they had undermined the logical structure of Darwin's central postulate. Nature provides no independent criterion of fitness; thus, natural selection is tautological.

Bethel then moves to two important corollaries of his major argument. First, if fitness only means survival, then how can natural selection be a "creative" force, as Darwinians insist. Natural selection can only tell us how "a given type of animal became more numerous"; it cannot explain "how one type of animal gradually changed into another." Secondly, why were Darwin and other eminent Victorians so sure that mindless nature could be compared with conscious selection by breeders? Bethell argues that the cultural climate of triumphant industrial capitalism had defined any change as inherently progressive. Mere survival in nature could only be for the good: "It is beginning to look as though what Darwin really discovered was nothing more than the Victorian propensity to believe in progress."

I believe that Darwin was right and that Bethell and his colleagues are mistaken: criteria of fitness independent of survival can be applied to nature and have been used consistently by evolutionists. But let me first admit that Bethell's criticism applies to much of the technical literature in evolutionary theory, especially to the abstract mathematical treatments that consider evolution only as an alteration in numbers, not as a change in quality. These studies do assess fitness only in terms of differential survival. What else can be done with abstract models that trace the relative successes of hypothetical genes A and B in populations that exist only on computer tape? Nature, however, is not limited by the calculations of theoretical geneticists. In nature, A's "superiority" over B will be *expressed* as differential survival, but it is not *defined* by it—or, at least, it better not be so defined, lest Bethell et al. triumph and Darwin surrender.

My defense of Darwin is neither startling, novel, nor profound. I merely assert that Darwin was justified in analogizing natural selection with animal breeding. In artificial selection, a breeder's desire represents a "change of environment" for a population. In this new environment, certain traits are superior a priori; (they survive and spread by our breeder's choice, but this is a *result* of their fitness, not a definition of it). In nature, Darwinian evolution is also a response to changing environments. Now, the key point: certain morphological, physiological, and behavioral traits should be superior a priori as designs for living in new environments. These traits confer fitness by an engineer's criterion of good design, not by the empirical fact of their survival and spread. It got colder before the woolly mammoth evolved its shaggy coat.

Why does this issue agitate evolutionists so much? OK, Darwin was right: superior design in changed environments is an independent criterion of fitness. So what? Did anyone ever seriously propose that the poorly designed shall triumph? Yes, in fact, many did. In Darwin's day, many rival evolutionary theories asserted that the fittest (best designed) must perish. One popular notion—the theory of racial life cycles—was championed by a former inhabitant of the office I now occupy, the great American paleontologist Alpheus Hyatt. Hyatt claimed that evolutionary lineages, like individuals, had cycles of youth, maturity, old age, and death (extinction). Decline and extinction are programmed into the history of species. As maturity yields to old age, the best-designed individuals die and the hobbled, deformed creatures of phyletic senility take over. Another anti-Darwinian notion, the theory of orthogenesis, held that certain trends, once initiated, could not be halted, even though they must lead to extinction caused by increasingly inferior design. Many nineteenth-century evolutionists (perhaps a majority) held that Irish elks became extinct because they could not halt their evolutionary increase in antler size; thus, they died—caught in trees or bowed (literally) in the mire. Likewise, the demise of saber-toothed "tigers" was often attributed to canine teeth grown so long that the poor cats couldn't open their jaws wide enough to use them.

Thus, it is not true, as Bethell claims, that any traits possessed by survivors must be designated as fitter. "Survival of the fittest" is not a tautology. It is also not the only imaginable or reasonable reading of the evolutionary record. It is testable. It had rivals that failed under the weight of contrary evidence and changing attitudes about the nature of life. It has rivals that may succeed, at least in limiting its scope.

If I am right, how can Bethell claim, "Darwin, I suggest, is in the process of being discarded, but perhaps in deference to the venerable old gentleman, resting comfortably in Westminster Abbey, next to Sir Isaac Newton, it is being done as discreetly and gently as possible with a minimum of publicity." I'm afraid I must say that Bethell has not been quite fair in his report of prevailing opinion. He cites the gadflies C. H. Waddington and H. J. Muller as though they epitomized a consensus. He never mentions the leading selectionists of our present generation—E. O. Wilson or D. Janzen, for example. And he quotes the architects of neo-Darwinism—Dobzhansky, Simpson, Mayr, and J. Huxley—only to ridicule their metaphors on the creativity of natural selection. (I am not claiming that Darwinism should be cherished because it is still popular; I am enough of a gadfly to believe that uncriticized consensus is a sure sign of impending trouble. I merely report that, for better or for worse, Darwinism is alive and thriving, despite Bethell's obituary.)

But why was natural selection compared to a composer by Dobzhansky; to a poet by Simpson; to a sculptor by Mayr; and to, of all people, Mr. Shakespeare by Julian Huxley? I won't defend the choice of metaphors, but I will uphold the intent, namely, to illustrate the essence of Darwinism—the creativity of natural selection. Natural selection has a place in all anti-Darwinian theories that I know. It is cast in a negative role as an executioner, a headsman for the unfit (while the fit arise by such non-Darwinian mechanisms as the inheritance of acquired characters or direct induction of favorable variation by the environment). The essence of Darwinism lies in its claim that natural selection creates the fit. Variation is ubiquitous and random in direction. It supplies the raw material only. Natural selection directs the course of evolutionary change. It preserves favorable variants and builds fitness gradually. In fact, since artists fashion their creations from the raw material of notes, words, and stone, the metaphors do not strike me as inappropriate. Since Bethell does not accept a criterion of fitness independent of mere survival, he can hardly grant a creative role to natural selection.

According to Bethell, Darwin's concept of natural selection as a creative force can be no more than an illusion encouraged by the social and political climate of his times. In the throes of Victorian optimism in imperial Britain, change seemed to be inherently progressive; why not equate survival in nature with increasing fitness in the nontautological sense of improved design.

I am a strong advocate of the general argument that "truth" as preached by scientists often turns out to be no more than prejudice inspired by prevailing social and political beliefs. I have devoted several essays to this theme because I believe that it helps to "demystify" the practice of science by showing its similarity to all creative human activity. But the truth of a general argument does not validate any specific application, and I maintain that Bethell's application is badly misinformed.

Darwin did two very separate things: he convinced the scientific world that evolution had occurred and he proposed the theory of natural selection as its mechanism. I am quite willing to admit that the common equation of evolution with progress made Darwin's first claim more palatable to his contemporaries. But Darwin failed in his second quest during his own lifetime. The theory of natural selection did not triumph until the 1940s. Its Victorian unpopularity, in my view, lay primarily in its denial of general progress as inherent in the workings of evolution. Natural selection is a theory of *local* adaptation to changing environments. It proposes no perfecting principles, no guarantee of general improvement; in short, no reason for general approbation in a political climate favoring innate progress in nature.

Darwin's independent criterion of fitness is, indeed, "improved design," but not "improved" in the cosmic sense that contemporary Britain favored. To Darwin, improved meant only "better designed for an immediate, local environment." Local environments change constantly: they get colder or hotter, wetter or drier, more grassy or more forested. Evolution by natural selection is no more than a tracking of these changing environments by differential preservation of organisms better designed to live in them: hair on a mammoth is not progressive in any cosmic sense. Natural selection can produce a trend that tempts us to think of more general progress—increase in brain size does characterize the evolution of group after group of mammals. But big brains have their uses in local environments; they do not mark intrinsic trends to higher states. And Darwin delighted in showing that local adaptation often produced "degeneration" in design—anatomical simplification in parasites, for example.

If natural selection is not a doctrine of progress, then its popularity cannot reflect the politics that Bethell invokes. If the theory of natural selection contains an independent criterion of fitness, then it is not tautological. I maintain, perhaps naively, that its current, unabated popularity must have something to do with its success in explaining the admittedly imperfect information we now possess about evolution. I rather suspect that we'll have Charles Darwin to kick around for some time.

MACRO-EVOLUTION

*STEPHEN JAY GOULD**

DARWINISM AND THE EXPANSION OF EVOLUTIONARY THEORY

Ben Sira, author of of the apocryphal book of Ecclesiasticus, paid homage to the heroes of Israel in a noted passage beginning, "let us now praise famous men." He glorified great teachers above all others, for their fame shall eclipse the immediate triumphs of kings and conquerors. And he argued that the corporeal death of teachers counts for nothing—indeed, it should be celebrated—since great ideas must live forever: "His name will be more glorious than a thousand others, and if he dies, that will satisfy him just as well." These sentiments express the compulsion we feel to commemorate the deaths of great thinkers; for their ideas still direct us today. Charles Darwin died 100 years ago, on April 19, 1882, but his name still causes fundamentalists to shudder and scientists to draw battle lines amid their accolades.

WHAT IS DARWINISM?

Darwin often stated that his biological work had embodied two different goals:[1] to establish the fact of evolution, and to propose natural selection as its primary mechanism. "I had," he wrote, "two distinct objects in view; firstly to show that species had not been separately created, and secondly, that natural selection had been the chief agent of change."[2]

---

*From Stephen Jay Gould, "Darwinism and the Expansion of Evolutionary Theory," *Science* 216, no. 4544 (April 1982): 380–87. Reprinted with permission of Rhonda Roland Shearer.

Although "Darwinism" has often been equated with evolution itself in popular literature, the term should be restricted to the body of thought allied with Darwin's own theory of mechanism, his second goal. This decision does not provide an unambiguous definition, if only because Darwin himself was a pluralist who granted pride of place to natural selection, but also advocated an important role for Lamarckian and other nonselectionist factors. Thus, as the nineteenth century drew to a close, G. J. Romanes and A. Weismann squared off in a terminological battle for rights to the name "Darwinian"—Romanes claiming it for his eclectic pluralism, Weismann for his strict selectionism.[3]

If we agree, as our century generally has, that "Darwinism" should be restricted to the world view encompassed by the theory of natural selection itself, the problem of definition is still not easily resolved. Darwinism must be more than the bare bones of the mechanics: the principles of superfecundity and inherited variation, and the deduction of natural selection therefrom. It must, fundamentally, make a claim for wide scope and dominant frequency; natural selection must represent the primary directing force of evolutionary change.

I believe that Darwinism, under these guidelines, can best be defined as embodying two central claims and a variety of peripheral and supporting statements more or less strongly tied to the central postulates; Darwinism is not a mathematical formula or a set of statements, deductively arranged.

(1) The creativity of natural selection. Darwinians cannot simply claim that natural selection operates since everyone, including Paley and the natural theologians, advocated selection as a device for removing unfit individuals at both extremes and preserving, intact and forever, the created type.[4] The essence of Darwinism lies in a claim that natural selection is the primary directing force of evolution, in that it creates fitter phenotypes by differentially preserving, generation by generation, the best adapted organisms from a pool of random variants[5] that supply raw material only, not direction itself. Natural selection is a creator; it builds adaptation step by step.

Darwin's contemporaries understood that natural selection hinged on the argument for creativity. Natural selection can only eliminate the unfit, his opponents proclaimed; something else must create the fit. Thus, the American Neo-Lamarckian E. D. Cope wrote a book with the sardonic title *The Origin of the Fittest*,[6] and Charles Lyell complained to Darwin that he could understand how selection might operate like two members of the "Hindoo triad"—Vishnu the preserver and Siva the destroyer—but not like Brahma the creator.[7]

The claim for creativity has important consequences and prerequisites that also become part of the Darwinian corpus. Most prominently, three constraints are imposed on the nature of genetic variation (or at least the evolutionarily significant portion of it). (i) It must be copious since selection makes nothing directly and requires a large pool of raw material. (ii) It must be small in scope. If new species characteristically arise all at once, then the fit are formed by the process of variation itself, and natural selection only plays the negative role of executioner for the unfit. True saltationist theories have always been considered anti-Darwinian on this basis. (iii) It must be undirected. If new environments can elicit heritable, adaptive variation, then creativity lies in the process of variation, and selection only eliminates the unfit. Lamarckism is an anti-Darwinian theory because it advocates directed variation; organisms perceive felt needs, adapt their bodies accordingly, and pass these modifications directly to offspring.

Two additional postulates, generally considered part and parcel of the Darwinian world view, are intimately related to the claim for creativity, but are not absolute prerequisites or necessary deductive consequences: (i) *Gradualism.* If creativity resides in a step-by-step process of selection from a pool of random variants, then evolutionary change must be dominantly continuous and descendants must be linked to ancestors by a long chain of smoothly intermediate phenotypes. Darwin's own gradualism precedes his belief in natural selection and has deeper roots;[8] it dominated his world view and provided a central focus for most other theories that he proposed, including the origin of coral atolls by subsidence of central islands, and formation of vegetable mold by earthworms.[9, 10] (ii) *The adaptationist program.* If selection becomes creative by superintending, generation by generation, the continuous incorporation of favorable variation into altered forms, then evolutionary change must be fundamentally adaptive. If evolution were saltational, or driven by internally generated biases in the direction of variation, adaptation would not be a necessary attribute of evolutionary change.

The argument for creativity rests on relative frequency, not exclusivity. Other factors must regulate some cases of evolutionary change—randomness as a direct source of modification, not only of raw material, for example. The Darwinian strategy does not deny other factors, but attempts to circumscribe their domain to few and unimportant cases.

(2) Selection operates through the differential reproductive success of individual organisms (the "struggle for existence" in Darwin's terminology). Selection is an interaction among individuals; there are no higher-order laws in nature, no statements about the "good" of species or ecosystems. If species survive longer, or if ecosystems appear to display harmony and balance, these features arise as a by-product of selection among individuals for reproductive success.

Although evolutionists, including many who call themselves Darwinians, have often muddled this point,[11] it is a central feature of Darwin's logic.[12] It underlies all his colorful visual imagery including the metaphor of the wedge or the true struggle that underlies an appearance of harmony: "we behold the face of nature bright with gladness," but. . . .[13] Darwin developed his theory of natural selection by transferring the basic argument of Adam Smith's economics into nature:[14] an ordered economy can best be achieved by letting individuals struggle for personal profits, thereby permitting a natural sifting of the most competitive (laissez-faire); an ordered ecology is a transient balance established by successful competitors pursuing their own Darwinian edge.

As a primary consequence, this focus upon individual organisms leads to reductionism, not to ultimate atoms and molecules of course, but of higher-order, or macroevolutionary, processes to the accumulated struggles of individuals. Extrapolationism is the other side of the same coin—the claim that natural selection within local populations is the source of all important evolutionary change.

DARWINISM AND THE MODERN SYNTHESIS

Although Darwin succeeded in his first goal, and lies in Westminster Abbey for his success in establishing the fact of evolution, his theory of natural selection did not triumph as an orthodoxy until long after his death. The Mendelian component to the modern, or Neo-Darwinian, theory only developed in our century. Moreover, and ironically, the first Mendelians emphasized macromutations and were non-Darwinians on the issue of creativity as discussed above.

The Darwinian resurgence began in earnest in the 1930s, but did not crystallize until the 1950s. At the last Darwinian centennial, in 1959 (both the 100th anniversary of the *Origin of Species* and the 150th of Darwin's birth), celebrations throughout the world lauded the "modern synthesis" as Darwinism finally triumphed.[15]

Julian Huxley, who coined the term,[16] defined the "modern synthesis" as an integration of the disparate parts of biology about a Darwinian core.[17] Synthesis occurred at two levels: (i) The Mendelian research program merged with Darwinian traditions of natural history, as Mendelians recognized the importance of micromutations and their correspondence with Darwinian variation, and as population genetics supplied a quantitative mechanics for evolutionary change. (ii) The traditional disciplines of natural history, systematics, paleontology, morphology, and classical botany, for example,[18] were integrated within the Darwinian core, or at least rendered consistent with it.

The initial works of the synthesis, particularly Dobzhansky's first (1937) edition of *Genetics and the Origin of Species*, were not firmly Darwinian (as defined above), and did not assert a dominant frequency for natural selection. They were more concerned with demonstrating that large-scale phenomena of evolution are consistent with the principles of genetics, whether Darwinian or not; and they therefore, for example, granted greater prominance to genetic drift than later editions of the same works would allow.

Throughout the late 1940s and 1950s, however, the synthesis hardened about its Darwinian core. Analysis of textbooks and, particularly, the comparison of first with later editions of the founding documents, demonstrates the emergence of natural selection and adaptation as preeminent factors of evolution. Thus, for example, G. C. Simpson redefined "quantum evolution" in 1953 as a limiting rate for adaptive phyletic transformation, not, as he had in 1944, as a higher-order analog of genetic drift, with a truly inadaptive phase between stabilized end points.[19] Dobzhansky removed chapters and reduced emphasis upon rapid modification and random components to evolutionary change.[20] David Lack reassessed his work on Darwin's finches and decided that minor differences among species are adaptive after all.[21] His preface to the 1960 reissue of his monograph features the following statement:[22]

> This text was completed in 1944 and . . . views on species-formation have advanced. In particular, it was generally believed when I wrote the book that, in animals, nearly all of the differences between subspecies of the same species, and between closely related species in the same genus, were without adaptive significance. . . . Sixteen years later it is generally believed that all or almost all subspecific and specific differences are adaptive. . . . Hence it now seems probable that at least most of the seemingly nonadaptive differences in Darwin's finches would, if more were known, prove to be adaptive.

Mayr's definition of the synthesis, offered without rebuttal at a conference of historians and architects of the theory reflects this crystallized version:

> The term "evolutionary synthesis" was introduced by Julian Huxley . . . to designate the general acceptance of two conclusions: gradual evolution can be explained in terms of small genetic changes ("mutations") and recombination, and the ordering of this genetic variation by natural selection; and the observed evolutionary phenomena, particularly macroevolutionary processes and speciation, can be explained in a manner that is consistent with the known genetic mechanisms.[24]

This definition restates the two central claims of Darwinism discussed in the last section: Mayr's first conclusion, with its emphasis on gradualism, small genetic change, and natural selection, represents the argument for creativity; while the second embodies the claim for reduction. I have been challenged for erecting a straw man in citing this definition of the synthesis,[24] but it was framed by a man who is both an architect and the leading historian of the theory, and it is surely an accurate statement of what I was taught as a graduate student in the mid-1960s. Moreover, these very words have been identified as the "broad version" of the synthesis (as opposed to a more partisan and restrictive stance) by White,[25] a leading evolutionist and scholar who lived through it all.

The modern synthesis has sometimes been so broadly construed usually by defenders who wish to see it as fully adequate to meet and encompass current critiques, that it loses all meaning by including everything. If, as Stebbins and Ayala claim, "'selectionist' and 'neutralist' views of molecular evolution are competing hypotheses within the framework of the synthetic theory of evolution,"[26] then what serious views are excluded? King and Jukes, authors of the neutralist theory, named it "non-Darwinian evolution" in the title of their famous paper.[27] Stebbins and Ayala have tried to win an argument by redefinition. The essence of the modern synthesis must be its Darwinian core. If most evolutionary change is neutral, the synthesis is severely compromised.

WHAT IS HAPPENING TO DARWINISM

Current critics of Darwinism and the modern synthesis are proposing a good deal more than a comfortable extension of the theory, but much less than a revolution. In my partisan view, neither of Darwinism's two central themes will survive in their strict formulation; in that sense, "the modern synthesis, as an exclusive proposition, has broken down on both of its fundamental claims."[28] However, I believe that a restructured evolutionary theory will embody the essence of the Darwinian argument in a more abstract, and hierarchically extended form. The modern synthesis is incomplete, not incorrect.

CRITIQUE OF CREATIVITY: GRADUALISM

At issue is not the general idea that natural selection can act as a creative force; the basic argument, in principle, is a sound one. Primary doubts center on the subsidiary claims—gradualism and the adaptationist program. If most evolutionary changes, particularly large-scale trends, include major nonadaptive components as primary directing or channeling features, and if they proceed more in an episodic than a smoothly continuous fashion, then we inhabit a different world from the one Darwin envisaged.

Critiques of gradualist thought proceed on different levels and have different import, but none are fundamentally opposed to natural selection. They are therefore not directed against the heart of Darwinian theory, but against a fundamental subsidiary aspect of Darwin's own world view—one that he consistently conflated with natural selection, as in the following famous passage: "If it could be demonstrated that any complex organ existed, which could not possibly have been formed by numerous, successive, slight modifications, my theory would absolutely break down."[29]

At the levels of microevolution and speciation, the extreme saltationist claim that new species arise all at once, fully formed, by a fortunate macromutation would be anti-Darwinian, but no serious thinker now advances such a view, and neither did Richard Goldschmidt,[30] the last major scholar to whom such an opinion is often attributed. Legitimate claims range from the saltational origin of key features by developmental shifts of dissociable segments of ontogeny[31] to the origin of reproductive isolation (speciation) by major and rapidly incorporated genetic changes that precede the acquisition of adaptive, phenotypic differences.[32]

Are such styles of evolution anti-Darwinian? What can one say except "yes and no." They do not deny a creative role to natural selection, but neither do they embody the constant superintending of each event, or the step-by-step construction of each major feature, that traditional views about natural selection have advocated. If new *Baupläne* often arise in an adaptive cascade following the saltational origin of a key feature, then part of the process is sequential and adaptive, and therefore Darwinian; but the initial step is not, since selection does not play a creative role in building the key feature. If reproductive isolation often precedes adaptation, then a major aspect of speciation is Darwinian (for the new species will not prosper unless it builds distinctive adaptations in the sequential mode), but its initiation, including the defining feature of reproductive isolation, is not.

At the macroevolutionary level of trends, the theory of punctuated equilibrium[33] proposes that established species generally do not change substantially in phenotype over a lifetime that may encompass many million years (stasis), and that most evolutionary change is concentrated in geologically instantaneous events of branching speciation. These geological instants, resolvable[34] in favorable stratigraphic circumstances (so that the theory can be tested for its proposed punctuations as well as for its evident periods of stasis), represent amounts of microevolutionary time fully consistent with orthodox views about speciation. Indeed, Eldredge and I originally proposed punctuated equilibrium as the expected geological consequence of Mayr's theory of peripatric speciation. The non-Darwinian implications of punctuated equilibrium lie in its suggestions for the explanation of evolutionary trends (see below), not in the tempo of individual speciation events. Although punctuated equilibrium is a theory for a higher level of evolutionary change, and must therefore be agnostic with respect to the role of natural selection in speciation, the world that it proposes is quite different from that traditionally viewed by paleontologists (and by Darwin himself) as the proper geological extension of Darwinism.

The "gradualist-punctuationalist debate," the general label often applied to this disparate series of claims, may not be directed at the heart of natural selection, but it remains an important critique of the Darwinian tradition. The world is not inhabited exclusively by fools, and when a subject arouses intense interest and debate, as this one has, something other than semantics is usually at stake. In the largest sense, this debate is but one small aspect of a broader discussion about the nature of change: Is our world (to construct a ridiculously oversimplified dichotomy) primarily one of constant change (with structure as a mere incarnation of the moment), or is structure primary and constraining, with change as a "difficult" phenomenon, usually accomplished rapidly when a stable structure is stressed beyond its buffering capacity to resist and absorb. It would be hard to deny that the Darwinian tradition, including the modern synthesis, favored the first view while "punctuationalist" thought in general, including such aspects of classical morphology as D'Arcy Thompson's theory of form,[35] prefers the second.

CRITIQUE OF CREATIVITY: ADAPTATION

The primary critiques of adaptation have arisen from molecular data, particularly from the approximately even ticking of the molecular clock, and the argument that natural populations generally maintain too much genetic variation to explain by natural selection, even when selection acts to preserve variation as in, for example, heterozygote advantage and frequency-dependent selection. To these phenomena, Darwinians have a response that is, in one sense, fully justified: Neutral genetic changes without phenotypic consequences are invisible to Darwinian processes of selection upon organisms and therefore represent a legitimate process separate from the subjects that Darwinism can treat. Still, since issues in natural history are generally resolved by appeals to relative frequency, the domain of Darwinism is restricted by these arguments.

But another general critique of the adaptationist program has been reasserted within the Darwinian domain of phenotypes.[36] The theme is an old one, and not unfamiliar to Darwinians. Darwin himself took it seriously, as did the early, pluralistic accounts of the modern synthesis. The later, "hard" version of the synthesis relegated it to unimportance or lip service. The theme is two-pronged, both arguments asserting that the current utility of a structure permits no assumption that selection shaped it. First, the constraints of inherited form and developmental pathways may so channel any change that even though selection induces motion down permitted paths, the channel itself represents the primary determinant of evolutionary direction. Second, current utility permits no necessary conclusion about historical origin. Structures now indispensable for survival may have arisen for other reasons and been "coopted" by functional shift for their new role.

Both arguments have their Darwinian versions. First, if the channels are set by past adaptations, then selection remains preeminent, for all major structures are either expressions of immediate selection, or channeled by a phylogenetic heritage of previous selection. Darwin struggled mightily with this problem. Ultimately, in a neglected passage that I regard as one of the most crucial paragraphs in the *Origin of Species*,[37] he resolved his doubts, and used this argument to uphold the great British tradition of adaptationism. Second, if coopted structures initially arose as adaptations for another function, then they too are products of selection, albeit in a regime not recorded by their current usage. We call this phenomenon preadaptation; as the primary solution to Mivart's taunt[38] about "the incipient stages of useful structures," it is a central theme of orthodox Darwinism.

But both arguments also have non-Darwinian versions, not widely appreciated but potentially fundamental. First, many features of organic architecture and developmental pathways have never been adaptations to anything, but arose as by-products or incidental consequences of changes with a basis in selection. Seilacher has suggested, for example, that the divaricate pattern of molluscan ornamentation may be nonadaptive in its essential design. In any case it is certainly a channel for some fascinating subsidiary adaptations.[39] Second, many structures available for cooptation did not arise as adaptations for something else (as the principle of preadaptation assumes) but were nonadaptive in their original construction. Evolutionary morphology now lacks a term for these coopted structures, and unnamed phenomena are not easily conceptualized. Vrba and I suggest that they be called exaptations,[40] and present a range of potential examples from the genitalia of hyenas to redundant DNA.

Evolutionists admit, of course, that all selection yields by-products and incidental consequences, but we tend to think of these nonadaptations as a sort of evolutionary frill, a set of small and incidental modifications with no major consequences. I dispute this assessment and claim that the pool of nonadaptations must be far greater in extent than the direct adaptations that engender them. This pool must act as a higher-level analog of genetic variation, as a phenotypic source of raw material for further evolution. Nonadaptations are not just incidental allometric and pleiotropic effects on other parts of the body, but multifarious expressions potentially within any adapted structure. No one doubts, for example, that the human brain became large for a set of complex reasons related to selection. But, having reached its unprecedented bulk, it could, as a computer of some sophistication, perform in an unimagined range of ways bearing no relation to the selective reasons for initial enlargement. Most of human society may rest on these nonadaptive consequences. How many human institutions, for example, owe their shape to that most terrible datum that intelligence permitted us to grasp—the fact of our personal mortality.

I do not claim that a new force of evolutionary change has been discovered. Selection may supply all immediate direction, but if highly constraining channels are built of nonadaptations, and if evolutionary versatility resides primarily in the nature and extent of nonadaptive pools, then "internal" factors of organic design are an equal partner with selection. We say that mutation is the ultimate source of variation, yet we grant a fundamental role to recombination and the evolution of sexuality—often as a prerequisite to multicellularity, the Cambrian explosion and, ultimately, us. Likewise, selection may be the ultimate source of evolutionary change, but most actual events may owe more of their shape to its nonadaptive sequelae.

IS EVOLUTION A PRODUCT OF SELECTION AMONG INDIVIDUALS?

Although arguments for a multiplicity of units of selection have been advanced and widely discussed,[41] evolutionists have generally held fast to the overwhelming predominance, if not exclusivity, of organisms as the objects sorted by selection—Dawkins'[42] attempt at further reduction to the gene itself notwithstanding. How else can we explain the vehement reaction of many evolutionists to Wynne-Edwards' theory of group selection for the maintenance of altruistic traits,[43] or the delight felt by so many when the same phenomena were explained, under the theory of kin selection, as a result of individuals pursuing their traditional Darwinian edge. I am not a supporter of Wynne-Edwards' particular hypothesis, nor do I doubt the validity and importance of kin selection; I merely point out that the vehemence and delight convey deeper messages about general attitudes.

Nonetheless, I believe that the traditional Darwinian focus on individual bodies, and the attendant reductionist account of macroevolution, will be supplanted by a hierarchical approach recognizing legitimate Darwinian individuals at several levels of a structural hierarchy, including genes, bodies, demes, species, and clades.

The argument may begin with a claim that first appears to be merely semantic, yet contains great utility and richness in implication, namely the conclusion advanced by Ghiselin and later supported by Hull that species should be treated as individuals, not as classes.[44] Most species function as entities in nature, with coherence and stability. And they display the primary characteristics of a Darwinian actor; they vary within their population (clade in this case), and they exhibit differential rates of birth (speciation) and death (extinction).

Our language and culture include a prejudice for applying the concept of individual only to bodies, but any coherent entity that has a unique origin, sufficient temporal stability, and a capacity for reproduction with change can serve as an evolutionary agent. The actual hierarchy of our world is a contingent fact of history, not a heuristic device or a logical necessity. One can easily imagine a world devoid of such hierarchy, and conferring the status of evolutionary individual upon bodies alone. If genes could not duplicate themselves and disperse among chromosomes, we might lack the legitimately independent level that the "selfish DNA" hypothesis establishes for some genes.[45] If new species usually arose by the smooth transformation of an entire ancestral species, and then changed continuously toward a descendant form, they would lack the stability and coherence required for defining evolutionary individuals. The theory of punctuated equilibrium allows us to individuate species in both time and space; this property (rather than the debate about evolutionary tempo) may emerge as its primary contribution to evolutionary theory.

In itself, individuation does not guarantee the strong claim for evolutionary agency: that the higher-level individual acts as a unit of selection in its own right. Species might be individuals, but their differential evolutionary success might still arise entirely from natural selection acting upon their parts, that is, upon phenotypes of organisms. A trend toward increasing brain size, for example, might result from the greater longevity of big-brained species. But big-brained species might prosper only because the organisms within them tend to prevail in traditional competition.

But individuation of higher-level units is enough to invalidate the reductionism of traditional Darwinism—for pattern and style of evolution depend critically on the disposition of higher-level individuals, even when all selection occurs at the traditional level of organisms. Sewall Wright, for example, has often spoken of "interdemic selection" in his shifting balance theory,[46] but he apparently uses this phrase in a descriptive sense and believes that the mechanism of change usually resides in selection among individual organisms, as when, for example, migrants from one deme swamp another. Still, the fact of deme structure itself—that is, the individuation of higher-level units within a species—is crucial to the operation of shifting balance. Without division into demes, and under panmixia, genetic drift could not operate as the major source of variation required by the theory.

We need not, however, confine ourselves to the simple fact of individuation as an argument against Darwinian reductionism. For the strong claim that higher-level individuals act as units of selection in their own right can often be made. Many evolutionary trends, for example, are driven by differential frequency of speciation (the analog of birth) rather than by differential extinction (the more usual style of selection by death). Features that enhance the frequency of speciation are often properties of populations, not of individual organisms, for example, dependence of dispersal (and resultant possibilities for isolation and speciation) on size and density of populations.

Unfortunately, the terminology of this area is plagued with a central confusion (some, I regret to say, abetted by my own previous writings). Terms like "interdemic selection" or "species selection"[47] have been used in the purely descriptive sense, when the sorting out among higher-level individuals may arise solely from natural selection operating upon organisms. Such cases are explained by Darwinian selection, although they are irreducible to organisms alone. The same terms have been restricted to cases of higher-level individuals acting as units of selection. Such situations are non-Darwinian, and irreducible on this strong criterion. Since issues involving the locus of selection are so crucial in evolutionary theory, I suggest that these terms only be used in the strong and restricted sense. Species selection, for example, should connote an irreducibility to individual organisms (because populations are acting as units of selection); it should not merely offer a convenient alternative description for the effects of traditional selection upon organisms.

The logic of species selection is sound, and few evolutionists would now doubt that it can occur in principle. The issue, again and as always in natural history, is one of relative frequency; how often does species selection occur, and how important is it in the panoply of evolutionary events. Fisher himself dismissed species selection because, relative to organisms, species are so few in number (within a clade) and so long in duration:[48]

> The relative unimportance of this as an evolutionary factor would seem to follow decisively from the small number of closely related species which in fact do come into competition, as compared to the number of individuals in the same species; and from the vastly greater duration of the species compared to the individual.

But Fisher's argument rests on two hidden and questionable assumptions. (i) Mass selection can almost always be effective in transforming entire populations substantially in phenotype. The sheer number of organisms participating in this efficient process would then swamp any effect of selection among species. But if stasis be prevalent within established species, as the theory of punctuated equilibrium asserts and as paleontological experience affirms (overwhelmingly for marine invertebrates, at least), then the mere existence of billions of individuals and millions of generations guarantees no substantial role for directional selection upon organisms. (ii) Species selection depends on direct competition among species. Fisher argues for differential death (extinction) as the mechanism of species selection. I suspect, however, that differential frequency of speciation (selection by birth) is a far more common and effective mode of species selection. It may occur without direct competition between species, and can rapidly shift the average phenotype within a clade in regimes of random extinction.

J. Maynard Smith has raised another objection against species selection: simply, that most features of organisms represent "things individual creatures do."[49] How, he asks, could one attribute the secondary palate of mammals to species selection? But the origin of a feature is one thing (and I would not dispute traditional selection among organisms as the probable mechanism for evolving a secondary palate), and the spread of features through larger clades is another. Macroevolution is fundamentally about the combination of features and their differential spread. These phenomena lie comfortably within the domain of effective species selection. Many features must come to prominence primarily through their fortuitous phyletic link with high speciation rates. Mammals represent a lineage of therapsids that may have survived (while all others died) as a result of small body sizes and nocturnal habits. Was the secondary palate a key to their success, or did it piggyback on the high speciation rates often noted (for other reasons) in small-bodied forms? Did mammals survive the Cretaceous extinction, thereby inheriting the world from dinosaurs, as a result of their secondary palate, or did their small size again preserve them during an event that differentially wiped out large creatures?

EVOLUTIONARY PATTERN BY INTERACTION BETWEEN LEVELS

The hierarchical model, with its assertion that selection works simultaneously and differently upon individuals at a variety of levels, suggests a revised interpretation for many phenomena that have puzzled people where they implicitly assumed causation by selection upon organisms. In particular, it suggests that negative interaction between levels might be an important principle in maintaining stability or holding rates of change within reasonable bounds.

The "selfish DNA" hypothesis, for example, proposes that much middle-repetitive DNA exists within genomes not because it provides Darwinian benefits to phenotypes, but because genes can (in certain circumstances) act as units of selection. Genes that can duplicate themselves and move among chromosomes will therefore accumulate copies of themselves for their own Darwinian reasons. But why does the process ever stop? The authors of the hypothesis suggest that phenotypes will eventually "notice" the redundant copies when the energetic cost of producing them becomes high enough to entail negative selection at the level of organisms. Stability may represent a balance between positive selection at the gene level and the negative selection it eventually elicits at the organism level.

All evolutionary textbooks grant a paragraph or two to a phenomenon called "overspecialization," usually dismissing it as a peculiar and peripheral phenomenon. It records the irony that many creatures, by evolving highly complex and ecologically constraining features for their immediate Darwinian advantage, virtually guarantee the short duration of their species by restricting its capacity for subsequent adaptation. Will a peacock or an Irish elk survive when the environment alters radically? Yet fancy tails and big antlers do lead to more copulations in the short run of a lifetime. Overspecialization is, I believe, a central evolutionary phenomenon that has failed to gain the attention it deserves because we have lacked a vocabulary to express what is really happening: the negative interaction of species-level disadvantage and individual-level advantage. How else can morphological specialization be kept within bounds, leaving a place for drab and persistent creatures of the world. The general phenomenon must also regulate much of human society, with many higher-level institutions compromised or destroyed by the legitimate demands of individuals (high salaries of baseball stars, perhaps).

Some features may be enhanced by positive interaction between levels. Stenotopy in marine invertebrates, for example, seems to offer advantages at both the individual level (when environments are stable) and at the species level (boosting rates of speciation by brooding larvae and enhancing possibilities for isolation relative to eurytopic species with planktonic larvae). Why then do eurytopic species still inhabit our oceans? Suppression probably occurs at the still higher level of clades, by the differential removal of stenotopic branches in major environmental upheavals that accompany frequent mass extinctions in the geological record.

If no negative effect from a higher level suppressed an advantageous lower-level phenomenon, then it might sweep through life. Sex in eukaryotic organisms may owe its prominence to unsuppressed positive interaction between levels. The advantages of sex have inspired a major debate among evolutionists during the past decade. Most authors seek traditional explanation in terms of benefit to organisms,[50] for example, better chance for survival of some offspring if all are not Xeroxed copies of an asexual parent, but the genetically variable products of two individuals. Some, however, propose a spread by species selection, for example, by vastly higher speciation rates in sexual creatures.[51]

The debate has often proceeded by mutual dismissal, each side proclaiming its own answers correct. Perhaps both are right, and sex predominates because two levels interact positively and are not suppressed at any higher level. No statement is usually more dull and unenlightening than the mediator's claim, "you're both right." In this case, however, we must adopt a different view of biological organization itself to grasp the mediator's wisdom—and the old solution, for once, becomes interesting in its larger implication. We live in a world with reductionist traditions, and do not react comfortably to notions of hierarchy. Hierarchical theories permit us to retain the value of traditional ideas, while adding substantially to them. They traffic in accretion, not substitution. If we abandoned the "either-or" mentality that has characterized arguments about units of selection, we would not only reduce fruitless and often acrimonious debate, but we would also gain a deeper understanding of nature's complexity through the concept of hierarchy.

A HIGHER DARWINISM?

What would a fully elaborated, hierarchically based evolutionary theory be called? It would neither be Darwinism, as usually understood, nor a smoothly continuous extension of Darwinism, for it violates directly the fundamental reductionist tradition embodied in Darwin's focus on organisms as units of selection.

Still, the hierarchical model does propose that selection operates on appropriate individuals at each level. Should the term "natural selection" be extended to all levels above and below organisms; there is certainly nothing unnatural about species selection. Some authors have extended the term, while others, Slatkin for example, restrict natural selection to its usual focus upon individual organisms: "Species selection is analogous to natural selection acting on an asexual population."[52]

Terminological issues aside, the hierarchically based theory would not be Darwinism as traditionally conceived; it would be both a richer and a different theory. But it would embody, in abstract form, the essence of Darwin's argument expanded to work at each level. Each level generates variation among its individuals; evolution occurs at each level by a sorting out among individuals, with differential success of some and their progeny. The hierarchical theory would therefore represent a kind of "higher Darwinism," with the substance of a claim for reduction to organisms lost, but the domain of the abstract "selectionist" style of argument extended.

Moreover, selection will work differently on the objects of diverse levels. The phenomena of one level have analogs on others, but not identical operation. For example, we usually deny the effectiveness of mutation pressure at the level of organisms. Populations contain so many individuals that small biases in mutation rate can rarely establish a feature if it is under selection at all. But the analog of mutation pressure at the species level, directed speciation (directional bias toward certain phenotypes in derived species), may be a powerful agent of evolutionary trends (as a macroevolutionary alternative to species selection). Directed speciation can be effective (where mutation pressure is not) for two reasons: first, because its effects are not so easily swamped (given the restricted number of species within a clade) by differential extinction; second, because such phenomena as ontogenetic channeling in phyletic size increase suggest that biases in the production of species may be more prevalent than biases in the genesis of mutations.

Each level must be approached on its own, and appreciated for the special emphasis it places upon common phenomena, but the selectionist style of argument regulates all levels and the Darwinian vision is extended and generalized, not defeated, even though Darwinism, strictly constructed, may be superseded. This expansion may impose a literal wisdom upon that famous last line of *Origin of Species*, "There is grandeur in this view of life."

Darwin, at the centenary of his death, is more alive than ever. Let us continue to praise famous men.

REFERENCES AND NOTES

1. I have argued [*Natural History* 91 (April 1982)] that a third and larger theme captures the profound importance and intellectual power of Darwin's work in a more comprehensive way: his successful attempt to establish principles of reasoning for historical science. Each of his so-called "minor" works (treatises on orchids, worms, climbing plants, coral reefs, barnacles, for example) exhibits both an explicit and a covert theme—and the covert theme is a principle of reasoning for the reconstruction of history. The principles can be arranged in order of decreasing availability of information, but each addresses the fundamental issue: how can history be scientific if we cannot directly observe a past process: (i) If we can observe present processes at work, then we should accumulate and extrapolate their results to render the past. Darwin's last book, on the formation of veg-

etable mold by earthworms (1881), is also a treatise on this aspect of uniformitarianism. (ii) If rates are too slow or scales too broad for direct observation, then try to render the range of present results as stages of a single historical process. Darwin's first book on a specific subject, the subsidence theory of coral atolls (1842), is (in its covert theme) a disquisition on this principle. (iii) When single objects must be analyzed, search for imperfections that record constraints of inheritance. Darwin's orchid book (1862), explicitly about fertilization by insects, argues that orchids are jury-rigged, rather than well built from scratch, because structures that attract insects and stick pollen to them had to be built from ordinary parts of ancestral flowers. Darwin used all three principles to establish evolution as well: (i) observed rates of change in artificial selection, (ii) stages in the process of speciation displayed by modern populations, and (iii) analysis of vestigial structures in various organisms. Thus, we should not claim that all Darwin's books are about evolution. Rather, they are all about the methodology of historical science. The establishment of evolution represents the greatest triumph of the method.

2. C. Darwin, *The Descent of Man*, ed. 2 (London: Murray 1889), p. 61.

3. G. J. Romanes, *Darwin, and After Darwin* (London: Longmans, Green and Co., 1900), pp. 1–36.

4. Failure to recognize that all creationists accepted selection in this negative role led Eiseley to conclude falsely that Darwin had "borrowed" the principle of natural selection from his predecessor E. Blyth [L. Eiseley, *Darwin and the Mysterious Mr. X* (New York: Dutton, 1979). The Reverend William Paley's classic work *Natural Theology*, published in 1803, also contains many references to selective elimination.

5. By "random" in this context, evolutionists mean only that variation is not inherently directed towards adaptation, not that all mutational changes are equally likely. The word is unfortunate, but the historical tradition too deep to avoid.

6. E. D. Cope, *The Origin of the Fittest* (New York: Appleton, 1887).

7. L. C. Wilson, ed., *Sir Charles Lyell's Scientific Journals on the Species Question* (New Haven, CT: Yale University Press, 1970), p. 369.

8. Darwin was convinced, for example, in part by reading a theological work arguing that extreme rapidity (as in the initial spread of Christianity) indicated a divine hand, that gradual and continuous change was the mark of a natural process [H. Gruber, *Darwin on Man* (New York: Dutton, 1974)].

9. C. Darwin, *The Structure and Distribution of Coral Reefs* (London: Smith, Elder and Co., 1842).

10. C. Darwin, *The Formation of Vegetable Mould, Through the Action of Worms* (London: Murray, 1881).

11. The following works have done great service in identifying and correcting this confusion: G. C. Williams, *Adaptation and Natural Selection* (Princeton, NJ: Princeton University Press, 1966); J. Maynard Smith, *The Evolution of Sex* (New York: Cambridge University Press, 1978).

12. A persuasive case for Darwin's active interest in this subject and for his commitment to individual selection has been recently made by M. Ruse, *Annals of Science* 37 (1980): 615.

13. Darwin, *On the Origin of Species* (London: Murray, 1859).

14. S. S. Schweber, *Journal of the History of Biology* 10 (1977): 229.

15. S. Tax, ed., *Evolution After Darwin*, vols. 1–3 (Chicago: University of Chicago Press, 1960).

16. J. Huxley, *Evolution, the Modern Synthesis* (London: Allen & Unwin, 1942).

17. For example: "The opposing factions became reconciled as the younger branches of biology achieved a synthesis with each other and with the classical disciplines: and the reconciliation converged upon a Darwinian center" (ibid., p. 25).

18. E. Mayr, *Systematics and the Origin of Species* (New York: Columbia University Press, 1942); C. C. Simpson, *Tempo and Mode in Evolution* (New York: Columbia University Press, 1944); B. Rensch, *Neuere Probleme der Abstammungslehre* (Stuttgart: Enke, 1947); G. L. Stebbins, *Variation and Evolution in Plants* (New York: Columbia University Press, 1950).

19. S. J. Gould, in *The Evolutionary Synthesis*, eds. E. Mayr and W. B. Provine (Cambridge, MA: Harvard University Press, 1980), p. 153.

20. Ibid., *Dobzhansky and the Modern Synthesis*, introduction to Th. Dobzhansky, *Genetics and the Origin of Species*, 1st ed. reprint (New York: Columbia University Press, 1982).

21. D. Lack, *Darwin's Finches* (New York: Harper Torchbook, 1960).

22. Ibid. This statement appears as the first paragraph in the preface.

23. Mayr, in *The Evolutionary Synthesis*, p. 1.

24. S. Orzack, (1981) *Paleobiology* 7:128.

25. M. J. D. White, *Paleobiology* 7:287.

26. C. L. Stebbins and F. J. Ayala, *Science* 213 (1981): 967.

27. J. L. King and T. H. Jukes, *Science* 164 (1969): 788.

28. Gould, *Paleobiology* 6 (1980): 119.

29. Darwin, *On the Origin of Species*, p. 189. On the day before publication of the *Origin of Species*, T. H. Huxley wrote to Darwin (letter of November 23, 1859): "You load yourself with an unnecessary difficulty in adopting *Natura non facit saltum* so unreservedly."

30. Gould, *The Uses of Heresy*, introduction to B. Goldschmidt, *The Material Basis of Evolution* (New Haven, CT: Yale University Press, 1982).

31. P. Alberch, *American Zoologist* 20 (1980): 653.

32. M. J. D. White, *Modes of Speciation* (San Francisco: Freeman, 1978); C. L. Bush, S. M. Case, A. C. Wilson, and J. L. Patton, *Proceedings of the National Academy of Sciences of the United States of America* 74 (1977): 3942.

33. N. Eldredge and S. J. Gould, in *Models in Paleobiology*, ed. T. J. M. Schopf (San Francisco: Freeman, Cooper, 1972), p. 82; S. J. Gould and N. Eldredge, *Paleobiology* 3 (1977): 115.

34. P. Williamson, *Nature* 293 (London, 1981): 437.

35. D'Arcy W. Thompson, *On Growth and Form* (New York: Cambridge University Press, 1942).

36. S. J. Gould and R. C. Lewontin, *Proceedings of the Royal Society of London. Series B, Containing Papers of a Biological Character* 205 (1979): 581; C. V. Lauder, *Paleobiology* 7 (1981): 430.

37. Darwin, *Origin of Species.* It is the concluding comment of chapter 6, and reads, in part: "It is generally acknowledged that all organic beings have been formed on two great laws—Unity of Type, and the Conditions of Existence. . . . Natural selection acts by either now adapting the varying parts of each being to its organic and inorganic conditions of life or by having adapted them during long past periods of time Hence in fact the law of the Conditions of Existence is the higher law as it includes through the inheritance of former adaptations, that of Unity of Type."

38. St. C. Mivart, *On the Genesis of Species* (London: Macmillan, 1871).

39. A. Seilacher, *Lethaia* 5 (1972): 325.

40. S. J. Gould and E. S. Vrba, *Paleobiology*, in press.

41. B. C. Lewontin, *Annual Review of Ecology and Systematics* 1 (1970): 1.

42. R. Dawkins, *The Selfish Gene* (New York: Oxford University Press, 1976).

43. V. C. Wynne-Edwards, *Animal Dispersion in Relation to Social Behavior* (Edinburgh: Oliver & Boyd, 1962).

44. M. Chiselin, *Systematic Zoology* 23 (1974): 536; D. L. Hull, *Annual Review of Ecology and Systematics* 11 (1980): 311.

45. W. F. Doolittle and C. Sapienza, *Nature* 284 (London, 1980): 601; L. E. Orgel and F. H. C. Crick, ibid., p. 604.

46. S. Wright, *Evolution and the Genetics of Populations*, vols. 1–4 (Chicago: University of Chicago Press, 1968–1978).

47. S. M. Stanley, *Macroevolution* (San Francisco: Freeman, 1979); *Proceedings of the National Academy of Sciences of the United States of America* 72 (1975): 646; also referenced in Eldredge and Gould, *Models in Paleobiology.*

48. R. A. Fisher, *The Genetical Theory of Natural Selection*, 2nd ed. (New York: Dover, 1958), p. 50.

49. J. Maynard Smith, personal communication.

50. C. C. Williams, *Sex and Evolution*, Monographs in Population Biology, no. 8 (Princeton, NJ: Princeton University Press, 1975).

51. S. M. Stanley, *Science* 190 (1975): 382.

52. M. Slatkin, *Paleobiology* 7 (1981): 421.

EDITORIAL NOTE

Gould makes reference to "saltationism." This is a view of the evolutionary process which supposes that change goes instantaneously from one form (say *fox)* to another form (say *dog).* Although critics have argued that Gould's theory of punctuated equilibrium is saltationary because he sees change as continuous (albeit rapid and spasmodic), he has always denied the charge.

He refers also to "genetic drift," which is the claim that in small populations accidents of sampling can lead to random, nonadaptive change. Most who accept drift also accept selection, although some biologists have argued that at the molecular level much change lies beneath selection and thus drifts. This is the so-called neutral theory of evolution.

*FRANCISCO J. AYALA**

BEYOND DARWINISM? THE CHALLENGE OF MACROEVOLUTION TO THE SYNTHETIC THEORY OF EVOLUTION

1. Evolution. An Unfinished Theory

The current theory of biological evolution (the "Synthetic Theory" or "Modern Synthesis") may be traced to Theodosius Dobzhansky's *Genetics and the Origin of Species*, published in 1937: a synthesis of genetic knowledge and Darwin's theory of evolution by natural selection. The excitement provoked by Dobzhansky's book soon became reflected in many important contributions which incorporated into the Modern Synthesis relevant fields of biological knowledge. Notable landmarks are Ernst Mayr's *Systematics and the Origin of Species* (1942), Julian S. Huxley's *Evolution: The Modern Synthesis* (1942), George Gaylord Simpson's *Tempo and Mode in Evolution* (1944), and G. Ledyard Stebbins' *Variation and Evolution in Plants* (1950). It seemed to many scientists that the theory of evolution was essentially complete and that all that was left was to fill in the details. This perception is reflected, for example, by Jacques Monod, in his widely known book, *Chance and Necessity* (1970, p. 139): "the elementary mechanisms of evolution have been not only understood in principle but identified with, precision . . . the problem has been resolved and evolution now lies well to this side of the frontier of knowledge."

There can be little doubt that some components of the Synthetic Theory are well established. In a nutshell, the theory proposes that mutation and sexual recombination furnish the raw materials for change; that natural selection fashions from these materials genotypes and gene pools; and that, in sexually repro-

*Peter D. Asquith and Thomas Nickles, eds., PSA 1982, Vol. 2 (East Lansing, MI: Philosophy of Science Association, 1983), pp. 275–91.

ducing organisms, the arrays of adaptive genotypes are protected from disintegration by reproductive isolating mechanisms (speciation). But Monod's unbridled optimism is unwarranted. Indeed,

> the causes of evolution, and the patterning of the processes that bring it about, are far from completely understood. We cannot predict the future course of evolution except in a few well-studied situations, and even then only short-range predictions are possible. Nor can we, again with a few isolated exceptions, explain why past evolutionary events had to happen as they did. A predictive theory of evolution is a goal for the future. Hardly any competent biologist doubts that natural selection is an important directing and controlling agency in evolution. Yet one current issue hotly debated is whether a majority or only a small minority of evolutionary changes are induced by selection. (Dobzhansky, Ayala, Stebbins, and Valentine, 1977, pp. 129–30)

Perhaps no better evidence can be produced of the unfinished status of the theory of evolution than pointing out some of the remarkable discoveries and theoretical developments of recent years. Molecular biology has been one major source of progress. One needs only mention the recently acquired ability to obtain quantitative measures of genetic variation in populations and of genetic differentiation during speciation and phyletic evolution; the discontinuous nature of the coding sequences of eukaryotic genes with its implications concerning the evolutionary origin of the genes themselves; the dynamism of DNA increases or decreases in amount and of changes in position of certain sequences; etc. But notable advances have occurred as well, and continue to take place, in evolutionary ecology, theoretical and experimental population genetics, and in other branches of evolutionary knowledge.

Any active field of systematic knowledge is likely to be beset by many unconfirmed hypotheses, unsettled issues, and controversies. Evolutionary theory is no exception. During the 1970s perhaps no other issue was more actively debated than the "neutrality" theory of molecular evolution—whether the evolution of informational macromolecules (nucleic acids and proteins) is largely governed by random changes in the frequency of adaptively equivalent variants, rather than by natural selection. The 1980s have started with another actively, and at times acrimoniously, contested problem: whether in the geological scale evolution is a more or less gradual process, or whether instead it is "punctuated." The model of punctuated equilibrium proposes that morphological evolution happens in bursts, with most phenotypic change occurring during speciation events, so that new species are morphologically quite distinct from their ancestors, but do not thereafter change substantially in phenotype over a lifetime that may encompass many millions of years. The punctuational model is contrasted with the gradualistic model, which sees morphological change as a more or less gradual process, not strongly associated with speciation events (fig. 4.1).

Two important issues are raised by the model of punctuated equilibrium. The first is scientific: whether morphological change as observed in the paleontological record is essentially always associated with speciation events, i.e., with the splitting of a lineage into two or more lineages. The second issue is epistemological: whether owing to the punctuated character of paleontological evolution, macroevolution is an autonomous field of study, independent from population genetics and other disciplines that study microevolutionary processes. I shall here examine the two issues; the second one at greater length than the first.

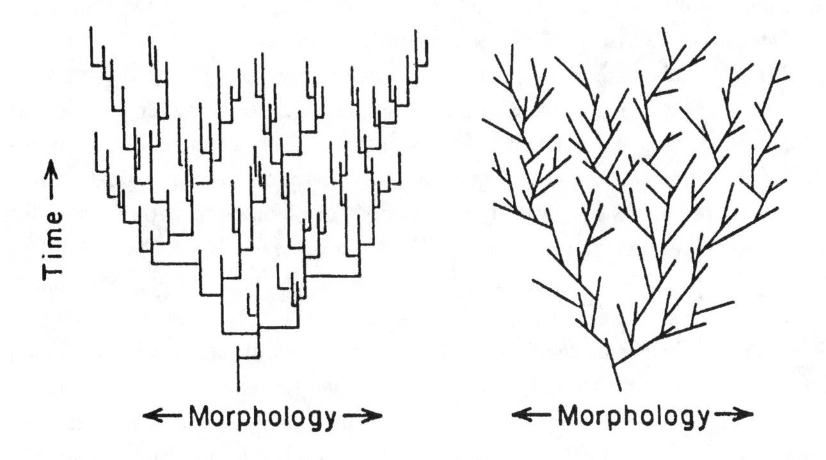

Figure 4.1. Simplified representation of two models of phenotypic evolution: punctuated equilibrium (left) and phyletic gradualism (right). According to the punctuated model, most morphological evolution in the history of life is associated with speciation events, which are geologically instantaneous. After their origin, established species generally do not change substantially in phenotype over a lifetime that may encompass many million years. According to the gradualist model morphological evolution occurs during the lifetime of a species, with rapidly divergent speciation playing a lesser role. The figures are extreme versions of the models. Punctualism does not imply that phenotypic change never occurs between speciation events. Gradualism does not imply that phenotypic change is occurring continuously at a more or less constant rate throughout the life of a lineage or that some acceleration does not take place during speciation but rather that phenotypic change may occur at any time throughout the lifetime of a species.

2. Phenotypic Change and Speciation

I note, first, that whether phenotypic change in macroevolution occurs in bursts or is more or less gradual, is a question to be decided empirically. Examples of rapid phenotypic evolution followed by long periods of morphological stasis are known in the fossil record. But there are instances as well in which phenotypic evolution appears to occur gradually within a lineage. The question is the relative frequency of one or the other mode; and paleontologists disagree in their interpretation of the fossil record. (Eldredge 1971, Eldredge and Gould 1972, Hallam 1978, Raup 1978, Stanley 1979, Gould 1980, and Vrba 1980, are among those who favor punctualism; whereas Kellogg 1975, Gingerich 1976, Levinton and Simon 1980, Schopf 1979 and 1981, Cronin, Boaz, Stringer, and Rak 1981, and Douglas and Avise 1982, favor phyletic gradualism.)

Whatever the paleontological record may show about the frequency of smooth, relative to jerky, evolutionary patterns, there is, however, one fundamental reservation that must be raised against the theory of punctuated equilibrium. This evolutionary model argues not only that most morphological change occurs in rapid bursts followed by long periods of phenotypic stability, but also that the bursts of change occur during the origin of new species. Stanley (1979, 1982), Gould (1982a, b), and other punctualists have made it clear that what is distinctive of the theory of punctuated equilibrium is this association between phenotypic change and speciation. One quotation should suffice: "Punctuated equilibrium is a specific claim about speciation and its deployment in geological time; *it should not be used as a synonym for any theory of rapid evolutionary change at any scale* . . . Punctuated equilibrium holds that accumulated speciation is the root of most major evolutionary change, and that what we have called anagenesis is usually no more than repeated cladogenesis (branching) filtered through the net of differential success at the species level" (Gould 1982a, pp. 84–85; italics added).

Species are groups of interbreeding natural populations that are reproductively isolated from any other such groups (Mayr 1963, Dobzhansky et al. 1977). Speciation involves, by definition, the development of reproductive isolation between populations previously sharing in a common gene pool. But it is no way apparent how the fossil record could provide evidence of the development of reproductive isolation. Paleontologists recognize species by their different morphologies as preserved in the fossil record. New species that are morphologically indistinguishable from their ancestors (or from contemporary species) go totally unrecognized. Sibling species are common in many groups of insects, in rodents, and in other well studied organisms (Mayr 1963, Dobzhansky 1970, Nevo and

Shaw 1972, Dobzhansky et al. 1977, White 1978, Benado, Aguilera, Reig, and Ayala 1979). Moreover, morphological discontinuities in a time series of fossils are usually interpreted by paleontologists as speciation events, even though they may represent phyletic evolution within an established lineage, without splitting of lineages.

Thus, when paleontologists use evidence of rapid phenotypic change in favor of the punctuational model, they are committing a definitional fallacy. Speciation as seen by the paleontologist always involves substantial morphological change because paleontologists identify new species by the eventuation of substantial morphological change. Stanley (1979, p. 144) has argued that "rapid change is concentrated in small populations and . . . that such populations are likely to be associated with speciation and unlikely to be formed by constriction of an entire lineage." But the two points he makes are arguable. First, rapid (in the geological scale) change may occur in populations that are not small. Second, bottlenecks in population size are not necessarily rare (again, in the geological scale) within a given lineage.

One additional point needs to be made. Punctualists speak of evolutionary change "concentrated in geologically instantaneous events of branching speciation" (Gould 1982b, p. 383). But events that appear instantaneous in the geological time scale may involve thousands, even millions of generations. Gould (1982a, p. 84), for example, has made operational the fuzzy expression "geologically instantaneous" by suggesting that "it be defined as 1 percent or less of later existence in stasis. This permits up to 100,000 years for the origin of a species with a subsequent life span of 10 million years." But 100,000 years encompasses one million generations of any insect such as *Drosophila*, and tens or hundreds of thousands of generations of fish, birds, or mammals. Speciation events or morphological changes deployed during thousands of generations may occur by the slow processes of allelic substitution that are familiar to the population biologist. Hence, the problem faced by microevolutionary theory is not how to account for rapid paleontological change, because there is ample time for it, but why lineages persist for millions of years without apparent morphological change. Although other explanations have been proposed it seems that stabilizing selection may be the process most often responsible for the morphological stasis of lineages (Stebbins and Ayala 1981, Charlesworth et al. 1982). Whether microevolutionary theory is sufficient to explain punctuated as well as gradual evolution is, however, a different question, to which I now turn.

3. *From Punctuated Evolution to the Autonomy of Macroevolution*

The second proposal made by the proponents of punctuated equilibrium is that macroevolution—the evolution of species, genera, and higher taxa—is an autonomous field of study, independent of microevolutionary theory (and the intellectual turf of paleontologists). This claim for autonomy has been expressed as a "decoupling" of macroevolution from microevolution (e.g., Stanley 1979, pp. x, 187, 193) or as a rejection of the notion that microevolutionary mechanisms can be extrapolated to explain macroevolutionary processes (e.g., Gould 1982b, p. 383).

The argument for the autonomy of macroevolution is grounded precisely on the claim that large-scale evolution is punctuated, rather than gradual. Two quotations shall suffice. "If rapidly divergent speciation interposes discontinuities between rather stable entities (lineages), and if there is a strong random element in the origin of these discontinuities (in speciation), then phyletic trends are essentially *decoupled* from phyletic trends within lineages. Macroevolution is decoupled from microevolution" (Stanley 1979, p. 187, italics in the original). "Punctuated equilibrium is crucial to the independence of macroevolution—for it embodies the claim that species are legitimate individuals, and therefore capable of displaying irreducible properties" (Gould 1982a, p. 94).

According to the proponents of punctuated equilibrium, phyletic evolution proceeds at two levels. First, there is change within a population that is continuous through time. This consists largely of allelic substitutions prompted by natural selection, mutation, genetic drift, and the other processes familiar to the population geneticist, operating at the level of the individual organism. Thus, most evolution within established lineages rarely, if ever, yields any substantial morphological change. Second, there is the process of origination and extinction of species. Most morphological change is associated with the origin of new species. Evolutionary trends result from the patterns of origination and extinction of species, rather than from evolution within established lineages. Hence, the relevant unit of macroevolutionary study is the species rather than the individual organism. It follows from this argument that the study of microevolutionary processes provides little, if any, information about macroevolutionary patterns, the tempo and mode of large-scale evolution. Thus, macroevolution is autonomous relative to microevolution, much in the same way as biology is autonomous relative to physics. Gould (1982b, p. 384) has summarized the argument: "Individuation of higher-level units is enough to invalidate the reductionism of traditional Darwinism—for pattern and style of evolution depend critically on the disposition of higher-level individuals [i.e., species]."

The question raised is the general issue of reduction of one branch of science to another. But as so often happens with questions of reductionism, the issue of whether microevolutionary mechanisms can account for macroevolutionary processes is muddled by confusion of separate issues. Identification of the issues involved is necessary in order to resolve them and to avoid misunderstanding, exaggerated claims, or unwarranted fears.

The issue "whether the mechanisms underlying microevolution can be extrapolated" to macroevolution involves, at least, three separate questions: (1) whether microevolutionary processes operate (and have operated in the past) throughout the organisms which make up the taxa in which macroevolutionary phenomena are observed. (2) Whether the microevolutionary processes identified by population geneticists (mutation, random drift, natural selection) are sufficient to account for the morphological changes and other macroevolutionary phenomena observed in higher taxa, or whether additional microevolutionary processes need to be postulated. (3) Whether theories concerning evolutionary trends and other macroevolutionary patterns can be derived from knowledge of microevolutionary processes.

The distinctions that I have made may perhaps become clearer if I state them as they might be formulated by a biologist concerned with the question whether the laws of physics and chemistry can be extrapolated to biology. The first question would be whether the laws of physics and chemistry apply to the atoms and molecules present in living organisms. The second question would be whether interactions between atoms and molecules according to the laws known to physics and chemistry are sufficient to account for biological phenomena, or whether the workings of organisms require additional kinds of interactions between atoms and molecules. The third question would be whether biological theories can be derived from the laws and theories of physics and chemistry.

The first issue raised can easily be resolved. It is unlikely that any biologist would seriously argue that the laws of physics and chemistry do not apply to the atoms and molecules that make up living things. Similarly, it seems unlikely that any paleontologist or macroevolutionist would claim that mutation, drift, natural selection, and other microevolutionary processes do not apply to the organisms and populations which make up the higher taxa studied in macroevolution. There is, of course, an added snarl—macroevolution is largely concerned with phenomena of the past. Direct observation of microevolutionary processes in populations of long-extinct organisms is not possible. But there is no reason to doubt that the genetic structures of populations living in the past were in any fundamental way different from the genetic structures of living populations. Nor is there any reason to believe that the processes of mutation, random drift and natural selection, or the nature of the interactions between organisms and the environment would have been different in nature for, say, Paleozoic trilobites or Mesozoic ammonites than for modern molluscs or fishes. Extinct and living populations—like different living populations—may have experienced quantitative differences in the relative importance of one or another process, but the processes could have hardly been different in kind. Not only are there reasons to the contrary lacking, but the study of biochemical evolution reveals a remarkable continuity and gradual change of informational macromolecules (DNA and proteins) over the most diverse organisms, which advocates that the current processes of population change have persisted over evolutionary history (Dobzhansky et al. 1977).

4. Are Microevolutionary Processes Sufficient?

The second question raised above is considerably more interesting than the first: Can the microevolutionary processes studied by population geneticists account for macroevolutionary phenomena or do we need to postulate new kinds of genetic processes? The large morphological (phenotypic) changes observed in evolutionary history, and the rapidity with which they appear in the geological record, is one major matter of concern. Another issue is "stasis," the apparent persistence of species, with little or no morphological change, for hundreds of thousands or millions of years. The dilemma is that microevolutionary processes apparently yield small but continuous changes, while macroevolution as seen by punctualists occurs by large and rapid bursts of change followed by long periods without change.

Goldschmidt (1940, p. 183) argued long ago that the incompatibility is real: "The decisive step in evolution, the first step toward macroevolution, the step from one species to another, requires another evolutionary method than that of sheer accumulation of micromutations." Goldschmidt's solution was to postulate "systemic mutations," yielding "hopeful monsters" that, on occasion, would find a new niche or way of life for which they would be eminently preadapted. The progressive understanding of the nature and organization of the genetic material acquired during the last forty years excludes the "systemic mutations" postulated by Goldschmidt, which would involve transformations of the genome as a whole.

Single-gene or chromosome mutations are known that have large effects on the phenotype because they act early in the embryo and their effects become magnified through development. Examples of such "macromutations" carefully analyzed in *Drosophila* are "bithorax" and the homeotic mutants that transform one body structure, e.g., antennae, into another, e.g., legs. Whether the kinds of morphological differences that characterize different taxa are due to such "macromutations" or to the accumulation of several mutations with small-effect, has been examined particularly in plants where fertile interspecific, and even intergeneric, hybrids can be obtained. The results do not support the hypothesis that the establishment of macromutations is necessary for divergence at the macroevolutionary level (Stebbins 1950, Clausen 1951, Grant 1971, see Stebbins and Ayala 1981). Moreover, Lande (1981 and references therein; see also Charlesworth et al. 1982) has convincingly shown that major morphological changes, such as in the number of digits or limbs, can occur in a geologically rapid fashion through the accumulation of mutations each with a small effect. The analysis of progenies from crosses between races or species that differ greatly (by as much as 30 phenotypic standard deviations) in a quantitative trait indicates that these extreme differences can be caused by the cumulative effects of no more than 5 to 10 independently segregating genes.

The punctualists' claim that mutations with large phenotypic effects must have been largely responsible for macroevolutionary change is based on the rapidity with which morphological discontinuities appear in the fossil record (Stanley 1979, Gould 1980). But the alleged evidence does not necessarily support the claim. Microevolutionists and macroevolutionists use different time scales. As pointed out earlier, the "geological instants" during which speciation and morphological shifts occur may involve expands as long as 100,000 years. There is little doubt that the gradual accumulation of small mutations may yield sizable morphological changes during periods of that length.

Anderson's (1973) study of body size in *Drosophila pseudoobscura* provides an estimate of the rates of gradual morphological change produced by natural selection. Large populations, derived from a single set of parents, were set up at different temperatures and allowed to evolve on their own. A gradual, genetically determined, change in body size ensued, with flies kept at lower temperature becoming, as expected, larger than those kept at higher temperatures. After 12 years, the mean size of the flies from a population kept at 16°C had become, when tested under standard conditions, approximately 10 percent greater than the size of flies kept at 27°C; the change of mean value being greater than the standard deviation in size at the time when the tests were made. Assuming 10 generations per year, the populations diverged at a rate of 8×10^{-4} of the mean value per generation.

Paleontologists have emphasized the "extraordinary high *net* rate of evolution that is the hallmark of human phylogeny" (Stanley 1979). Interpreted in terms of the punctualist hypothesis, human phylogeny would have occurred as a succession of jumps, or geologically instantaneous saltations, interspersed by long periods without morphological change. Could these bursts of phenotypic evolution be due to the gradual accumulation of small changes? Consider cranial capacity, the character undergoing the greatest relative amount of change. The fastest rate of net change occurred between 500,000 years ago, when our ancestors were represented by *Homo erectus* and 75,000 years ago, when Neanderthal man had acquired a cranial capacity similar to that of modern humans. In the intervening 425,000 years, cranial capacity evolved from about 900 cc in Peking man to about 1,400 cc in Neanderthal people. Let us assume that the increase in brain size occurred in a single burst at the rate observed in *D. pseudoobscura* of 8×10^{-4} of the mean value per generation. The change from 900 cc to 1,400 cc could have taken place in 540 generations or, assuming generously 25 years per generation, in 13,500 years. Thirteen thousand years are, of course, a geological instant. Yet, this evolutionary "burst" could have taken place by gradual accumulation of mutations with small effects at rates compatible with those observed in microevolutionary studies.

The known processes of microevolution can, then, account for macroevolutionary change, even when this occurs according to the punctualist model—i.e., at fast rates concentrated on geologically brief time intervals. But what about the problem of stasis? The theory of punctuated equilibrium argues that after the initial burst of morphological change associated with their origin, "species generally do not change substantially in phenotype over a lifetime that may encompass many million years" (Gould 1982b, p. 383). Is it necessary to postulate new processes, yet unknown to population genetics, in order to account for the long persistence of lineages with apparent phenotypic change? The answer is "no."

The geological persistence of lineages without morphological change was already known to Darwin, who wrote in the last edition of *The Origin of Species* (1872, p. 375). "Many species once formed never undergo any further change; . . . and the periods, during which species have undergone modification, though long as measured by years, have probably been short in comparison with the periods during which they retain the same form." A successful morphology may remain unchanged for extremely long periods of time, even though successive speciation events—as manifested, e.g., by the existence of sibling species, which in many known instances have persisted for millions of years (Stebbins and Ayala 1981).

Evolutionists have long been aware of the problem of paleontological stasis and have explored a number of alternative hypotheses consistent with microevolutionary principles and sufficient to account for the phenomenon. Although the issue is far from definitely settled, the weight of the evidence favors stabilizing selection as the primary process responsible for morphological stasis of lineages through geological time (Stebbins and Ayala 1981, Charlesworth et al. 1982).

5. Emergence and Hierarchy Versus Reduction

Macroevolution and microevolution are *not* decoupled in the two senses so far expounded: identity at the level of events and compatibility of theories. First, the populations in which macroevolutionary patterns are observed are the same populations that evolve at the microevolutionary level. Second, macroevolutionary phenomena can be accounted for as the result of known microevolutionary processes. That is, the theory of punctuated equilibrium (as well as the theory of phyletic gradualism) is consistent with the theory of population genetics. Indeed, any theory of macroevolution that is correct must be compatible with the theory of population genetics, to the extent that this is a well established theory.

Now, I pose the third question raised earlier: can macroevolutionary theory be derived from microevolutionary knowledge? The answer can only be "no." If macroevolutionary theory were deducible from microevolutionary principles, it would be possible to decide between competing macroevolutionary models simply by examining the logical implications of microevolutionary theory. But the theory of population genetics is compatible with both, punctualism and gradualism; and, hence, logically it entails neither. Whether the tempo and mode of evolution occur predominantly according to the model of punctuated equilibria or according to the model of phyletic gradualism is an issue to be decided by studying macroevolutionary patterns, not by inference from microevolutionary processes. In other words, macroevolutionary theories are not reducible (at least at the present state of knowledge) to microevolution. Hence, macroevolution and microevolution are decoupled in the sense (which is epistemologically most important) that macroevolution is an autonomous field of study that must develop and test its own theories.

Punctualists have claimed autonomy for macroevolution because species—the units studied in macroevolution—are higher in the hierarchy of organization of the living world than individual organisms. Species, they argue, have therefore "emergent" properties, not exhibited by, nor predictable from, lower-level entities. In Gould's (1980, p. 121) words, the study of evolution embodies "a concept of hierarchy—a world constructed not as a smooth and seamless continuum, permitting simple extrapolation from the lowest level to the highest, but as a series of ascending levels, each bound to the one below it in some ways and independent in others . . . 'emergent' features not implicit in the operation of processes at lower levels, may control events at higher levels." Although I agree with the thesis that macroevolutionary theories are not reducible to microevolutionary principles, I shall argue that it is a mistake to ground this autonomy on the hierarchical organization of life, or on purported emergent properties exhibited by higher-level units.

The world of life is hierarchically structured. There is a hierarchy of levels that go from atoms, through molecules, organelles, cells, tissues, organs, multicellular individuals and populations, to communities. Time adds another dimension of the evolutionary hierarchy, with the interesting consequence that transitions from one level to another occur: as time proceeds the descendants of a single species may include separate species, genera, families, etc. But hierarchical differentiation of subject matter is neither necessary nor sufficient for the autonomy of scientific disciplines. It is not necessary, because entities of a given hierarchical level can be the subject of diversified disciplines: cells are appropriate subjects of study for cytology, genetics, immunology, and so on. Even a single event can be the subject matter of several disciplines. My writing of this paragraph can be studied by a physiologist interested in the workings of muscles and nerves, by a psychologist concerned with thought processes, by a philosopher interested in the epistemological question at issue, and so on. Nor is the hierarchical differentiation of subject matter a sufficient condition for the autonomy of scientific disciplines: relativity theory obtains all the way from subatomic particles to planetary motions and genetic laws apply to multicellular organisms as well as to cellular and even subcellular entities.

One alleged reason for the theoretical independence of levels within a hierarchy is the appearance of "emergent" properties, which are "not implicit in the operation of lesser levels, [but] may control events at higher levels." The question of emergence is an old one, particularly in discussions on the reducibility of biology to the physical sciences. The issue is, for example, whether the functional properties of the kidney are simply the properties of the chemical constituents of that organ. In the context of macroevolution, the question is: do species exhibit properties different from those of the individual organisms of which they consist? I have argued elsewhere (Dobzhansky et al. 1977, ch. 16), that questions about the emergence of properties are ill-formed, or at least unproductive, because they can only be solved by definition. The proper way of formulating questions about the relationship between complex systems and their component parts is by asking whether the properties of complex systems can be inferred from knowledge of the properties that their components have in isolation. The issue of emergence cannot be settled by discussions about the "nature" of things or their properties, but it is resolvable by reference to our *knowledge* of those objects.

Consider the following question: Are the properties of common salt, sodium chloride, simply the properties of sodium and chlorine when they are associated according to the formula NaCl? If among the properties of sodium and chlorine I include their association into table salt and the properties of the latter, the answer is "yes"; otherwise, the answer is "no." But the solution, then, is simply a matter of definition; and resolving the issue by a definitional maneuver contributes little to understanding the relationships between complex systems and their parts.

Is there a rule by which one could decide whether the properties of complex systems should be listed among the properties of their component parts? Assume that by studying the components in isolation we can infer the properties they will have when combined with other component parts in certain ways. In such a case, it would seem reasonable to include the "emergent" properties of the whole among the properties of the component parts. (Notice that this solution to the problem implies that a feature that may seem emergent at a certain time, might not appear as emergent any longer at a more advanced state of knowledge.) Often, no matter how exhaustively an object is studied in isolation, there is no way to ascertain the properties it will have in association with other objects. We cannot infer the properties of ethyl alcohol, proteins, or human beings from the study of hydrogen, and thus it makes no good sense to list their properties among those of hydrogen. The important point, however, is that the issue of emergent properties is spurious and that it needs to be reformulated in terms of propositions expressing our knowledge. It is a legitimate question to ask whether the statements concerning the properties of organisms (but not the properties themselves) can be logically deduced from statements concerning the properties of their physical components.

The question of the autonomy of macroevolution, like other questions of reduction, can only be settled by empirical investigation of the logical consequences of propositions, and not by discussions about the "nature" of things or their properties. What is at issue is not whether the living world is hierarchically organized—it is; or whether higher level entities have emergent properties—which is a spurious question. The issue is whether in a particular case, a set of *propositions* formulated in a defined field of knowledge (e.g., macroevolution) can be derived from another set of propositions (e.g., microevolutionary theory). Scientific theories consist, indeed, of propositions about the natural world. Only the investigation of the logical relations between propositions can establish whether or not one theory or branch of science is reducible to some other theory or branch of science. This implies that a discipline which is autonomous at a given stage of knowledge may become reducible to another discipline at a later time. The reduction of thermodynamics to statistical mechanics became possible only after it was discovered that the temperature of a gas bears a simple relationship to the mean kinetic energy of its molecules. The reduction of genetics to chemistry could not take place before the discovery of the chemical nature of the hereditary material (I am not, of course, intimating that genetics can now be fully reduced to chemistry, but only that a partial reduction may be possible now, whereas it was not before the discovery of the structure and mode of replication of DNA).

Nagel (1961, see also Ayala 1968) has formulated the two conditions that are necessary and, jointly, sufficient to effect the reduction of one theory or branch of science to another. These are the condition of derivability and the condition of connectability.

The *condition of derivability* requires that the laws and theories of the branch of science to be reduced be derived as logical consequences from the laws and theories of some other branch of science. The *condition of connectability* requires that the distinctive terms of the branch of science to be reduced be redefined in the language of the branch of science to which it is reduced—this redefinition of terms is, of course, necessary in order to analyze the logical connections between the theories of the two branches of science.

Microevolutionary processes, as presently known, are compatible with the two models of macroevolution—punctualism and gradualism. From microevolutionary knowledge, we cannot infer which one of those two macroevolutionary patterns prevails, nor can we deduce answers for many other distinctive macroevolutionary issues, such as rates of morphological evolution, patterns of species extinctions, and historical factors regulating taxonomic diversity. The condition of derivability is not satisfied: the theories, models, and laws of macroevolution cannot be logically derived, at least at the present state of knowledge, from the theories and laws of population biology.

In conclusion, then, macroevolutionary processes are underlain by microevolutionary phenomena and are compatible with microevolutionary theories, but macroevolutionary studies require the formulation of autonomous hypotheses and models (which must be tested using macroevolutionary evidence). In this (epistemologically) very important sense, macroevolution is decoupled from microevolution: macro-evolution is an autonomous field of evolutionary study.

6. Whence Macroevolution and Paleontology?

I have argued that distinctive macroevolutionary issues, such as rates of morphological evolution or patterns of species origination and extinction, cannot presently be derived from microevolutionary theories and principles. It deserves notice, however, that the discipline of (morphological) macroevolution is notoriously lacking in theoretical constructs of great deductive import and generality.

There is little by way of hypotheses, theories, and models that would allow one to make inferences about what should be the case in one or other group of organisms and that would lead to the corroboration or falsification of a given general hypothesis, theory, or model. There are, of course, models such as punctuated equilibrium, species selection (Stanley 1979), and the Red Queen Hypothesis (Van Valen 1973). But this amounts to considerably less than a substantial body of deductive theory, such as it exists in population genetics, population ecology, and other microevolutionary disciplines. Awareness of the autonomy of macroevolutionary studies may perhaps help paleontologists in developing their science into a mature theoretical discipline.

There is one sector of macroevolutionary problems where considerable advances have occurred in the last decade yielding propositions of great generality and import: the field of molecular evolution. General theoretical constructs include the hypothesis of the molecular clock and related concepts concerning rates of amino acid and nucleotide substitution; the constraints imposed upon fundamental biochemical processes (transcription, translation, the genetic code, . . ." and structures (ribosomes, membranes, mitotic apparatus . . .); and so on. These have led to notable advances, for example, in the reconstruction of the topology and timing of phylogenetic history, including the identification of a new phylum (the archeobacteria).

It may seem invidious to call attention to the successes of molecular biology in solving macroevolutionary issues. But it does corroborate my claim that hierarchical distinction of the objects studied is no grounds for the autonomy of scientific theories. Molecular macroevolution is not an autonomous field of study, because it can be reduced to microevolutionary theories, even though it concerns macroevolutionary entities. The hypothesis of the molecular clock of evolution, for example, is founded on concepts—such as mutation rate, population size, and genetic drift—distinctive of the field of population genetics.

REFERENCES

Anderson, W. W. "Genetic Divergence in Body Size Among Experimental Populations of *Drosophila Pseudoobscura* Kept at Different Temperatures." *Evolution* 27 (1973): 278–84.

Ayala, F. J. "Biology as an Autonomous Science." *American Scientist* 56 (1968): 207–21.

Ayala, F. J., G. L. Stebbins, and J. W. Valentine. *Evolution*. San Francisco, CA: W. H. Freeman and Co., 1977.

Benado, M., M. Aguilera, D. A. Reig, and F. J. Ayala, "Biochemical Genetics of Venezuelan Spiny Rats of the *Proechimys Guainae* and *Proechimys Trinitatis* Superspecies." *Genetica* 50 (1979): 89–97.

Charlesworth, B., R. Lande, and M. Slatkin. "A Neo-Darwinian Commentary on Macroevolution." *Evolution* 36 (1982): 474–98.

Clausen, J. *Stages in the Evolution of Plant Species*. Ithaca, NY: Cornell University Press, 1951.

Cronin, J. E., N. T. Boaz, C. B. Stringer, and Y. Rak. "Tempo and Mode in Hominid Evolution." *Nature* 292 (1981): 113–22.

Darwin, C. R. 1872. *The Origin of Species*, 6th ed. Repr., New York: Random House, 1936.

Dobzhansky, Th. *Genetics and the Origin of Species*. Columbia Biological Series, no. XI. New York: Columbia University Press, 1937.

———. *Genetics of the Evolutionary Process*. New York: Columbia University Press, 1970.

Douglas, M. E., and J. C. Avise. "Speciation Rates and Morphological Divergence in Fishes. Tests of Gradual Versus Rectangular Modes of Evolutionary Change." *Evolution* 36 (1982): 224–32.

Eldredge, N. "The Allopatric Model and Phylogeny in Paleozoic Invertebrates." *Evolution* 25 (1971): 156–67.

Eldredge, N., and S. J. Gould. "Punctuated Equilibria: An Alternative to Phyletic Gradualism." In *Models in Paleobiology*, edited by T. J. M. Schopf, 82–115. San Francisco: Freeman, Cooper and Co., 1972.

Gingerich, P. D. "Paleontology and Phylogeny: Patterns of Evolution at the Species Level in Early Tertiary Mammals." *American Journal of Science* 276 (1976): 1–28.

Goldschmidt, R. B. *The Material Basis of Evolution*. New Haven, CT: Yale University Press, 1940.

Gould, S. J. (1980). "Darwinism and the Expansion of Evolutionary Theory." *Science* 216 (1982b): 380–87.

———. "Is a New and General Theory of Evolution Emerging?" *Paleobiology* 6:119–30.

———. "The Meaning of Punctuated Equilibrium and Its Role in Validating a Hierarchical Approach to Macroevolution." In *Perspectives on Evolution*, edited by R. Milkman, 83–104. Sunderland, MA: Sinauer Press, 1982a.

Grant, V. *Plant Speciation*. New York: Columbia University Press, 1971.

Hallam, A. "How Rare Is Phyletic Gradualism and What Is Its Evolutionary Significance? Evidence from Jurassic Bivalves." *Paleobiology* 4 (1978): 16–25.

Huxley, J. S. *Evolution: The Modern Synthesis*. New York: Harper, 1942.

Kellogg, D. E. "The Role of Phyletic Change in the Evolution of *Pseudocubus Vema* (Radiolaria)." *Paleobiology* 1 (1975): 359–70.

Lande, R. "The Minimum Number of Genes Contributing to Quantitative Variation Between and Within Populations." *Genetics* 99 (1981): 541–53.

Levinton, J. S., and C. M. Simon. "A Critique of the Punctuated Equilibria Model and Implications for the Detection of Speciation in the Fossil Record." *Systematic Zoology* 29 (1980): 130–42.

Lewin, R. "Evolution Theory Under Fire." *Science* 210 (1980): 883–87.

Mayr, E. *Animal Species and Evolution*. Cambridge, MA: Harvard University Press, 1963.

———. *Systematics and the Origin of Species*. New York: Columbia University Press, 1942.

Monod, J. *Le hasard et la nécessité*, translated by A. Wainhouse. Paris: Editions du Seuil, 1970. Repr. in *Chance and Necessity*. New York: Vintage Books, 1972.

Nagel, E. *The Structure of Science*. New York: Harcourt, Brace and World, 1961.

Nevo, E., and C. R. Shaw. "Genetic Variation in a Subterranean Mammal, *Spalax Ehrenbergi*." *Biochemical Genetics* 7 (1972): 235–41.

Raup, D. M. "Cohort Analysis of Generic Survivorship." *Paleobiology* 4 (1978): 1–15.

Schopf, T. J. M. "Evolving Paleontological Views on Deterministic and Stochastic Approaches." *Paleobiology* 5 (1979): 337–52.

———. "Punctuated Equilibrium and Evolutionary Stasis." *Paleobiology* 7 (1981): 156–66.

Simpson, G. C. *Tempo and Mode in Evolution*. New York: Columbia University Press, 1944.

Stanley, S. M. "Macroevolution and the Fossil Record." *Evolution* 36 (1982): 460–73.

———. *Macroevolution: Pattern and Process*. San Francisco, CA: W. H. Freeman, Co., 1979.

Stebbins, G. L. *Variation and Evolution in Plants*. New York: Columbia University Press, 1950.

Stebbins, G. L., and F. J. Ayala. "Is a New Evolutionary Synthesis Necessary?" *Science* 213 (1981): 967–71.

Van Valen, L. "A New Evolutionary Law." *Evolutionary Theory* 1 (1973): 1–30.

Vrba, E. S. "Evolution, Species, and Fossils: How Does Life Evolve?" *South African Journal of Science* 76 (1980): 61–84.

White, M. J. D. *Modes of Speciation*. San Francisco, CA: W. H. Freeman, Co., 1978.

DAVID W. E. HONE AND MICHAEL J. BENTON

THE EVOLUTIONOF LARGE SIZE: HOW DOES COPE'S RULE WORK?

Cope's Rule is the tendency for organisms in evolving lineages to increase in size over time. The concept is detailed in many textbooks, but has rarely been demonstrated. Many suggestions of the benefits of large body size exist, but none has yet been confirmed empirically. Using a large-scale analysis of recent studies, Kingsolver and Pfennig have now shown how size benefits survival, mating success and fecundity, and they provide convincing arguments for a mechanism that is capable of driving Cope's Rule.

Cope's Rule is the tendency for evolutionary lineages to increase in size over time[1] and, in spite of being over a century old,[2] is still poorly understood. Studies across a wide range of taxa have supported,[3] and rejected it,[4] with one recent study suggesting that previous analyses had been tested at too high a taxonomic level to be effective.[5] Cope's Rule has also been explained as a statistical artefact of increasing variance in size in a clade that arises from small ancestors.[6] For Cope's Rule to operate, large body size must provide an increase in fitness. A new study by Kingsolver and Pfennig[7] appears to demonstrate this for the first time.

Using a recent study of data assembled from natural populations,[8] Kingsolver and Pfennig assessed information about the effects of size change on fitness.[9] Requirements for data to be included in this new study were: (i) natural variation of quantitative traits; (ii) measurements of fitness; and (iii) estimation of selection differentials or gradients (providing a standardized measure of directional selection, enabling cross-study comparisons).

The authors split characteristic traits into "overall size traits" (e.g. total length, mass, etc.), or "other morphological traits." This "other trait" category included size-dependent characteristics, such as wingspan, flower size, and so on, and therefore the new results are likely to be conservative as many of these traits will be tied to body mass. This splitting of traits gave a total of 854 trait estimates: 91 for size and 763 for other traits, spread over 39 species from 42 studies, covering vertebrates, invertebrates and plants. The selection gradients were then plotted for both size and other traits according to their effects on "overall fitness" and three key "fitness" criteria: survival, fecundity and mating success, thus covering the effects of both size and other traits on both natural selection and sexual selection.

SIZE AND FITNESS

The results are quite startling (figure 5.1): size increase produces a marked increase in survival, fecundity and mating success, whereas the effects of other traits were neutral. The results overall were highly significant: 79 percent gave a positive effect with larger size (with a median of 0.15), but for other traits this was 50 percent (with a median of 0.02), an almost neutral effect because half produced positive results and half negative. When broken down into the three components, this positive effect is seen across all three analyses, and also across all of the taxonomic groups within them, so the results are unlikely to be an artefact of methodology or taxonomic bias. Furthermore, the patterns are inconsistent with alternating selection, which has been suggested for some findings of Cope's Rule, and negative selection associated with large size was rare. Although few studies had the appropriate data, the authors tested, where possible, for stabilizing selection and showed that individual selection on size was not the primary mechanism preventing evolution towards large size.

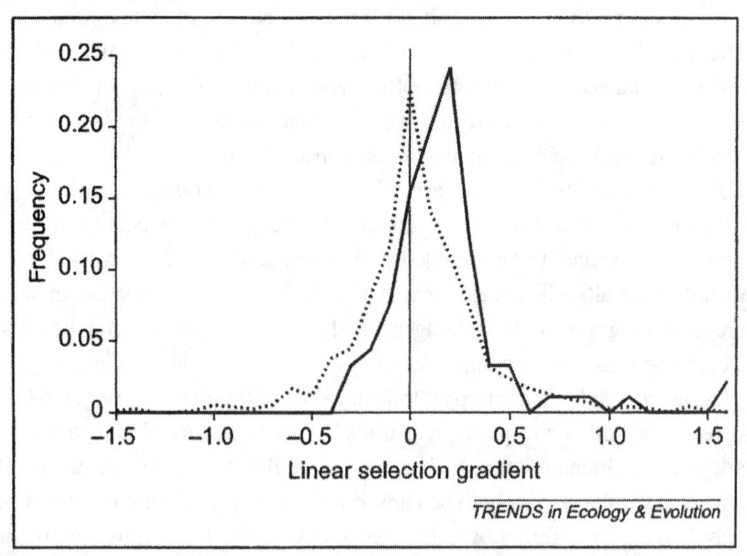

Figure 5.1. The distribution of linear selection gradients (β) for body size (solid line) and for other morphological traits (dashed line). Body size is a positive result, whereas the other traits provide only neutral selection effects. Reproduced with permission from J. G. Kingsolver and D. W. Pfenning.[12]

EFFECTS OF LARGE SIZE

An increase in body size is supposed to convey many selective advantages on an organism, but also presents new problems.[10] For Cope's Rule to operate, any benefits must outweigh these problems. However, benefits and costs operate at various levels: a small increase in size might confer an advantage in interspecific competition for an individual against its rivals, but to evolve gigantohomeothermy would require many, many generations of size increase and would have little short-term benefit (for this trait). Below are some of the most common arguments both for and against large size.[11]

Benefits

- Increased defense against predation (might be more vulnerable in some circumstances)
- Increase in predation success
- Greater range of acceptable foods
- Increased success in mating and intraspecific competition
- Increased success in interspecific competition
- Extended longevity
- Increased intelligence (with increased brain size)
- At very large size, the potential for thermal inertia (e.g. sauropod dinosaurs and tuna)
- Survival through lean times and resistance to climatic variation and extremes

Problems

- Increased development time (both pre- and post-natal)
- Increased requirement of food, water, and so on
- Susceptibility to extinction: a longer generation time gives a slower rate of evolution and, consequently, a reduced ability to adapt to sudden change
- Lower fecundity: switch from r-selection to a K-selection strategy produces fewer offspring with a high parental investment in each over many "poorer" offspring tested, where possible, for stabilizing selection and showed that individual selection on size was not the primary mechanism preventing evolution toward large size.

Several biases were present and identified in the article: (i) the studies were weighted toward birds, plants and insects, which might be unusual groups in terms of size change and fitness.[13] However, Cope's Rule has also been found in several mammal groups[14] and the dinosaurs[15] so this potential bias might be less significant than Kingsolver and Pfennig suggest; (ii) only one component of fitness was analysed for most studies—a more comprehensive coverage might reveal counter selection; (iii) the analysis was based only on published studies, and these are unlikely to have reported marginal or non-significant results. There might be an additional bias: positive selection might be observed because small individuals are fundamentally unhealthy and die more easily (being "the runt of the litter") rather than owing to active selection for large size. However, it is difficult to determine the extent of these possible biases, particularly in the face of such a positive set of results.

ENDLESS TRENDS TO GIGANTISM

In the light of these results, why do species not continue to increase in size infinitely? The selective advantages and the measured size increases over just a few generations are several orders of magnitude higher than is required for the size increases shown in palaeontological studies of Cope's Rule. If unopposed, these trends could lead to a macroevolutionary pattern of size change.

Within the study, development time was shown to correlate positively with size increase;[16] that is, increased size leads to an increase in development time (although few studies were appropriate for this and these were mainly for plants). Therefore, an increase in developmental time is an obvious limitation on rapid size increase, but cannot account for the observed lack of stabilizing selection. Whereas a recent study of tyrannosaurid dinosaurs showed that development need not be a limiting factor on large size,[17] this appears to be the exception rather than the rule. Even with the accelerated development seen in *Tyrannosaurus rex*, it still developed more slowly in absolute terms than did its close relatives among the tyrannosaurs.

Kingsolver and Pfennig also suggest that mass extinctions account for the cap on size increase,[18] as large organisms are more vulnerable to environmental crises.[19] These opposing trends and the variation demonstrated might explain why Cope's Rule is found in some groups and not in others.

However, there are probably additional caps at the species or generational level that could prevent the large short-term size increases predicted by these results. First, there is the issue of morphological constraints on current size: for example, the giraffe *Giraffa cameleopardis* might not be able to grow larger than it is as it could not achieve the necessary arterial pressure for the blood to reach a head on a longer neck; also, the pied kingfisher *Ceryle rudis* would no longer be capable of hovering if it became heavier, preventing it from feeding. Second, there are the probable effects of niche overlap. Most organisms are restricted ecologically to an "n-dimensional morphospace," enclosed on all sides by the morphospaces of other species. If an organism becomes too large, it will encroach on the morphospaces of other species, leading to increased competition. The species increasing in size will be less adapted to competition in this niche than is the incumbent species and will be outcompeted and forced to return to its existing niche. There is also the issue of the "evolutionary clock," which limits the ability of large organisms to adapt to severe changes because of their long generation times. Small organisms can adapt or even speciate to survive times of mass extinction, terminating larger species.[20] Finally, an extreme case of size increase might ultimately lead to divergence and speciation (as shown in the Galapagos finches *Geospiza*),[21] although this would be impossible to determine over such a short time period.

PROSPECTS

Cope's Rule appears to be alive and well, but where do we go from here? To test Cope's Rule fully, we need to cover the intermediate ground of multiple generations in extant lineages, and across shorter paleontological timescales. Alroy's recommendation of using lower taxonomic levels for tests of Cope's Rule has yet to be followed on a larger scale,[22] with only a few studies being carried out since.[23] There is still skepticism among many paleobiologists about the reality of Cope's Rule. It would be ironic if Cope's Rule turned out to be a statistical artifact,[24] and was not supported by Cope's original macroevolutionary observation,[25] and yet the traditional explanation for this supposed trend (the fitness advantage of large size) turned out to be a valid evolutionary driving force.[26]

ACKNOWLEDGMENTS

The authors thank Mark Pagel, Anthony Arnold and two anonymous referees for their helpful comments on an earlier version of this article.

REFERENCES

1. M. J. Benton, "Cope's Rule," in *Encyclopeadia of Evolution*, ed. M. Pagel (New York: Oxford University Press, 2002), pp. 185–86.

2. E. D. Cope, *The Primary Factors of Organic Evolution* (Chicago: Open Court, 1896).

3. J. Alroy, "Cope's Rule and the Dynamics of Body Mass Evolution in North American Fossil Mammals," *Science* 280 (1998): 731–34; D. W. McShea, "Trends, Tools, and Terminology," *Paleobiology* 26 (2000): 330–33.

4. D. Jablonski, "Body Size and Macroevolution," in *Evolutionary Paleobiology*, ed. by D. Jablonski, et al. (Chicago: University of Chicago Press, 1996), pp. 256–89; D. Jablonski, "Body-size Evolution in Cretaceous Molluscs and the Status of Cope's Rule," *Nature* 385 (1997): 250–52; A. J. Arnold, et al., "Causality and Cope's Rule: evidence from the Planktonic-Foramininifera," *Journal of Paleontology* 69 (1995): 203–10.

5. Cope, *The Primary Factors of Organic Evolution.*

6. S. J. Gould, "Cope's Rule as Psychological Artifact," *Nature* 385 (1997): 199–200.

7. J. G. Kingsolver and D. W. Pfennig, "Individual-level Selection as a Cause of Cope's Rule of Phyletic Size Increase," *Evolution* 58 (2004): 1608–12.

8. J. G. Kingsolver, et al., "The Strength of Phenotypic Selection in Natural Populations," *The American Naturalist* 157 (2001): 245–61.

9. Kingsolver and Pfennig, "Individual-level Selection as a Cause of Cope's Rule of Phyletic Size Increase."

10. K. Schmidt-Nielsen, *Scaling: Why is Animal Size so important* (New York: Cambridge University Press, 1984).

11. Benton, "Cope's Rule"; Schmidt-Nielsen, *Scaling: Why is Animal Size so important.*

12. Kingsolver and Pfennig, "Individual-level Selection as a Cause of Cope's Rule of Phyletic Size Increase."

13. Ibid.

14. Alroy, "Cope's Rule and the Dynamics of Body Mass Evolution in North American Fossil Mammals"; B. J. McFadden, *Fossil Horses: Systematics, Palaeobiology and Evolution of the Family Equidae* (New York: Cambridge University Press, 1992).

15. D. W. E. Hone, MSc Thesis, Imperial College, London, 2001.

16. Kingsolver and Pfennig, "Individual-level Selection as a Cause of Cope's Rule of Phyletic Size Increase."

17. Cope, *The Primary Factors of Organic Evolution.*

18. Kingsolver and Pfennig, "Individual-level Selection as a Cause of Cope's Rule of Phyletic Size Increase."

19. M. L. McKinney, "Extinction Vulnerability and Selectivity: Combining Ecological and Paleontological Views," *Annual Review of Ecology and Systematics* 28 (1997): 495–516.

20. Arnold, et al., "Causality and Cope's Rule: Evidence from the Planktonic-Foramininifera."

21. B. R. Grant, "Evolution in Darwin's Finches: A Review of a Study on Isla

Daphne Major in the Galapagos Archipelago," *Zoology* 106 (2003): 255–59.

22. Alroy, "Cope's Rule and the Dynamics of Body Mass Evolution in North American Fossil Mammals."

23. J. R. Knouft and L.M. Page, "The Evolution of Body Size in Extant Groups of North American Freshwater Fishes: Speciation, Size Distributions and Cope's Rule," *The American Naturalist* 161 (2003): 413–21.

24. D. Jablonski, "Body Size and Macroevolution"; Arnold, et al., "Causality and Cope's Rule: Evidence from the planktonic-foramininifera"; Gould, "Cope's rule as psychological artifact."

25. Cope, *The Primary Factors of Organic Evolution.*

26. Kingsolver and Pfennig, "Individual-level Selection as a Cause of Cope's Rule of Phyletic Size Increase."

DEREK TURNER

BEYOND DETECTIVE WORK: EMPIRICAL TESTING IN PALEOBIOLOGY

1. EMPIRICAL TESTING WITHOUT EXPERIMENT?

Like all scientists, paleobiologists (those who study ancient life) need to subject their hypotheses and theories to rigorous empirical tests. In many areas of science, empirical testing involves experimentation: researchers carry out a series of trials, while varying the initial conditions a little bit each time in order to see how those changes affect the outcomes. For example, Charles Darwin became interested in the question whether seeds could germinate after prolonged exposure to saltwater. He wanted to understand more about seed dispersal: How could a plant species which started out on the mainland end up flourishing on an island hundreds of miles away? To test the hypothesis that seeds could have survived transportation by ocean currents, he placed large numbers of seeds from many different plant species in small bottles of saltwater for varying periods of time (two weeks, three weeks, etc.). Then he planted the seeds to see which ones sprouted. Darwin knew that if none of the seeds sprouted, he would need to find some other explanation of seed dispersal. This type of biological experimentation—and indeed much of the research that biologists do today—involves manipulation and control. Although it may sound so obvious as to be hardly worth pointing out, *no one can manipulate or experiment with the past.* This simple fact creates a major methodological problem for paleobiologists: How can they test hypotheses about prehistoric life without being able to experiment on the things they wish to study?

Contrary to the impression that one might get from textbook presentations of the scientific method, it really is possible to carry out rigorous empirical tests without actually performing any experimental manipulations on the objects of interest. In what follows, I describe two techniques that paleobiologists have come to rely on over the last few decades. The first involves *virtual experiments*: if you cannot actually manipulate the past, then the next best thing is to use a computer model that you can manipulate. A second technique, which Michael Ruse (1999) has called *crunching the fossils*, involves the compilation of huge amounts of data from the fossil record in order to test hypotheses about large-scale evolutionary processes. The development of these techniques for carrying out non-experimental empirical tests is a major—and largely unsung—scientific achievement. However, there are still important theoretical problems in paleobiology which these techniques may not be able to resolve. I argue that both the empirical methods and the theoretical problems of paleobiology go beyond historical detective work.

2. VIRTUAL EXPERIMENTS

In his entertaining and influential book, *The Dinosaur Heresies* (1986), the paleontologist Robert Bakker sought to overthrow the traditional view of dinosaurs as cold-blooded, dim-witted, slow-moving giants. He deployed a series of biomechanical arguments in defense of the view that *Tyrannosaurus rex* was an agile predator, capable of pouncing, lunging, and running down prey at speeds up to 45 miles per hour. Bakker's view caught on, and in the movie *Jurassic Park*, a tyrannosaur runs fast enough to keep up with a speeding jeep. How could scientists possibly test the hypothesis that *T. rex* could run at 40–45 mph? After all, no one can observe a living tyrannosaur in action or experiment with a creature that has been extinct for 65 million years. Was the portrayal of *T. rex* in *Jurassic Park* based on solid, well-tested science, or merely on educated (but untested) guesswork?

Bakker knew that one way to test his hypothesis would be to look for tyrannosaur tracks. There is a well-established mathematical technique for estimating the speed of an animal based on its footprints (Alexander 1976). All you need to know is the stride length (which can be acquired from a fossilized trackway) and the height of the animal's hip (which can be determined from skeletons). There are, however, two problems with this approach. First, no one has ever found any well-preserved trackways that can be assigned to *T. rex*. Second, this technique only tells scientists how fast an animal was going when it happened to make a given set of footprints; it cannot reveal the animal's maximum speed. Since animals do not run at maximum speed very often, it is likely that if we ever did find a *T. rex* track way, the footprints would indicate that the animal was moving much more slowly than the maximum speed proposed by Bakker. When *Jurassic Park* was released in theaters, the hypothesis that *T. rex* could have kept up with a speeding jeep was simply not well tested.

Since then, other scientists have brought computing technology to bear on the problem, and at the moment things are not looking good for Bakker's five-ton sprinting dinosaurs. A team of scientists from Stanford University have taken a computer program designed for modeling the musculoskeletal systems of living organisms, including humans, and used it to create three dimensional computer models of the musculoskeletal system of *T. rex* (Hutchinson et al. 2005). In essence, they created a virtual tyrannosaur, which they then proceeded to experiment on. They placed their virtual dinosaur in a variety of different poses in order to study the forces that would act on the animal's joints. On the basis of these virtual experiments, they argue that *T. rex* could never have hit 40 mph, as Bakker suggested, and that its maximum speed was probably much slower.

Although this is the first study involving a computer model of a dinosaur's musculoskeletal system, so-called "numerical experiments" involving computer models of evolutionary, geological and climatic processes have for some time been a mainstay of research in paleobiology and earth science. Numerical experiments involving computer models have enabled scientists to test their ideas even in cases where they cannot manipulate the things they really want to study (Huss 2004). Geologists, for example, have used computer models to study the ebb and flow of glaciers, and such models have also been indispensable to scientists who are trying to understand the complexities of global climate change.

3. COPE'S RULE

Few paleobiologists spend much of their time wondering how fast dinosaurs could run. Much work in paleobiology involves testing ideas about large-scale evolutionary trends. One example of a possible evolutionary trend is *size increase*. According to "Cope's Rule," named in honor of the nineteenth-century American paleontologist, E. D. Cope, "the size of individuals tends to increase in most evolutionary groups" (McShea 1998, p. 306). Cope is perhaps most famous today for his feud with the Yale paleontologist and rival fossil hunter, O.C. Marsh. He is somewhat less well known for his neo-Lamarckian views about evolution. Cope himself, though he did make some tantalizing claims about body size and evolution, never quite asserted that which has come to bear his name (Polly 1998). He came close, though. In a discussion of "the phylogeny of the horse," Cope did remark that "the species have all the while been growing gradually larger" (1896, p. 148). Indeed for a long time, horses served as the classic illustration of Cope's Rule. Existing horses and zebras evolved from *Hyracotherium*, a small animal that was only about 15–20 inches high at the shoulder, and that lived during the Eocene, about 55–57 million years ago. Horses really do seem to have gotten bigger as they have evolved.

Today, Cope's Rule has some ardent and able defenders within the paleontological community, such as John Alroy (1998). Yet there are also some well-documented exceptions to it. The story of the evolution of horses turns out to be rather complicated, with animals in some lineages getting bigger and bigger over time, but animals in a few (now extinct) lineages getting smaller (McFadden 1986; Gould and McFadden 2004). The horse genus *Nannipus*, which went extinct in North America several million years ago, was much smaller than today's horses, and evolved from earlier horses that were much larger than it was. Exceptions like this help explain why no scientists these days are willing to talk anymore about "Cope's law" (which is what it used to be called). A law of nature is supposed to be a generalization that holds true at all times and places, without any exceptions.

Still, even if we demote the idea from a law to a rule, why would a prominent scientist such as Alroy persist in defending Cope's Rule in the face of such clear counterexamples? The answer to this question reveals a lot about how paleobiological research is done today. Alroy and others have put Cope's Rule to the test by *crunching the fossils*—that is, by searching for interesting patterns in huge collections of fossils. For some time, paleobiologists have relied on large databases to facilitate this work.

In his 1998 study, for instance, Alroy looked at 1,534 different species of mammals that have lived in North America during the Cenozoic Era (roughly, the last 65 million years). He then asked a very straightforward question: if you examine *pairs* of species that belong to the same genus, and one of which shows up in the fossil record earlier than the other, will the organisms in the newer species be bigger than those in the older one? He found that on average, the newer species are about 9.1 percent larger than the older species in the same genus. Notice that this is perfectly compatible with there being some cases, like that of *Nannipus*, in which the older species are actually bigger. But Alroy takes his results to provide a strong confirmation of "the most narrow and deterministic interpretation of Cope's Rule; namely, that there are directional trends within lineages" (1998, p. 732).

In recent years, a number of other scientists have carried out fossil crunching tests of Cope's Rule in different groups of animals. In one of the most important such studies, David Jablonski (1996; 1997) found that the rule does not hold up very well for marine mollusks that lived during the Cretaceous period. Other recent fossil-crunching studies have focused on dinosaurs (Hone et al. 2005), the earliest reptiles (Laurin 2004), deep sea ostracodes (Hunt and Roy 2006), as well as fish (Knouft and Page 2003). It is not too much of a stretch to see this as a larger research program using fossil crunching techniques to determine where Cope's Rule applies and where it doesn't.

4. NO SMOKING GUNS

As these two sketches show, scientists really can test predictions without actually experimenting on the things they want to study. But that is not the only reason why these two cases involving virtual experiments and fossil crunching tests are interesting.

It is natural to suppose that since paleobiologists cannot directly experiment with the past, their historical research will have a lot in common with detective work. An investigator may visit a crime scene in search of a clue—the smoking gun, the testimony of an eyewitness, a DNA sample, anything—that will indicate whether Suspect A or Suspect B committed the crime. Similarly, you might think that paleontologists begin by formulating different hypotheses about whodunit—say, about what caused the extinction of the dinosaurs. Then they head out into the field in search of some fossil or geological clue that will solve the mystery. At least one philosopher of science has recently argued that this is exactly how historical scientists typically proceed (Cleland 2002).

Cleland is right that paleobiologists often do proceed in this way. However, one remarkable feature of the cases I have described here is that there simply are no proverbial smoking guns—no particular clues in the fossil record that answer the scientists' questions. In the first case, the problem is that even if *T. rex* could have run 40 mph, it is highly unlikely that anyone would ever discover the smoking gun—a set of footprints made by a tyrannosaur that was running that fast. *T. rex* tracks are hopelessly rare, and animals seldom run at their top speeds. In the second case, Cope's Rule is not even the kind of scientific hypothesis that could be supported or disconfirmed by any particular clue from the fossil record. Not only do these cases show that paleobiologists have devised ways of testing hypotheses in spite of the obvious fact that they cannot experiment on the past, but they also show that research in paleobiology often goes beyond detective work.

5. THEORETICAL PROBLEMS IN PALEOBIOLOGY

Virtual experiments and fossil crunching tests have enabled scientists to squeeze information from the fossil record that no one previously suspected was there.

But these methods also have their limitations. A good deal of the recent discussion of Cope's Rule, for example, has focused on conceptual and theoretical issues that do not seem to lend themselves to resolution by empirical testing. I want to conclude by describing two of these theoretical problems which the new methods may not be able to resolve completely.

If you ask *why* average size in mammals tends to increase within lineages over time, as Alroy found, one plausible answer is that natural selection generally favors bigger organisms. There are a number of different ways in which size might give one an edge in the Darwinian struggle for existence: bigger animals could be better able to intimidate rivals, more desirable as mates, more effective predators, more difficult for predators to kill, better able to regulate their body temperatures, and so on (Hone and Benton 2005). We might call this the "bigger is fitter" explanation of Cope's Rule.

In an important and widely cited paper published in 1973, the paleontologist Steven Stanley argued that there is another equally good explanation of Cope's Rule. Stanley's elegant reasoning has inspired a great deal of further work on what have come to be known as "passive diffusion" explanations of evolutionary trends (McShea 1998; for a highly accessible introduction, see Gould 1996). For present purposes, though, I will focus on just one line of argument that can be found in Stanley's original paper. He contrasted the following two ways of thinking about natural selection and body size:

1. Traditional "bigger is fitter" view:
   Smaller size → Larger size

2. Alternative "bigger is not necessarily fitter" view:
   Starting size → Optimum size

Rather than supposing that natural selection always favors larger size, Stanley invites us to suppose that selection will drive a lineage from its starting size (whatever that happens to be) in the direction of the optimum size for whatever ecological niche it happens to occupy.

Stanley then borrows an idea from Cope himself, which is that smaller species are likelier to survive mass extinction events. Cope wrote:

> Changes of climate and food consequent on disturbances of the earth's crust have rendered existence impossible to many plants and animals [including the dinosaurs] . . . Such changes have been often especially severe in their effects on large species, which required food in large quantities. The results have been degeneracy or extinction. On the other hand, plants and animals of unspecial-

> ized habits [mammals, perhaps?] have survived . . . species of small size would survive a scarcity of food, while large ones would perish (1896, pp. 173–74).

Cope was an avid collector of dinosaur fossils, and it is a good bet that he had the extinction of the dinosaurs in mind when he wrote this passage. If this reasoning is correct, it would explain why most lineages happen to start out small. The average size of the organisms within each lineage would then increase over time as a result of natural selection—until, that is, the organisms within the lineage reach the optimum size (whatever that happens to be). According to this view, we need not suppose that larger body size necessarily confers any advantage in the struggle for existence. If lineages happened to start out bigger than their optimum size, then natural selection would in that case lead to size *decrease*. Thus, Stanley argued that if you want to explain why Cope's Rule holds true, the important thing is to explain why lineages typically start out small. Cope's Rule "is more fruitfully viewed as describing evolution from small size rather than toward large size" (1973, p. 22).

Although these two strategies for explaining Cope's Rule may seem very similar, there is one extremely important difference between the "bigger is fitter" view and Stanley's "bigger is not necessarily fitter" view. Cope's Rule is an example of what's known as a macroevolutionary trend, where the "macro" just means that the trend has to do with evolutionary processes at the largest scale. According to the "bigger is fitter" view, the best way to explain this macroevolutionary trend is by looking at *microevolution*: what sort of fitness advantage would a larger body size confer on individual organisms in particular environments? But suppose we take Stanley's route and ask instead why lineages typically start out small. In order to answer that question, we would have to say something about why some lineages (i.e. the ones with the larger body size) are more prone to extinction than others. Notice that here we are still talking about *macroevolution*—the focus is on differences between entire evolutionary lineages, rather than on differences between individual organisms in a population. Should scientists try to explain Cope's Rule, a macroevolutionary trend, in terms of microevolutionary processes ("the bigger is fitter" view) or in terms of macroevolution (as Stanley proposed)? Do scientists even have to choose between these two explanations? (Grantham 1999 suggests that they might not. See also Sterelny 1996.)

I have sketched two closely related theoretical problems: first, is Cope's trend best thought of as evolution toward larger size or as evolution away from smaller size? Second, what is the best way to think about the relationship between macro and microevolutionary processes? It is by no means clear whether the fossil record contains answers to questions like these. Certainly detective work will not help scientists to address them. No smoking gun—no single fossil clue—will tell scientists which explanation of Cope's trend is the correct one, or how macroevolution is related to microevolution. These theoretical problems are not historical whodunits like the question what caused the extinction of the dinosaurs. Detectives search for evidence to try to determine what caused a *particular event* (e.g., "Did Colonel Mustard or Professor Plum commit the murder in the drawing room?"). But the theoretical problems described here do not involve any particular event at all. At issue, rather, is how much scientists can infer about evolution from patterns in the fossil record.

Paleobiologists cannot experiment with the past. However, it would be a mistake to suppose that the only alternative to experimental manipulation is historical detective work. Both the methods of empirical testing (including virtual experiments as well as fossil crunching) and the theoretical problems of paleobiology go beyond the search for the smoking gun.

REFERENCES

Alexander, R. McN. "Estimates of Speeds of Dinosaurs." *Nature* 261 (1976): 129–30.

Alroy, J. "Cope's Rule and the Dynamics of Body Mass Evolution in North American Fossil Mammals." *Science* 280 (1998): 731–34.

Bakker, R. T. *The Dinosaur Heresies*. New York: Kensington, 1986.

Cleland, C. E. "Methodological and Epistemic Differences Between Historical and Experimental Science." *Philosophy of Science* 69 (2002): 474–96.

Cope, E. D. *The Primary Factors of Organic Evolution*. Chicago: Open Court, 1896.

Gould, G. C., and B. J. McFadden. "Gigantism, Dwarfism, and Cope's Rule: 'Nothing in Evolution Makes Sense Without a Phylogeny.'" *Bulletin of the American Museum of Natural History* 285 (2004): 219–37.

Gould, S. J. *Full House: The Spread of Excellence from Plato to Darwin*. New York: Harmony Books, 1996.

Grantham, T. "Explanatory Pluralism in Paleobiology." *Philosophy of Science* 66 (1999): S223–36.

Hone, D. W. E., and M. J. Benton. "The Evolution of Large Size: How Does Cope's Rule Work?" *Trends in Ecology and Evolution* 20 (2005): 4–6.

Hone, D. W. E., T. M. Keesey, D. Pisanis, and A. Purvis. "Macroevolutionary Trends in the Dinosauria: Cope's Rule." *Journal of Evolutionary Biology* 18 (2005): 587–95.

Hunt, G., and K. Roy. "Climate Change, Body Size Evolution, and Cope's Rule in Deep-Sea Ostracodes." *Proceedings of the National Academy of Sciences* 103 (2006): 1347–52.

Huss, J. E. Experimental Reasoning in Non-experimental Science: Case Studies from Paleobiology. PhD diss., University of Chicago, 2004. UMI no. 3125619.

Hutchinson, J. R., F. C. Anderson, S. S. Blemker, and S. J. Delp. "Analysis of Hindlimb Muscle Moment Arms in Tyrannosaurus rex Using a Three-dimensional Musculoskeletal Computer Model: Implications for Stance, Gait, and Speed." *Paleobiology* 31 (2005): 676–701.

Jablonski, D. "Body Size and Macroevolution." In *Evolutionary Paleobiology*, ed. D. Jablonski, et al., 256–289. Chicago: University of Chicago Press, 1996.

———. "Body-size Evolution in Cretaceous Molluscs and the Status of Cope's Rule." *Nature* 385 (1997): 250–52.

McFadden, B. J. "Fossil Horses from 'Eohippus' (Hyracotherium) to Equus: Scaling, Cope's Law, and the Evolution of Body Size." *Paleobiology* 12 (1986).

McShea, D. W. "Mechanisms of Large-Scale Evolutionary Trends." *Evolution* 48 (1994): 1747–63.

———. "Possible Largest-Scale Trends in Organismal Evolution: Eight 'Live Hypotheses.'" *Annual Review of Ecology and Systematics* 29 (1998): 293–318.

Polly, P. D. "Cope's Rule." *Science* 282 (1998): 51.

Ruse, M. *Mystery of Mysteries: Is Evolution a Social Construction*? Cambridge, MA: Harvard University Press, 1999.

Stanley, S. M. "An Explanation for Cope's Rule." *Evolution* 27 (1973): 1–26.

Sterelny, K. "Explanatory Pluralism in Evolutionary Biology." *Biology and Philosophy* 11 (1996): 193–214.

CLASSIFICATION

*ERNST MAYR**

SPECIES CONCEPTS AND THEIR APPLICATION

Darwin's choice of title for his great evolutionary classic, *On the Origin of Species*, was no accident. The origin of new "varieties" within species had been taken for granted since the time of the Greeks. Likewise the occurrence of gradations, of "scales of perfection" among "higher" and "lower" organisms, was a familiar concept, though usually interpreted in a strictly static manner. The species remained the great fortress of stability, and this stability was the crux of the anti-evolutionist argument. "Descent with modification," true biological evolution, could be proved only by demonstrating that one species could originate from another. It is a familiar and often-told story how Darwin succeeded in convincing the world of the occurrence of evolution and how—in natural selection—he found the mechanism that is responsible for evolutionary change and adaptation. It is not nearly so widely recognized that Darwin failed to solve the problem indicated by the title of his work. Although he demonstrated the modification of species in the time dimension, he never seriously attempted a rigorous analysis of the problem of the multiplication of species, of the splitting of one species into two. I have examined the reasons for this failure (Mayr 1959a) and found that foremost among them was Darwin's uncertainty about the nature of species. The same can be said of those authors who attempted to solve the problem of speciation by saltation or other heterodox hypotheses. They all failed to find solutions that are workable in the light of the modern appreciation of the population structure of species. An understanding of the nature of species, then, is an indispensable prerequisite for the understanding of the evolutionary process.

*"Species Concepts and Their Application," reprinted by permission of the publisher from *Populations, Species, and Evolution, An Abridgment of Animal Species and Evolution* by Ernst Mayr, pp. 10–20, Cambridge, MA: Belknap Press of Harvard University Press, Copyright © 1963 by the President and Fellows of Harvard College.

SPECIES CONCEPTS

The term *species* is frequently used to designate a class of similar things to which a name has been attached. Most often this term is applied to living organisms, such as birds, fishes, flowers, or trees, but it has also been used for inanimate objects and even for human artifacts. Mineralogists speak of species of minerals, physicists of nuclear species; interior decorators consider tables and chairs species of furniture. The application of the same term both to organisms and to inanimate objects has led to much confusion and an almost endless number of species definitions (Mayr 1963, 1969); these, however, can be reduced to three basic species concepts. The first two, mainly applicable to inanimate objects, have considerable historical significance, because their advocacy was the cause of much past confusion. The third is the species concept now prevailing in biology.

1. The Typological Species Concept

The typological species concept, going back to the philosophies of Plato and Aristotle (and thus sometimes called the essentialist concept), was the species concept of Linnaeus and his followers (Cain, 1958). According to this concept, the observed diversity of the universe reflects the existence of a limited number of underlying "universals" or types (*eidos* of Plato). Individuals do not stand in any special relation to one another, being merely expressions of the same type. Variation is the result of imperfect manifestations of the idea implicit in each species. The presence of the same underlying essence is inferred from similarity, and morphological similarity is, therefore, the species criterion for the essentialist. This is the so-called morphological species concept. Morphological characteristics do provide valuable clues for the determination of species status. However, using degree of morphological difference as the primary criterion for species status is completely different from utilizing morphological evidence together with various other kinds of evidence in order to determine whether or not a population deserves species rank under the biological species concept. Degree of morphological difference is not the decisive criterion in the ranking of taxa as species. This is quite apparent from the difficulties into which a morphological—typological species concept leads in taxonomic practice. Indeed, its own adherents abandon the typological species concept whenever they discover that they have named as a separate species something that is merely an individual variant.

2. *The Nominalistic Species Concept*

The nominalists (Occam and his followers) deny the existence of "real" universals. For them only individuals exist; species are man-made abstractions. (When they have to deal with a species, they treat it as an individual on a higher plane.) The nominalistic species concept was popular in France in the eighteenth century and still has adherents today. Bessey (1908) expressed this viewpoint particularly well: "Nature produces individuals and nothing more . . . species have no actual existence in nature. They are mental concepts and nothing more . . . species have been invented in order that we may refer to great numbers of individuals collectively."

Any naturalist, whether a primitive native or a trained population geneticist, knows that this is simply not true. Species of animals are not human constructs, nor are they types in the sense of Plato and Aristotle; but they are something for which there is no equivalent in the realm of inanimate objects.

From the middle of the eighteenth century on, the inapplicability of these two medieval species concepts (1 and 2 above) to biological species became increasingly apparent. An entirely new concept, applicable only to species of organisms, began to emerge in the later writings of Buffon and of many other naturalists and taxonomists of the nineteenth century (Mayr 1968).

3. *The Biological Species Concept*

This concept stresses the fact that species consist of populations and that species have reality and an internal genetic cohesion owing to the historically evolved genetic program that is shared by all members of the species. According to this concept, then, the members of a species constitute (1) *a reproductive community.* The individuals of a species of animals respond to one another as potential mates and seek one another for the purpose of reproduction. A multitude of devices ensures intraspecific reproduction in all organisms. The species is also (2) *an ecological unit* that, regardless of the individuals composing it, interacts as a unit with other species with which it shares the environment. The species, finally, is (3) *a genetic unit* consisting of a large intercommunicating gene pool, whereas an individual is merely a temporary vessel holding a small portion of the contents of the gene pool for a short period of time. These three properties raise the species above the typological interpretation of a "class of objects" (Mayr 1963, p. 21). The species definition that results from this theoretical species concept is: *Species are groups of interbreeding natural populations that are reproductively isolated from other such groups.*

The development of the biological concept of the species is one of the earliest manifestations of the emancipation of biology from an inappropriate philosophy based on the phenomena of inanimate nature. The species concept is called biological not because it deals with biological taxa, but because the definition is biological. It utilizes criteria that are meaningless as far as the inanimate world is concerned.

When difficulties are encountered, it is important to focus on the basic biological meaning of the species: A species is a protected gene pool. It is a Mendelian population that has its own devices (called isolating mechanisms) to protect it from harmful gene flow from other gene pools. Genes of the same gene pool form harmonious combinations because they have become coadapted by natural selection. Mixing the genes of two different species leads to a high frequency of disharmonious gene combinations; mechanisms that prevent this are therefore favored by selection. Thus it is quite clear that the word "species" in biology is a relational term. A is a species in relation to B or C because it is reproductively isolated from them. The biological species concept has its primary significance with respect to sympatric and synchronic populations (existing at a single locality and at the same time), and these—the "nondimensional species"—are precisely the ones where the application of the concept faces the fewest difficulties. The more distant two populations are in space and time, the more difficult it becomes to test their species status in relation to each other, but also the more irrelevant biologically this becomes.

The biological species concept also solves the paradox caused by the conflict between the fixity of the species of the naturalist and the fluidity of the species of the evolutionist. It was this conflict that made Linnaeus deny evolution and Darwin the reality of species (Mayr 1957). The biological species combines the discreteness of the local species at a given time with an evolutionary potential for continuing change.

THE SPECIES CATEGORY AND SPECIES TAXA

The advocacy of three different species concepts has been one of the two major reasons for the "species problem." The second is that many authors have failed to make a distinction between the definition of the species category and the delimitation of species taxa (for fuller discussion see Mayr 1969).

A *category* designates a given rank or level in a hierarchic classification. Such terms as "species," "genus," "family," and "order" designate categories. A category, thus, is an abstract term, a class name, while the organisms placed in these categories are concrete zoological objects.

Organisms, in turn, are classified not as individuals, but as groups of organisms. Words like "bluebirds," "thrushes," "songbirds," or "vertebrates" refer to such groups. These are the concrete objects of classification. Any such group of populations is called a *taxon* if it is considered sufficiently distinct to be worthy of being formally assigned to a definite category in the hierarchic classification. *A taxon is a taxonomic group of any rank that is sufficiently distinct to be worthy of being assigned to a definite category.*

Two aspects of the taxon must be stressed. A taxon always refers to specified organisms. Thus *the* species is not a taxon, but any given species, such as the Robin (*Turdus migratorius*) is. Second, the taxon must be formally recognized as such, by being described under a designated name.

Categories, which designate a rank in a hierarchy, and taxa, which designate named groupings of organisms, are thus two very different kinds of phenomena. A somewhat analogous situation exists in our human affairs. Fred Smith is a concrete person, but "captain" or "professor" is his rank in a hierarchy of levels.

THE ASSIGNMENT OF TAXA TO THE SPECIES CATEGORY

Much of the task of the taxonomist consists of assigning taxa to the appropriate categorical rank. In this procedure there is a drastic difference between the species taxon and the higher taxa. Higher taxa are defined by intrinsic characteristics. Birds is the class of feathered vertebrates. Any and all species that satisfy the definition of "feathered vertebrates" belong to the class of birds. An essentialist (typological) definition is satisfactory and sufficient at the level of the higher taxa. It is, however, irrelevant and misleading to define species in an essentialistic way because the species is not defined by intrinsic, but by *relational* properties.

Let me explain this. There are certain words that indicate a relational property, like the word "brother." Being a brother is not an inherent property of an individual, as hardness is a property of a stone. An individual is a brother only with respect to someone else. The word "species" likewise designates such a relational property. A population is a species with respect to all other populations with which it exhibits the relationship of reproductive isolation—noninterbreeding. If only a single population existed in the entire world, it would be meaningless to call it a species.

Noninterbreeding between populations is manifested by a gap. It is this gap between populations that coexist (are sympatric) at a single locality at a given time which delimits the species recognized by the local naturalist. Whether one studies birds, mammals, butterflies, or snails near one's home town, one finds each species clearly delimited and sharply separated from all other species. This demarcation is sometimes referred to as the species delimitation *in a nondimensional system* (a system without the dimensions of space and time).

Anyone can test the reality of these discontinuities for himself, even where the morphological differences are slight. In eastern North America, for instance, there are four similar species of the thrush genus *Catharus* (Table 1), the Veery (*C. fuscescens*), the Hermit Thrush (*C. guttatus*), the Olive-Backed or Swainson's Thrush (*C. ustulatus*) , and the Gray-Cheeked Thrush (*C. minimus*). These four species are sufficiently similar visually to confuse not only the human observer, but also silent males of the other species. The species-specific songs and call notes, however, permit easy species discrimination, as observationally substantiated by Dilger (1956). Rarely do more than two species breed in the same area, and the overlapping species, *f*+ *g*, *g* + *u*, and *u* + *m*, usually differ considerably in their foraging habits and niche preference, so that competition is minimized with each other and with two other thrushes, the Robin (*Turdus migratorius*) and the Wood Thrush (*Hylocichia mustelina*), with which they share their geographic range and many ecological requirements. In connection with their different foraging and migratory habits the four species differ from one another (and from other thrushes) in the relative length of wing and leg elements and in the shape of the bill. There are thus many small differences between these at first sight very similar species. Most important, no hybrids or intermediates among these four species have ever been found. Each is a separate genetic, behavioral, and ecological system, separated from the others by a complete biological discontinuity, a gap.

Table 1. Characteristics of Four Eastern North American Species of Catharus (from Dilger 1956)

Characteristics compared	*C. fuscescens*	*C. guttatus*	*C. ustulatus*	*C. minimus*
Breeding range	Southernmost	More	Boreal northerly	Arctic
Wintering area	No. South America Bottomland	So. United States, Coniferous	C. America to to Arentina, mixed or pure	No. South America, Stunted
Breeding habitat	Woods with lush undergrowth	Woods mixed with deciduous	Tall Coniferous forests	Northern fir and spruce forests

Foraging	Ground and arboreal (forest interior)	Ground (inner forest edges)	Largely arboreal (forest interior)	Ground (forest inerior)
Nest	Ground	Ground	Trees	Trees
Spotting on eggs	Rare	Rare	Always	Always
Relative wing length	Medium	Short	Very long	Medium
Hostile call	*veer pheu*	*chuck seeeep*	*peep chuck-burr*	*beer*
Song	Very distinct	Very distinct	Very distinct	Very distinct
Flight song	Absent	Absent	Absent	Present

DIFFICULTIES IN THE APPLICATION OF THE BIOLOGICAL SPECIES CONCEPT

The practicing taxonomist often has difficulties when he endeavors to assign populations to the correct rank. Sometimes the difficulty is caused by a lack of information concerning the degree of variability of the species with which he is dealing. Helpful hints on the solution of such practical difficulties are given in the technical taxonomic literature (Mayr 1969).

More interesting to the evolutionist are the difficulties that are introduced when the dimensions of time and space are added. Most species taxa do not consist merely of a single local population but are an aggregate of numerous local populations that exchange genes with each other to a greater or lesser degree. The more distant that two populations are from each other, the more likely they are to differ in a number of characteristics. I show elsewhere (Mayr 1963, ch. 10 and 11) that some of these populations are incipient species, having acquired some but not all characteristics of species. One or another of the three most characteristic properties of species taxa—reproductive isolation, ecological difference, and morphological distinguishability—is in such cases only incompletely developed. The application of the species concept to such incompletely speciated populations raises considerable difficulties. There are six wholly different situations that may cause difficulties.

(1) *Evolutionary continuity in space and time.* Widespread species may have terminal populations that behave toward each other as distinct species even though they are connected by a chain of interbreeding populations. Cases of reproductive isolation among geographically distant populations of a single species are discussed in Mayr (1963, ch. 16).

(2) *Acquisition of reproductive isolation without corresponding morphological change.* When the reconstruction of the genotype in an isolated population has resulted in the acquisition of reproductive isolation, such a population must be considered a biological species. If the correlated morphological change is very slight or unnoticeable, such a species is called a sibling species (Mayr 1963, ch. 3).

(3) *Morphological differentiation without acquisition of reproductive isolation.* Isolated populations sometimes acquire a degree of morphological divergence one would ordinarily expect only in a different species. Yet some such populations, although as different morphologically as good species, interbreed indiscriminately where they come in contact. The West Indian snail genus *Cerion* illustrates this situation particularly well (fig. 6.1).

(4) *Reproductive isolation dependent on habitat isolation.* Numerous cases have been described in the literature in which natural populations acted toward each other like good species (in areas of contact) as long as their habitats were undisturbed. Yet the reproductive isolation broke down as soon as the characteristics of these habitats were changed, usually by the interference of man. Such cases of secondary breakdown of isolation are discussed in Mayr (1963, ch. 6).

(5) *Incompleteness of isolating mechanisms.* Very few isolating mechanisms are all-or-none devices (see Mayr 1963, ch. 5). They are built up step by step, and most isolating mechanisms of an incipient species are imperfect and incomplete. Species level is reached when the process of speciation has become irreversible, even if some of the secondary isolating mechanisms have not yet reached perfection (see Mayr 1963, ch. 17).

(6) *Attainment of different levels of speciation in different local populations.* The perfecting of isolating mechanisms may proceed in different populations of a polytypic species (one having several subspecies) at different rates. Two widely overlapping species may, as a consequence, be completely distinct at certain localities but may freely hybridize at others. Many cases of sympatric hybridization discussed in Mayr (1963, ch. 6) fit this characterization (see Mayr 1969 for advice on handling such situations).

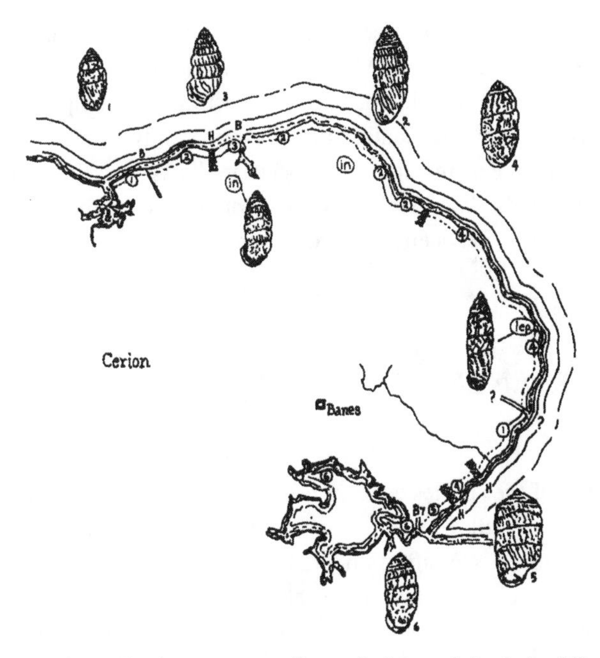

Figure 6.1. The distribution pattern of populations of the halophilous land snail *Cerion* on the Banes Peninsula in eastern Cuba. Numbers refer to distinctive races or "species." Where two populations come in contact (with one exception) they hybridize (*H*), regardless of degree of difference. In other cases contact is prevented by a barrier (*B*). *In* = isolated inland population.

These six types of phenomena are consequences of the gradual nature of the ordinary process of speciation (excluding polyploidy; see Mayr 1963, p. 254). Determination of species status of a given population is difficult or arbitrary in many of these cases.

DIFFICULTIES POSED BY UNIPARENTAL REPRODUCTION

The task of assembling individuals into populations and species taxa is very difficult in most cases involving uniparental (asexual) reproduction. Self-fertilization, parthenogenesis, pseudogamy, and vegetative reproduction are forms of uniparental reproduction. The biological species concept, which is based on the presence or absence of interbreeding between natural populations, cannot be applied to groups with obligatory asexual reproduction because interbreeding of populations is nonexistent in these groups. The nature of this dilemma is discussed in more detail elsewhere (Mayr 1963, 1969). Fortunately, there seem to be rather well-defined discontinuities among most kinds of uniparentally reproducing organisms. These discontinuities are apparently produced by natural selection from the various mutations that occur in the asexual lines (clones). It is customary to utilize the existence of such discontinuities and the amount of morphological difference between them to delimit species among uniparentally reproducing types.

THE IMPORTANCE OF A NONARBITRARY DEFINITION OF SPECIES

The clarification of the species concept has led to a clarification of many evolutionary problems as well as, often, to a simplification of practical problems in taxonomy. The correct classification of the many different kinds of varieties (phena), of polymorphism (Mayr 1963, ch. 7), of polytypic species (ibid., ch. 12), and of biological races (ibid., ch. 15) would be impossible without the arranging of natural populations and phenotypes into biological species. It was impossible to solve, indeed even to state precisely, the problem of the multiplication of species until the biological species concept had been developed. The genetics of speciation, the role of species in large-scale evolutionary trends, and other major evolutionary problems could not be discussed profitably until the species problem was settled. It is evident then that the species problem is of great importance in evolutionary biology and that the growing agreement on the concept of the biological species has resulted in a uniformity of standards and a precision that is beneficial for practical as well as theoretical reasons.

THE BIOLOGICAL MEANING OF SPECIES

The fact that the organic world is organized into species seems so fundamental that one usually forgets to ask why there are species, what their meaning is in the scheme of things There is no better way of answering these questions than to try to conceive of a world without species. Let us think, for instance, of a world in which there are only individuals, all belonging to a single interbreeding community. Each individual is in varying degrees different from every other one, and each individual is capable of mating with those others that are most similar to it. In such a world, each individual would be, so to speak, the center of a series of concentric rings of increasingly more different individuals. Any two mates would be on the average rather different from each other and would produce a vast array of genetically different types among their offspring. Now let us assume that one of these recombinations is particularly well adapted for one of the available niches. It is prosperous in this niche, but when the time for mating comes, this superior genotype will inevitably be broken up. There is no mechanism that would prevent such a destruction of superior gene combinations, and there is, therefore, no possibility of the gradual improvement of gene combinations. The significance of the species now becomes evident. The reproductive isolation of a species is a protective device that guards against the breaking up of its well-integrated, coadapted gene system. Organizing organic diversity into species creates a system that permits genetic diversification and the accumulation of favorable genes and gene combinations without the danger of destruction of the basic gene complex. There are definite limits to the amount of genetic variability that can be accommodated in a single gene pool without producing too high a proportion of inviable recombinants. Organizing genetic diversity into protected gene pools—that is, species—guarantees that these limits are not overstepped. This is the biological meaning of species.

REFERENCES

Bessey, C. E. "The Taxonomic Aspect of the Species." *American Naturalist* 42 (1908): 218–24.

Cain, A. J. "Logic and Memory in Linnaeus's System of Taxonomy." *Proceedings of the Linnean Society of New South Wales* 169 (1958): 144–63.

Dilger, W. C. "Hostile Behavior and Reproductive Isolating Mechanisms in the Avian Genera *Catharus* and *Hylocichla*." *Auk* 73 (1956): 313–53.

Mayr, E. "Species Concepts and Definitions." In *The Species Problem*, ed. E. Mayr. Washington, DC: American Association for the Advancement of Science publ. no. 50, 1957: 1–22.

———. "Darwin and the Evolutionary Theory in Biology." In *Evolution and Anthropology: A Centennial Approach*. Washington, DC: Anthropological Society of America, 1959a.

———. *Animal Species and Evolution*. Cambridge: Harvard University Press, 1963.

———. "Illiger and the Biological Species Concept." *Journal of the History of Biology* 1 (1968): 163–78.

———. *Principles of Systematic Zoology*. New York: McGraw-Hill, 1969.

RICHARD A. RICHARDS

SOLVING THE SPECIES PROBLEM: KITCHER AND HULL ON SETS AND INDIVIDUALS

A dairy farmer notices a dramatic decrease in output from his milk cows. After trying everything he can think of to solve the problem, he turns in despair to the smartest person he knows—a friend who also happens to be a mathematician. The farmer explains the problem to his friend, who listens intently, scribbles down some notes and leaves. After a week of intense deliberation the mathematician friend returns, clears his throat and says with some authority: "Consider a perfectly spherical cow . . ."[1]

I. THE SPECIES PROBLEM

Ask anyone what species are, and if he or she knows at least some biology, the answer you will most likely get is something like "species are groups of actually or potentially interbreeding organisms." Unfortunately, this standard way of conceiving species, the *biological species concept*, is obviously inadequate, excluding all those organisms that reproduce asexually—some vertebrates such as whiptail lizards (genus *Cnemidophorus*), the agamosperms in the plant kingdom, and numerous invertebrates. Systematists have proposed additional species concepts to rectify this and similar problems, until there are now, on some counts, over 20 species concepts in circulation. The problem here is that first, each of these species concepts seems inadequate—limited in applicability and scope; second, different concepts divide biodiversity up in different ways; third, there is no obvious reason to prefer one concept over another. This is *the species problem*: There are multiple, inconsistent ways to conceive of species, that divide up biodiversity in different and inconsistent ways without any obvious resolution.

The species problem contributes to a disparity in species counts. Given any group of organisms, systematists typically disagree—sometimes wildly—about the number of species. *Metriaclima*, a genus of fish species in Lake Malawi, was proposed to replace a single species *Pseudotropheus zebra*, on the basis of additional samples, but systematists disagree about how many species are represented in the samples, some counting two, others ten. Conventional classifications count 9,000 birds in the world, but some ornithologists believe there are 20,000. On one count of Mexican bird species, there are 101, on another there are 249 (Hey 2001, 20). Some of this disparity is due to what systematists call the "lumper" and "splitter" tendencies of individual systematists. But the disparity in counts is also due to the fact that systematists employ different species concepts and these concepts lump and split organisms differently. This is not just a theoretical problem for systematists; the application of endangered species law and the preservation of biodiversity depend on how we identify and individuate species.

Systematists and philosophers of biology have worried greatly about the proliferation of species concepts. Two philosophical issues have come to be central to the species problem. First is the monism-pluralism question: is there a single way to conceive species, or are we forced into recognizing multiple species concepts? Second is the ontological question: what basic kinds of things are species—sets or individuals? The views of two prominent philosophers, Philip Kitcher and David Hull, have come to represent the fundamental philosophical positions here. Kitcher argues for pluralism, and that species should be conceived as sets. Hull accepts monism as a worthy goal, and argues that species should be conceived as historical individuals. My intentions here are first, to lay out in rough outline these competing philosophical positions, and second, to suggest how the disputes can be resolved. I will propose a hierarchical approach to species concepts—a theoretical monism coupled with an operational pluralism; and then argue for a modest, pragmatic preference for Hull's species as individuals thesis based on theoretical, metaphorical and heuristic value. To better understand the issues here, a brief survey of the history of the species debate is useful.

2. SPECIES CONCEPTS

Swedish botanist Carolus Linnaeus, the best known of pre-Darwinian systematists, created the modern taxonomic method, employing binomial nomenclature and a hierarchical taxonomy of species, genera, family, class and order. Linnaeus is also associated with the view that species have unchanging essences that are passed on in reproduction (even though late in his career he believed that it was higher taxa that have the essences). Darwin challenged the fixity of species, arguing for the gradualistic production of new species through the operation of natural selection on populations. This commitment to gradualism and population thinking caused him to doubt that there really were such things as species. After all, if there is variability extending throughout populations, and these populations change gradually over time to produce new forms, then it is unclear where to draw the lines demarcating species.

Darwin's evolutionary theory provided the resources for the many species concepts that have since found application. Ernst Mayr, one of the architects of the Modern Synthesis, focused on the reproductive cohesion found in species in his *biological species concept* discussed in the first sentences of this essay (Mayr 1982, 273). After all, members of species reproduce with each other, but not with members of other species. Following Mayr's example, systematists have proposed multiple concepts based on important evolutionary processes. There are the *ecological species concept*, based on fit within an adaptive zone; the *recognition species concept,* based on fertilization systems; and the *genetic species concept*, based on genetic cohesion. Other systematists focus on the similarities found among members of species with the *morphological species concept*, based on specific morphological similarities and differences; the *phenetic species concept*, based on overall similarity, the *genotypic cluster concept*, based on genotypical similarity and clustering; and *the polythetic species concept* that conceives of species in terms of clusters of statistically covarying characters. Yet other systematists focus on the fact that species are extended in time. Here, we have the *evolutionary species concept,* which conceives of species as evolutionary lineages; the *successional species concept, chronospecies concept*, and *paleospecies concept*, which conceive of species as segments of a changing lineage; while the *cladistic species concept* and *phylogenetic species concept* are based on the branching of lineages (Mayden 1997).

3. PLURALISM AND MONISM

This multiplicity of species concepts begs the question: is there a single concept that is adequate, or must we employ multiple concepts? In other words: is there a single kind of species thing, or are there multiple kinds of species things? Philip Kitcher gives a pluralistic answer:

> Species are sets of organisms related to one another by complicated, biological interesting relations. There are many such relations which could be used to

> delimit species tax. However, there is no unique relation which is privileged in that the species taxa it generates will answer to the needs of all biologists and will be applicable to all groups of organisms. (Kitcher 1992, 317)

On this approach, dubbed "pluralistic realism" by Kitcher, biologists can legitimately divide up a single group of organisms into different sets on the basis of a variety of factors—similarity, reproductive isolation, adaptive niche or phylogeny. This pluralism is apparent in his taxonomy of species concepts reproduced below (see fig. 7.1). The groups identified are "real" in the sense that the division is based on real structural and historical factors. But how they are divided up ultimately depends on the interests and needs of the researcher. In effect, what counts as a species is pragmatic and varies according to the needs of researchers.

If we adopt Kitcher's suggestion, then there is no single thing "species" by which we can classify all organisms. Geneticists, systematists, developmental biologists and ecologists will all likely focus on different processes and features, and group on those different grounds, as will the researchers who focus on particular kinds of organisms. This seems subjective (you group on your grounds, I group on mine), but whatever science is, surely it cannot endorse differences of outcome based on the interests of the researcher. On the other hand, the basis for the different groupings is objective in the sense that it is not *just* preference that operates here. The structural features and historical processes are all real. It is

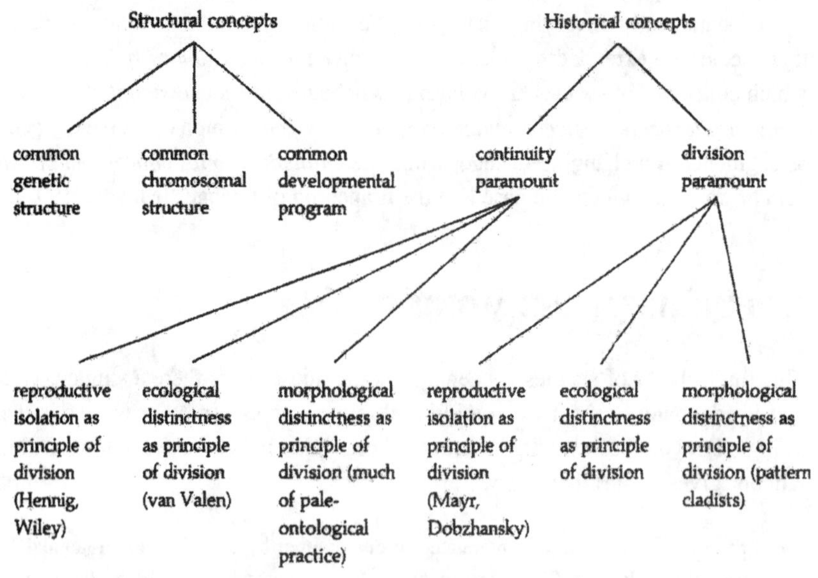

Figure 7.1. The taxonomy of species concepts

only the choice of criteria that is subjective. After all, we make these pragmatic classifications all the time. We classify carnivores together, separate from herbivores, on the basis of what they eat. We distinguish sexually reproducing organisms from the asexual on the basis of reproduction. We even classify green vegetables together on the basis of nutritive functioning for humans.

We can grant the usefulness of these pluralistic, pragmatic classifications, however, without being pluralists and subjectivists. The real question for Kitcher is: do we want to call any or all of these sets of organisms "species"? In dividing biodiversity into species, most systematists think that they are somehow "dividing nature at its joints"—picking out *uniquely* important entities, not just pragmatically significant categories. Kitcher concedes that his pluralism does not tell us which sets should count as species (Kitcher 1989, 186). Unfortunately, we do not get much further guidance from Kitcher here about which sets might count as species, except that further "specification is needed" (Kitcher 1989, 187). Kitcher's concession raises the obvious objection: might not further specification indicate that some *single* kind of set should count as species? In other words, isn't Kitcher giving up too quickly on a monistic account of species? One reason to worry that Kitcher's pluralism is premature is that there seems to be a developing consensus about some aspects of the species category.

One widely accepted view is that species are just those things that evolve. One way of understanding this claim is that species are things that are "born" in speciation change over a period of time, and "die" in extinction or new speciation. This is an idea that has become associated with a species concept advocated by G. G. Simpson, who argued that a species is "a lineage (an ancestral-descendent sequence of populations) evolving separately from others and with its own unitary evolutionary role and tendencies" (Mayden 1997, 395). Simpson's idea has been widely adopted, not only by those who advocate a historical conception of species, but also by those who have advocated process or similarity based concepts. Most systematists now agree that whatever species are, they are historical lineages that originate in speciation events, change over time and go extinct. This is true whether a species is sexual or asexual, and whatever its morphological, behavioral and genetic properties are. If so, we might ask whether the "further specification" acknowledged by Kitcher might focus on lineages.

4. DIVISION OF LABOR SOLUTION

One of the standard fallacies covered in informal logic classes is the fallacy of "false dichotomy"—the incorrect assumption that there are only two alternative positions, and the inference that one must be correct because the other is false. In much of the debate over monism and pluralism lurk two assumptions. The first assumption is the dichotomy between simple monism and simple pluralism: either one species concept will turn out to be correct and the others incorrect, or there is an irreducible plurality of kinds of species. The second assumption is that all species concepts are equivalent. Both of these assumptions are problematic. One plausible solution to the species problem begins by rejecting these assumptions.

Standard attempts to solve the species problem often begin by comparing each species concept against a standard set of criteria: Does the concept have wide scope? Is it theoretically significant? Is it operational (Hull 1997)? As useful as this approach is, perhaps some species concepts should instead be judged against only one criterion, such as theoretical significance, while other species concepts should be judged against another criterion, such as operationality. In other words, perhaps there is a conceptual division of labor. This idea is behind Richard Mayden's proposal that we should distinguish the *theoretical concepts* that function to tell us what these species things are, from the *operational concepts* that tell us how to recognize them. Mayden follows Simpson here in proposing for a *theoretical concept* that species are lineages with beginnings, ends and unique fates, as asserted by the *evolutionary species concept.* He argues that competing species concepts are really operational concepts designed to help us recognize species in nature (Mayden 1997). Mayr's *biological species concept*, for instance, helps us recognize *some* lineages because of the patterns of sexual reproduction that occur within the lineage and the reproductive isolation from other lineages. For asexual species we might turn to another operational principle based on the similarities found within the lineage and differences between lineages. But which operational principle will be relevant would depend on the nature of the organisms under investigation, so no single operational principle will suffice.

Mayden's distinction between theoretical and operational concepts echoes an earlier debate in the philosophy of science. Rudolf Carnap criticized the tendency to equate operational concepts with definitions (by P. W. Bridgman and others), then he proposed his own way of making this distinction. According to Carnap, there are theoretical principles that can function as definitions, telling us what a thing is, but they get connected to observation via *correspondence rules* that tell us how to identify or observe a thing (Carnap 1966, 233–36). We could, for instance, have a theoretical concept telling us what sort of thing an electron is, and then a set of rules telling us how to "observe" an electron. Adopting Carnap's way of making this distinction, we could in principle be a monist relative to a theoretical species concept, and a pluralist relative to correspondence rules. So we can accept that species have all sorts of different properties that may be theoretically significant in a variety of ways (as Kitcher does), but still endorse a single theoretical concept—perhaps the lineage based *evolutionary species concept*. The bottom line is that we can grant Kitcher's claim that there are many important relations, but reject his conclusion that there is therefore "no unique relation which is privileged."

5. SETS OR INDIVIDUALS

Suppose Simpson's suggestion that species are historical lineages continues to gain widespread support. Can we reconcile this with Kitcher's claim above that species are sets? Lurking here is an ontological question: whichever species concept we adopt, what basic, fundamental kind of thing are species? Traditionally this question has been whether species are natural kinds or not, but recent debate has focused on the related question whether species are sets or individuals. David Hull (along with Michael Ghiselin) has argued that while species have traditionally been conceived as spatio-temporally unrestricted classes or sets with members, there is another, better way to conceive of species "mereologically"—as spatio-temporally restricted individuals with parts. Hull explains:

> By "individuals" I mean spatiotemporally localized cohesive and continuous entities (historical entities). By "classes" I intend spatiotemporally unrestricted classes, the sorts of things which can function in traditionally defined laws of nature. The contrast is between Mars and planets, the Weald and geological strata, between Gargantua and organisms. (Hull 1992, 294)

According to this *species as individuals* thesis, individuality is minimally constituted by spatio-temporal restriction—species individuals have a beginning and an end, and spatial boundaries. But it is often understood to require something more—a cohesion constituted by gene flow, similar selection pressures and outcomes and even causal integration.

The *species as individuals* thesis has seemed counterintuitive to many. Clearly species are made up of individual organisms and do not have the same kind of cohesion we see in individual organisms. Individual organisms have cellular structure, and often have extensive causal integration. In vertebrates, for instance, lungs (or gills) and hearts interact in the cardiovascular system, muscular systems are supported by skeletal systems, and gastrointestinal systems provide chemical nutrients to all the other systems. One standard objection to the *species as individuals* thesis is that this kind of causal interaction is necessary for the existence of the individual, but not so obviously for a species (Ruse 1992). Kitcher endorses this commonsense objection when he claims that the difference between individuals and species is that if we separate the parts (cells) of an individual we destroy it, but if we separate the parts of a species (individual members), the species can still exist. We can bring the parts of the species back together for reproduction (Kitcher 1989, 186). Therefore, Kitcher concludes, species are not individuals in the same sense organisms are.

Many philosophers have found this *cohesion argument* both compelling and decisive, but systematists are less convinced. Examination of another of Kitcher's arguments reveals why this is the case. This argument is that the dispute about sets and individuals is not really an ontological issue because "all our discourse about evolution can be reconstructed equally well within set theory or within mereology" (Kitcher 1989, 185). We could treat species—even if construed as historical lineages—as either sets or individuals. Kitcher's claim here could be interpreted superficially as just an issue about lexicon: we can use either set or individual (mereological) vocabulary and it does not really make any difference. If it were just about lexicon, Kitcher would surely be right, but there are other philosophical issues lurking.

One standard tool in the philosopher's toolbox is conceptual analysis, used to get a clarified understanding of what we mean in our use of terms. But as Hull objects, the relevant terms here get used in a variety of ways and there is no single meaning to uncover and clarify.

> The terms "gene," "organism," and "species" have been used in a wide variety of ways in a wide variety of contexts. Anyone who attempts merely to map this diversity is presented with a massive and probably pointless task. (Hull 1992, 293)

More to the point here, the term "individual" undergoes analysis in both a philosophical and biological context, and there are significant differences in the contexts that affect understanding of the term. Hull explains:

> Differences between these two analyses have three sources: first, philosophers have been most interested in individuating persons, the hardest case of all, while biologists have been content to individuate organisms; second, when philosophers have discussed the individuation of organisms, they have usually limited themselves to adult mammals, while biologists have attempted to develop a notion of organism adequate to handle the wide variety of organisms which exist in nature; and finally, philosophers have felt free to resort to hypothetical science fiction examples to test their conceptions, while biologists rely on actual cases. In each instance, I prefer the biologists' strategy. A clear notion of an individual organism seems an absolute prerequisite for any adequate notion of a person,

> and this notion should be applicable to all organisms, not just a miniscule fraction. But most importantly, real examples tend to be much more detailed and bizarre than those made up by philosophers. Too often the example is constructed for the sole purpose of supporting the preconceived intuitions of the philosophers and has no life of its own. It cannot force the philosopher to improve his analysis the way that real examples can. Biologists are in the fortunate position of being able to test their analyses against a large stock of extremely difficult, extensively documented actual cases. (Hull 1992, 301)

One of Hull's concerns in this long passage is that when philosophers think about individuals, they typically think about a narrow range of organisms—primarily humans, with some consideration of large mammals or vertebrates. We focus on them for obvious reasons: first, these are the kind of individual organisms we encounter all the time in everyday life; second, these are the kinds of organisms we can observe without any special assistance from either microscopes or biological theory. As Hull warns, though, we can be misled by this perspective:

> Given our relative size, period of duration, and perceptual acuity, organisms appear to be historical entities, species appear to be classes of some sort, and genes cannot be seen at all. However, after acquainting oneself with the various entities which biologists count as organisms and the roles which organisms and species play in the evolutionary process, one realizes exactly how problematic our commonsense notions are. (Hull 1992, 295)

The biologist can avoid these mistaken commonsense notions by adopting a much broader perspective, based not just on large mammals and vertebrates, but on microscopic organisms, fungi, bacteria, viruses, slime molds, jellyfish and more. What is important here is that since we are here discussing biological species in general, we must base our analysis on the wide range of organisms we find in nature—not just the humans and a narrow range of vertebrates we first think of when we think of individuals.

Furthermore when we observe a broad range of organisms, the cohesion we find in vertebrate individuals is often lacking. There is, in fact, a wide range of "individuality" in biodiversity that makes it difficult to draw the kind of lines presupposed by the cohesion argument, when it asserts that individual organisms cease to exist if their parts are separated. Michael Ghiselin asks us to consider cases where cohesiveness is lacking to varying degrees among individual organisms:

> A situation in which an organism breaks up into component parts that never get back together again is familiar even to the lay person. Propagation by cuttings,

> budding, and fission of some animals such as starfish are good examples. There are fewer examples of organisms that break up, then fuse back together, but slime-molds are an example. These fungi, which form lineages produced by asexual reproduction, forage independently on organic materials. Later in their life cycle they come together to form a single mass, complete with reproductive organs that give rise to spores. Sometimes they are called "social amoebae," and the term aptly compares them to societies that are united only from time to time. (Ghiselin 1997, 55)

One could provide additional examples like those cited by Ghiselin, where individual organisms seem to survive the separation of the parts, and where what may seem like groups of organisms come together to form a cohesive whole joined by causal interaction. But the important point here is that a survey of biodiversity reveals that the cohesiveness found in ourselves and vertebrate individuals in general is not universal, nor is there a clear line between what we would identify as individual organisms and mere groups of cells. We would be making a mistake if we ignore the vast number of intermediate cases in evaluating our *general* biological concepts. The conclusion we might want to draw from our investigation of biodiversity is that individuality really comes in degrees: some organisms are highly cohesive with extensive causal interaction among their parts, while other organisms cohere much more loosely. Similarly, those lineages we might want to call species might also have degrees of individuality, with some cohering to a greater degree than others. If so, it may also be that being a species comes in degrees, and some organisms form lineages that are more fully species than others.

6. EVOLUTIONARY THEORY AND ONTOLOGY

The preceding defense of the *species as individuals* thesis is mostly negative in that it can only establish that species *could be* conceived as individuals, not that they *should be*. We might wonder then, if there are positive reasons to treat species as individuals. Hull gives a reason, arguing that it is more consistent with evolutionary theory to treat species as individuals than as sets. At minimum, there is the claim that it is species that evolve—by speciation, change over time and extinction. This change over time is a product of natural selection operating on a population of organisms that are connected by ancestor-descendent relationships and reproduction, and vary as a result of imperfect copying in reproduction. What this requires, according to Hull, is that we conceive of species as a changing lineage.

> The relevant organismal units in evolution are not sets of organisms defined in

> terms of structural similarity but lineages formed by the imperfect copying processes of reproduction. Organisms can belong to the same lineage even though they are structurally different from other organisms of that lineage. What is more, continued changes in structure can take place indefinitely. If evolution is to occur, not only can such indefinite structural variation take place within organisms lineages but also it must . . . . (Hull 1992, 298)

This, by itself, is not sufficient to recommend the individuality thesis. After all, Kitcher recognizes the possibility of "historically connected sets," where set inclusion is based on the historical relation found in a lineage (Kitcher 1992, 320). But there are further considerations that might favor the individuality thesis.

In a changing lineage what is important is not just the bare fact that there is a historical relation between all the members of the lineage (as required for set inclusion). The temporal ordering of the members and the pattern of change are important as well. Furthermore, because there is some cohesion, the factors that affect cohesion are important—reproduction, mate recognition systems, social structures, etc. An historical set based only on historical relations cannot obviously represent all these factors and the many more that might be relevant. What the individuality conception can do is cause us to focus on those factors that are important to individuality—cohesion, speciation and extinction processes. If these are important factors in evolutionary processes, then the individuality thesis is valuable heuristically—to help us understand processes, and see how theory can be developed and extended.

Moreover, this heuristic function can extend beyond the narrow concern with species. We can regard genes as individuals, and extend that to include gene lineages (Hull 1992, 294). Similarly, we can regard colonies as individuals, much as we would species and individual organisms. And once we do that we can ask the same set of "individuality" questions about origins, change, cohesion factors and ends. In these ways, individuality seems to function more broadly as a governing metaphor, guiding the many ways we think about biological entities and processes. Whether or not the *individual* approach functions better within evolutionary theory than the *set* approach in these various ways cannot be decided on the basis of such limited analysis from some philosopher. The theoretical issues are numerous and complex, and the suitability of each approach will likely be determined over time by the biologists engaged in research and theoretical development who use the various concepts. (As pragmatically minded philosophers are fond of saying, "the proof is in the pudding.") For now, trends among evolutionists seem to favor the *species as individuals* approach.

7. CONCLUSION

Given the species problem—that there are multiple species concepts that lead to divergent species counts—we might be tempted to follow Kitcher, and adopt a pluralistic account of species. If so, there would be no single kind of species thing. But surely we do not want to give in to this pluralism and its attendant subjectivity unless necessary. The division of labor solution gives us an alternative: while there may be multiple operational concepts (better described as *correspondence rules*), that tell us how to identify species things, there can still be a single theoretical concept that tells us what sort of thing species are. We can have our cake and eat it too! And if we follow biologists in accepting that species are lineages, then we might also regard species as individuals rather than sets.

In the joke that introduces this paper, we are amused (presumably) by the response of the mathematician. When confronted with a dairy cow that has reduced milk output, he asks us to consider a perfectly spherical cow. What is amusing is that he is adopting a strategy that is perfectly appropriate for certain mathematical problems, but not for the *physiological* problem of milk production. The problem is not that it is impossible to use the "discourse" of mathematics in discussing milk production, but in this case it seems to miss the point—the shape of the cow is irrelevant. What I would like to suggest here is that the attempt to treat species as sets is like the approach to the milk production problem that uses the simplifying assumption of spherical shape. It can be done, but looks to be the wrong approach, and if so, that will be born out in the efforts to regain full milk production.

SOURCES

Carnap, Rudolf. *An Introduction to the Philosophy of Science*, edited by M. Gardner. New York: Basic Books, 1966.

Darwin, Charles. 1859. *On the Origin of Species*. Facsimile of the 1st ed. Cambridge, MA: Harvard University Press, 1964.

Ghiselin, Michael T. *Metaphysics and the Origin of Species*. Albany: State University of New York Press, 1997.

Hey, Jody. *Genes, Categories and Species: the Evolutionary and Cognitive Causes of the Species Problem.* Oxford: Oxford University Press, 2001.

Hull, David L. "A Matter of Individuality." In *The Units of Evolution: Essays on the Nature of Species*, edited by M. Ereshefsky. Cambridge, MA: Bradford Books, 1992.

———. "The Ideal Species Concept—and Why We Can't Get It." In *Species: the Units of Biodiversity*, edited by M. F. Claridge, H. A. Dawah, and M. R. Wilson. London: Chapman and Hall, 1997.

Kitcher, Philip. "Some Puzzles About Species." In *What the Philosophy of Biology Is: Essays Dedicated to David Hull*, edited by Michael Ruse. Dordrecht, Netherlands: Kluwer Academic Books, 1989.

———. "Species." In *The Units of Evolution: Essays on the Nature of Species*, edited by M. Ereshefsky. Cambridge, MA: Bradford Books, 1992.

Mayden, Richard L. "A Hierarchy of Species Concepts: the Denouement in the Saga of the Species Problem." In Claridge et al., *Species*.

Mayr, Ernst. *The Growth of Biological Thought*. Cambridge, MA: Belknap Press, 1982.

Ruse, Michael. "Biological Species: Natural Kinds, Individuals, or What?" In Ereshefsky, *The Units of Evolution*.

NOTE

1. I am indebted to Peter Achinstein for this scenario, who uses it in discussion of formal models.

MARK RIDLEY*

PRINCIPLES OF CLASSIFICATION

What is the proper relation of the theory of evolution and the classification of living things? The strongest possible relation would be one of practical necessity, if classification were practically impossible without the theory of evolution. The facts of history alone show that this relation does not hold. People had successfully classified animals and plants for two millennia before evolution was ever accepted. The simplest act of classification, indeed, requires no theory at all, let alone the theory of evolution; it merely requires that groups be recognized, defined, and named. A group, in this simple sense, is a collection of organisms that share a particular defining trait; the group Chordata for instance contains all animals that possess a notochord, a hollow dorsal nerve chord, and segmented muscles. Classification, as the definition and naming of groups, is in principle easy, but it is also important. It is even essential. Biologists could not communicate or check their discoveries if their specimens had not been classified into publicly recognized groups.

If communication were the only purpose of classification it would not matter what groups were defined, provided that the definitions were agreed upon. Chordata happens to be a group that is generally recognized, but by the same method we could define other groups that are not normally recognized. We might, for instance, define the Ocellata as the group of all living things that possess eyes. It would contain most vertebrates, many insects and crustaceans, some molluscs and worms, and some other odd invertebrates. The Ocellata has not, so far as I know, ever been considered as a taxonomic group; but, if classification is only a matter of defining and naming groups, we might ask why it is not.

That question brings us to the fundamental problem of classification. Different traits define different groups. We could just accept some groups and not others. We could agree that Ocellata is intrinsically as good a group as Chordata,

---

but declare that it just so happens that we have decided to recognize Chordata but not Ocellata. Classification would then be subjective. And if that satisfied us, evolution would not only be practically unnecessary but completely unnecessary, for there would be no more to classification than its practice.

Most biologists, however, are not so easily satisfied. They would prefer the choice of groups to be principled rather than subjective. Then, if we do recognize Chordata but not Ocellata, it must be because some principle shows that the Chordata are an acceptable group, but the Ocellata are not. A perfect principle would unambiguously show whether any group was acceptable, and admit no conflict between acceptable groups. No groups would then be chosen subjectively: they would all be chosen by reference to the principle. Even in the absence of a perfect principle, a principle might still be useful if it narrowed down the number of groups that were acceptable. Once the need for a principle of the choice of traits is recognized a third relation of the theory of evolution and classification is opened up between practical necessity and complete dispensability. If evolution supplied the only valid principle of choosing traits it would be philosophically necessary for classification or if merely one among several principles philosophically desirable. We have seen that evolution is practically unnecessary; from now on we shall be concerned with whether it is philosophically necessary, philosophically desirable, or completely unnecessary.

We can assume that the classification of life will at all events be hierarchical. A hierarchical classification is one whose groups are contained completely within more inclusive groups with no overlap; humans (for example) are contained within the genus *Homo*, which is contained within the order of primates, which is contained within the class Mammalia, which is contained within the sub phylum Vertebrata, which is within the phylum Chordata, which is within the kingdom Animalia. In principle biological classifications might not be hierarchical, but in practice they nearly all are. We are not ignoring a contentious practical issue.

What then could supply a principle for the hierarchical classification of life? Two kinds of answers have been offered: a *phenetic* hierarchy, or a *phylogenetic* one. A phenetic hierarchy is one of the similarity of form of the groups being classified; it is defined by any traits, such as leg length, skin color, number of spines on back, or some collection of them. A phylogenetic hierarchy is one of the pattern of evolutionary descent; groups are formed according to recency of common ancestry.

The phenetic and phylogenetic principles may agree or disagree, according to the species considered. Figure 8.1 shows the phenetic and phylogenetic relations of three sets of three species. The phenetic and phylogenetic classifications are the same if the rate of evolution is approximately constant and its direction is divergent, as is probably true of a human, a chimp, and a rabbit (fig. 8.1a); the human and the chimp share a more recent common ancestor and resemble each other more closely than does either with the rabbit. The two principles disagree when there is convergence or differential rates of divergent evolution. A barnacle, a limpet, and a lobster illustrate the case of convergence (fig. 8.1b); the barnacle and limpet are phenetically closer, but the barnacle has a more recent common ancestor with the lobster. The barnacle has converged, during evolution, on to the molluscan form. The salmon, lungfish, and cow illustrate the other source of disagreement (fig. 8.1c); a lungfish is phenetically more like a salmon than a cow; but it shares a more recent common ancestor with a cow than with a salmon. The evolutionary line leading from lungfish to cows has changed so rapidly that cows now look utterly different from their piscine ancestors. Lungfish indeed have hardly changed at all in 400 million years; they are often called living fossils.

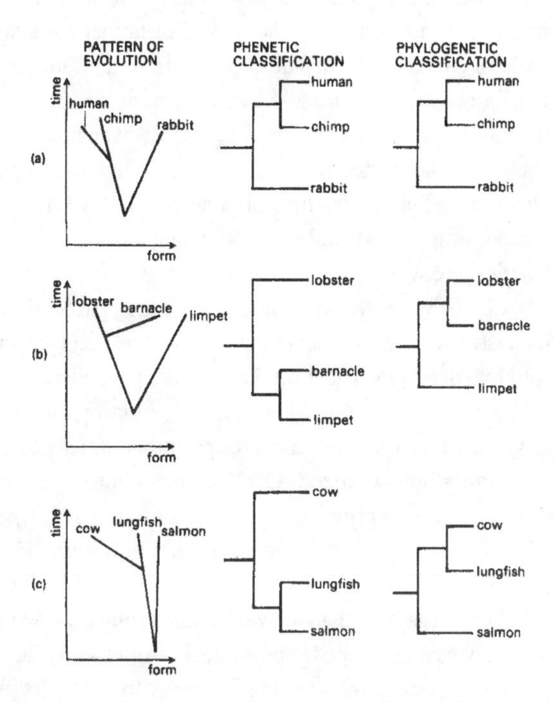

Figure 8.1. Phenetic and phylogenetic classification. The pattern of evolution (left), phenetic classification (center), and phylogenetic classification (right) for three cases. Phenetic and phylogenetic classification agree in the case of (a) human, chimp, and rabbit; but disagree when there is convergence, as in the case of (b) barnacle, limpet, and lobster; or when there is differential divergence, as in the case of (c) salmon, lungfish, and cow.

Clearly, evolution is a necessary assumption of phylogenetic classification: if organisms did not have evolutionary relations, we could not classify according to them. But evolution is not an assumption of the phenetic system: we could classify organisms by their similarity of appearance whether they shared a common ancestor or had been separately created. In the phenetic system, classification is by similarity of appearance, not evolution. If the phylogenetic principle is invalid, evolution will be completely unnecessary in classification; if both principles are valid, evolution will be philosophically desirable; if only the phylogenetic principle is valid, evolution will be philosophically necessary. If neither principle is valid, we shall have to fall back on the kind of subjectivity that we aimed to escape from. Such is the significance of the question of whether the two principles are valid.

Phenetic and phylogenetic classification have each grown into a whole school, complete with its own philosophical self-justifications, techniques, and advocates. Phenetic classification is advocated by the school of numerical taxonomy; phylogenetic by the school called cladism. There is another important school of classification as well. It is a school often called evolutionary taxonomy, whose practitioners are more numerous than either numerical or cladistic taxonomy. But despite its importance, we need not concern ourselves with it here. Its hierarchical classifications mix phenetic and phylogenetic components; the "evolutionary" classification of the barnacle, limpet, and lobster is phylogenetic, but of lungfish, salmon, and cow is phenetic. In covering the pure extremes, we shall cover the arguments of the mixed school too. What we want to know are the justifications of numerical taxonomy, which should justify phenetic classification, and of cladism, which should justify phylogenetic classification. Let us take phenetic classification first.

We have already met the difficulty of classification by an arbitrarily chosen trait. It is that some traits define some groups, other traits other groups; eyes defined the (customarily unrecognized) Ocellata, notochords the (customarily recognized) Chordata. If our only principle is to pick traits and define groups by them, we are left with a subjective choice among conflicting groups like Ocellata and Chordata. The numerical taxonomic school, which flourished from the late fifties into the sixties, believed that it had an answer to this problem. It would classify not by single traits, but by as many traits as possible. It would study dozens, even hundreds, of traits, which it would then average in order to define its groups. It came with a repertory of statistical procedures designed to realize that end. The general kind of statistic is what is called a multivariate cluster statistic. Given many measurements of many traits in the units to be classified, the cluster statistic averages all the measurements, to form groups (or "clusters") of units according to their similarity in all the traits. The groups in the classification are said to be defined by their "overall morphological similarity." It was believed that if many traits were used, the groups discerned by the cluster statistic would be less arbitrary. Whereas groups defined by a few different traits may contain very different members, as did Chordata and Ocellata, groups defined by a large number of traits (it was thought) would have more consistent memberships.

We must consider a cluster statistic in more detail. The clusters are formed according to what is called the "distance" between the units being classified. A distance in this sense is not the distance from one place to another, such as five miles, but is the difference between the values of a trait in two groups. Suppose that we are classifying two species. If the legs of one species are on average six inches long and those of the other four inches, the distance between the two species with respect to leg length is two inches. Numerical taxonomy, however, does not operate with only one trait. It uses dozens. The distance between the groups is therefore measured as the average for all the traits. If we had also measured skin color in the two species and the distance between the colors had been 0.3 units, then the average distance for the leg length and skin color combined is $(2 + 0.3)/2 = 1.15$ units. This figure is called the "mean trait distance"; it would also be possible to use the Euclidean distance, which is measured, in two dimensions, by Pythagoras's theorem. The method can be applied for an indefinitely large number of traits and species (or whatever unit). The cluster statistic can then set to work.

The cluster statistic forms groups (or "clusters") by successively aggregating the units with the shortest distances to each other. It forms a hierarchy of clusters as more and more distant units are added in. The numerical taxonomy, or phenetic classification, of the species will then be exactly defined. The classification *is* the hierarchic output of the statistic.

The advantages which numerical taxonomy claims for itself are its objectivity and repeatability. Any taxonomist could take the same group of animals or plants, measure many traits quantitatively, feed the measurements into the cluster statistic, and the same classification would always emerge. Numerical taxonomists claimed that by contrast the methods used to reconstruct phylogenetic trees were hopelessly vague and woolly. We have not yet come to those techniques; but we can look now at how well the method of numerical taxonomy stands up by its own criterion. Is numerical taxonomy, and its resulting phenetic classification, really objective and repeatable?

The answer is that it is not. The reason is abstract, but too important to ignore. It was pointed out most powerfully by an Australian entomologist, L. A. S. Johnson, in 1968. When I wrote above that the cluster statistic simply forms a hierarchy by adding in turn the next least distant group, I ignored a problem.

There is more than one cluster statistic, because there is more than one way of recognizing the "nearest" group. As we shall see, these different cluster statistics define different groups. If numerical taxonomy is truly objective, its own principle must dictate which cluster statistic should be used and which classificatory groups recognized. If it does not, its own claim to repeatability will be exploded. It will be hoist on its own petard.

We can illustrate the point by two different statistics, called a nearest neighbor statistic and an average neighbor statistic. These are just two among many, which makes the real problem even worse than what follows; but we can use only two statistics to illustrate the nature, if not the extent, of the problem. Nearest neighbor statistics form successively more inclusive groups by combining the sub-groups with the nearest neighbor to each other. We can see it in figure 7.2, along with the average neighbor statistic, which forms more inclusive groups not from those subgroups with the nearest *nearest* neighbor, but from those with the nearest *average* neighbor.

In figure 8.2, the nearest neighbor and average neighbor cluster statistics produce different hierarchies. In many cases the two statistics will produce hierarchies of the same shape, even if they do differ quantitatively. But sometimes they will not. The two statistics then give different classifications. (The figure only has two dimensions, which might be, say, leg length and skin color; as we have seen, numerical taxonomists rely on many more than two traits. But that simplification is only to fit the printed page; it does not matter for the general point. Indeed the problem grows worse as more dimensions are introduced.)

The nearest neighbor statistic joins it to the group with the nearest *nearest* neighbor: it compares the two distances labeled nearest neighbor: the one to species 5 is shorter; species 4 is classified with species 5, 6, 7 (classification [a]). The average neighbor statistic joins it to the group with the nearest average neighbor. The two points marked x are the average distances from species 4 to each group. The average neighbor statistic compares the two distances labeled average neighbor: the distance to the group of species 1, 2, and 3 is shorter, and species 4 is therefore classified with them (classification [b]).

The principle of numerical taxonomy provides no guidance among the different cluster statistics. It implies no criterion by which to choose among the different hierarchies produced by different statistics. The principle of numerical taxonomy is to classify according to "overall morphological similarity," but overall morphological similarity can only be measured by a cluster statistic. There is no higher measure of overall morphological similarity against which the different cluster statistics can be compared. When different statistics conflict, the practical numerical taxonomist has to decide which one he prefers. He can make a choice, of course; but it will have to be subjective.

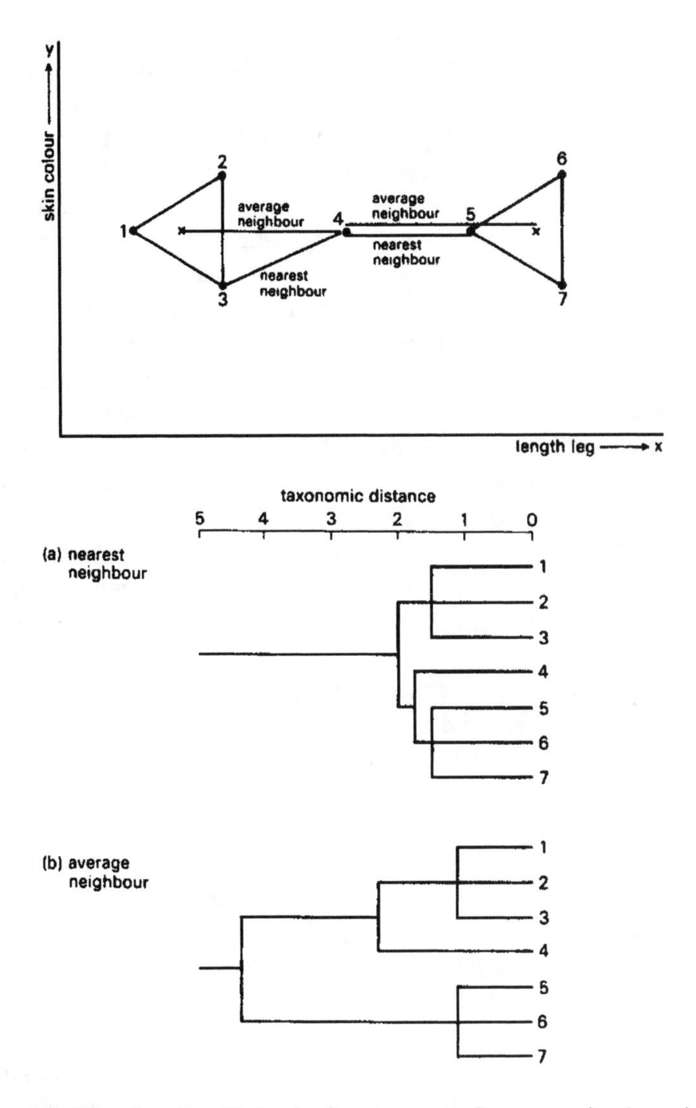

Figure 8.2. Two cluster statistics in disagreement. Seven species (nos. 1–7) have been measured for two traits, leg length and skin color, and plotted in a two-dimensional space, with their leg length on the x-axis and their skin color on the y-axis. Two cluster statistics, a nearest neighbor statistic and an average neighbor statistic, have been used to classify the seven species. The resulting classifications are shown below: (a) by nearest neighbor (b) by average neighbor. Both statistics first recognize the two groups, one of species 1, 2, 3, and the other of species 5, 6, 7. They disagree about the classification of species 4.

The principle of phenetic classification, therefore, is a failure. Numerical taxonomy successfully removed subjectivity from the choice of traits, but only to see it pop up again (in a less obvious but equally destructive form) in the choice of cluster statistic. If phenetic relations cannot provide a valid principle for the hierarchical classification of living things, what about phylogeny?

Here there is more hope. Unlike the hierarchy of phenetic resemblance (or overall morphological similarity), the phylogenetic hierarchy does exist independently of our techniques to measure it. The phylogenetic tree is a unique hierarchy. It really is true, of any two species, that they either do or do not share a more recent common ancestor with each other than with another species. In phylogenetic classification, there is no problem of subjective choice among different possible hierarchies. There is only one correct phylogeny. If the evidence suggests that one classification is more like it than another, that is the classification to choose. The Chordata are allowed, but the Ocellata are not, because the group of chordates share a common ancestor but the group of species that possess eyes do not. Sometimes there will not be enough evidence to say whether one classification is more phylogenetic than another: then we should either have to make a provisional subjective choice, or refuse to choose until more evidence becomes available. The advantage of the phylogenetic principle is that it does possess a higher criterion to compare the evidence with: the phylogenetic hierarchy. This advantage had been vaguely understood by many evolutionary taxonomists, but it was first thoroughly thought through by the German entomologist Hennig. Cladism is sometimes called Hennigian classification.

Now that we have solved the philosophical problem we are left with the practical one. How can phylogenetic relations be discovered? Evolution took place in the past. Unlike phenetic relations, phylogenetic relations cannot be directly observed. They have to be inferred. But how?

For each species, we need to know which other species it shares its most recent common ancestor with, for that is the species with which it should be classified. How can we discover it? The method proposed by Hennig (and previously applied by many others) is to look for traits that are evolutionary innovations. During evolution, traits change from time to time. According to whether a particular trait is an earlier or a later evolutionary stage, it can be called a primitive or a derived trait. Most traits pass through several stages in evolution, and whether a particular stage is primitive or derived depends on which other stage it is being compared with; it is primitive with respect to later stages, but derived with respect to earlier ones. Consider as an example the evolution of the vertebrate limb. The most primitive stage is its absence; then, in fish, it appears as a fin; in amphibians the fin evolves into the tetrapodan pentadactyl (five-digit) limb; it has stayed like that in most vertebrates, but in some lizards and independently in some ungulates the number of digits on the limb has been reduced from five to four, three, two, or even (in horses) one. If we compare amphibians with fish, the pentadactyl limb is the derived state and fins the primitive state of this trait; but if we compare an amphibian with a horse, the five-toed state becomes primitive and the one-toed state in the horse is now the derived. Similarly if we compare the hand of a human with the front foot of a horse, the pentadactyl human hand is in the primitive state relative to the single-toed equine foot.

Such is the meaning of primitive and derived states. The distinction is necessary, in Hennig's system, in order that the derived traits can be selected for use in classification. The derived traits are selected because shared derived traits indicate common ancestry, whereas shared primitive traits do not. Let us stay with the same example. Suppose that we wish to classify a five-toed lizard, a horse, and an ape in relation to each other. The ape and the lizard share the trait "five-toed," but this does not indicate that they share a more recent common ancestor than either does with the horse: the trait is primitive and does not indicate common ancestry within the group of horse, ape, lizard. Whenever there is an evolutionary innovation, it is retained (until the next evolutionary change) by the species descended from the innovatory species: shared derived traits do indicate common ancestry. That is why they are used to discover phylogenetic relations.

We can now move a further step in the search for a method. The problem has now become to distinguish primitive from derived traits. In the case of the vertebrate limb we assumed that the course of evolution was known. But how could it be discovered to begin with? There are several techniques. We need not consider them all. Let us look at one in detail to demonstrate that the distinction can be made. Let us consider outgroup comparison. The simplest case has one trait, with two states, in two species; lactation and its absence, for instance, in a horse and a toad. The problem is whether lactation in horses is derived with respect to its absence in toads, or its absence in toads is derived with respect to its presence in horses. The solution, by outgroup comparison, is obtained by examining the state of some related species, called the outgroup. The outgroup should be a species which is not more closely related to one of the two species than the other, that is why it is an *out*group: it is separated from the species under consideration. In this case any fish or invertebrate would do but a cow would not, because it is more closely related to the horse than it is to the toad. By the method of outgroup comparison, that trait is taken to be primitive which is found in the outgroup. Whether we took a species of fish or an invertebrate, the answer would be the same. The outgroup lacks lactation: lactation is the derived state.

The result of outgroup comparison is uncertain. It will be wrong whenever there is unrecognized convergence. Thus if we compared the trait "body shape" in dolphins and dogs with some such outgroup as a fish we should determine that the dog had the derived shape. Actually the dolphin does. No one would be mistaken in the case of dolphins and fish; but other more subtle cases of convergence surely exist which are not as easy to recognize, and in them outgroup comparison will be misleading. But although it can go wrong, it is probably better than nothing. Shared traits are probably more often due to common ancestry than to convergence: but only more often, not always.

Advocates of phenetic classification often remark that phylogenetic classification is impossible because its techniques are circular. In the case of outgroup comparison, for instance, in order to apply the technique we needed to know that the outgroup (the fish) was less related to the toad and the horse than either to each other. It appears that we need to know the classification before we can apply the techniques; which would be quite a problem since the technique is supposed to be used to discover the classification. The problem, however, is not as destructive as it appears.

The argument of outgroup comparison is not circular. It works by what is often called successive approximation. It is the method by which theories are developed in all sciences. As new facts are collected and considered, they are examined in the light of the present theory. If they fit it, confidence in the theory is increased. If they do not, they may suggest a new theory, which can be used in considering yet further evidence. There is a continual reexamination of the theory in the light of new evidence, and when the theory is changed, our interpretation of all previous evidence should change too. This is not circular reasoning: it is testing a theory. In outgroup comparison, we can start with some crude idea of which species is an outgroup; if further evidence fits the crude idea, the hypothesis is (tentatively) confirmed, and it can then be used in interpreting further facts. Let us consider a hypothetical example.

Let us suppose that we have six species, and we suspect that one of them is less closely related than the other five. That one can be used as an outgroup. Comparison with it can be used to classify the other five, whose relations are not yet known. We first examine a trait in all six species. We take the state in the outgroup to be primitive for the group of five. Figure 8.3 shows the procedure. There are two points to notice. One is that the procedure can start from a very vague starting point; we do not need a firm classification to apply outgroup comparison. The other is that, if further evidence demands it, we can modify our initial ideas. If one trait after another suggests that species 6 is not separate from 1–5 then we can modify the classification and put 6 in its appropriate place. All the previous steps would then have to be reconsidered. As the analysis proceeds any error at the beginning, it has a decreasing effect; the initial errors are gradually discovered, and their damaging consequences removed. Such is the method of successive approximation. It is the common method of scientific theory-building: only sciences that completely lack theories can do without the feedback between the interpretation of facts and the testing of theories.

Outgroup comparison is not the only cladistic method. Another method supposes that, as the organism develops, the evolutionarily derived stages appear after the primitive stages. The backbone of vertebrates is a derived state relative to its absence in invertebrates; and the backbone develops after its absence in a vertebrate embryo. Like outgroup comparison, the embryological criterion is imperfect but better than nothing. Another method is to look at the order in which the traits appear in the fossil record. The most powerful technique is to take all these methods together, and use all the evidence available. I do not wish to give the impression that the techniques of reconstructing phylogeny are perfect. They are far from that. Many problems remain, especially that of how to reconcile conflicting information from different traits. But although the system has difficulties, it is probably not altogether impractical. The cladistic evidence suggests that humans share a more recent common ancestor with chimps than with butterflies, and few biologists would deny that the evidence is correct in this case.

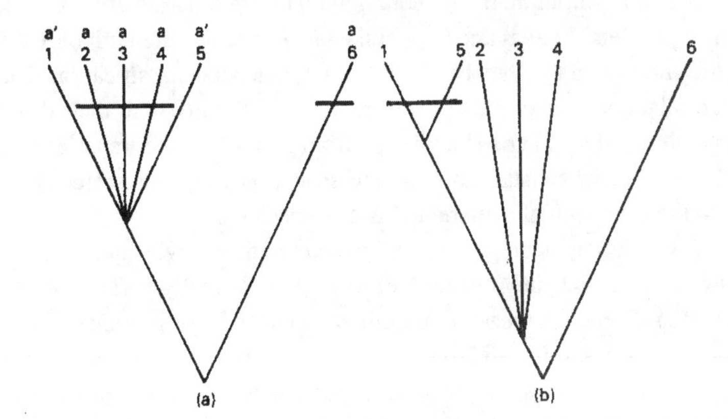

Figure 8.3. Outgroup comparison by successive approximation. (a) Five species (nos. 1–5) are to be classified, and it is thought that another species (6) is less related to them than any of the five to each other. Trait *A* is compared in all six species: species 2, 3, and 4 have it in state *a*, as does species 6, but species 1 and 5 have another state *á*; by outgroup comparison *á* is reasoned to be a derived state. (b) Species 1 and 5 are classified together, for they share a derived trait. The relations of species 2, 3, and 4 remain unknown; but the procedure can be repeated when evidence for new traits becomes available.

Derived traits, therefore, can be distinguished from primitive ones. Groups can be defined by shared derived traits. The cladistic system of phylogenetic classification is workable. But although cladistic groups can be recognized and defined, and although the classification (so far as it is phylogenetic) will be philosophically sound, it will only be valid for any one time. Evolution is continually going on; the traits of lineages continually change. What will define a group at one time may not define it at another. The traits defining groups are temporarily contingent, not essential. Biological groups do not have Aristotelian "essences." The phylogenetic group Vertebrata may happen to be defined, at present, by the possession of bones (or cartilage), but that may no longer be true in a million years' time. A descendant of a vertebrate species may lack bones, and the trait will cease to define the evolutionary group. The traits that define groups are not eternal and inevitable; they just happen to be useful sometimes.

For the same reason, difficulties arise when species from different times (particularly fossils and present-day species) are put in the same classification. It can be done, but it is often awkward. Because traits are continually changing, groups with members from more than one geological period cannot be defined by constant traits. Furthermore, because (as we shall see in the next chapter) species may vary in space, the traits used in a classification strictly only apply to one place. But spatial variation, being more limited than temporal variation, is less of a practical problem. Spatial and temporal changes in traits are both only difficulties in the practice, not in the philosophy, or phylogenetic classification. The phylogenetic relations of a species remain real. The difficulty is in their discovery; phylogenetic relations have to be inferred from shared derived traits. But because traits do not remain constant in time and space, they are not perfectly reliable guides to the true phylogenetic relations of a species.

Because phylogenetic classifications are defined by shared traits, it is tempting to think that phylogenetic classification is really a form of phenetic classification. There is a measure of truth in this idea. Phylogenetic classification is defined by shared traits, and to that extent is phenetic. But it could not be otherwise. All classifications are phenetic in this sense. The proper description of a classificatory system as phylogenetic or phenetic should not be according to the techniques that it uses, but the hierarchy that it seeks to represent. There is an utterly different philosophy behind the two systems. One tries to represent the branching pattern of evolution; the other tries to represent the pattern of morphological similarity.

Moreover, different *kinds* of traits define phylogenetic and phenetic classifications. Phylogenetic classification supposes that some traits—shared derived traits in particular—are better indicators of phylogeny than are others. Phylogenetic classification, at least of the cladistic variety, uses only shared derived traits. Phenetic classification, by contrast, indiscriminately uses both primitive and derived traits. Both for reasons of philosophy and technique, therefore, it is misleading to call phylogenetic classification a form of phenetic classification.

Phylogenetic classification is philosophically preferable to its only competitor, phenetic classification. It is also a practical possibility. Its main problems are in its techniques, which are (as yet) far from perfect. But the techniques of cladism have been improved, even within their short history, and further work, particularly on molecules, should improve them further. The difficulty in phenetic classification is more fundamental. Its claim to preference, which is its objectivity, is a false claim. It is left with little to recommend itself by. Phenetic classification should, I think, be avoided whenever possible. Classifications, if they are to be objective, must represent phylogeny.

If this conclusion is correct, evolution and classification are closely related. The relation is one of philosophical necessity. Evolution is required to justify the kind of classification that is practiced. Evolution is not merely desirable, but necessary, because phylogeny is the only known principle of classification. If we were content with merely subjective classification, evolution would be unnecessary. But if we are not—if we seek a principled classification—evolution becomes essential. It underwrites the entire philosophy of phylogenetic classification. Without evolution, phylogenetic classification, and its method of searching for derived traits, would be as subjective as any other technique. Only because we can assume that evolution is true can we even begin to think about phylogenetic classification. Then we need techniques to detect phylogeny. The source of those techniques is our understanding of how traits change in evolution. The theory of evolution, therefore, not only guarantees the philosophy of classification: it is also the breeding ground of taxonomic techniques.

HUMAN NATURE

EDWARD O. WILSON*

HEREDITY

We live on a planet of staggering organic diversity. Since Carolus Linnaeus began the process of formal classification in 1758, zoologists have cataloged about one million species of animals and given each a scientific name, a few paragraphs in a technical journal, and a small space on the shelves of one museum or another around the world. Yet despite this prodigious effort, the process of discovery has hardly begun. In 1976 a specimen of an unknown form of giant shark, fourteen feet long and weighing sixteen hundred pounds, was captured when it tried to swallow the stabilizing anchor of a United States Naval vessel near Hawaii. About the same time entomologists found an entirely new category of parasitic flies that resemble large reddish spiders and live exclusively in the nests of the native bats of New Zealand. Each year museum curators sort out thousands of new kinds of insects, copepods, wireworms, echinoderms, priapulids, pauropods, hypermastigotes, and other creatures collected on expeditions around the world. Projections based on intensive surveys of selected habitats indicate that the total number of animal species is between three and ten million. Biology, as the naturalist Howard Evans expressed it in the title of a recent book, is the study of life "on a little known planet."

Thousands of these species are highly social. The most advanced among them constitute what I have called the three pinnacles of social evolution in animals: the corals, bryozoans, and other colony-forming invertebrates; the social insects, including ants, wasps, bees, and termites; and the social fish, birds, and mammals. The communal beings of the three pinnacles are among the principal objects of the new discipline of sociobiology, defined as the systematic study of the biological basis of all forms of social behavior, in all kinds of organisms,

*Reprinted by permission of the publisher from "Heredity" in *On Human Nature, With a New Preface* by Edward O. Wilson, pp. 15–25, Cambridge, MA: Harvard University Press,

including man. The enterprise has old roots. Much of its basic information and some of its most vital ideas have come from ethology, the study of whole patterns of behavior of organisms under natural conditions. Ethology was pioneered by Julian Huxley, Karl von Frisch, Konrad Lorenz, Nikolaas Tinbergen, and a few others and is now being pursued by a large new generation of innovative and productive investigators. It has remained most concerned with the particularity of the behavior patterns shown by each species, the ways these patterns adapt animals to the special challenges of their environments, and the steps by which one pattern gives rise to another as the species themselves undergo genetic evolution. Increasingly, modern ethology is being linked to studies of the nervous system and the effects of hormones on behavior. Its investigators have become deeply involved with developmental processes and even learning, formerly the nearly exclusive domain of psychology, and they have begun to include man among the species most closely scrutinized. The emphasis of ethology remains on the individual organism and the physiology of organisms.

Sociobiology, in contrast, is a more explicitly hybrid discipline that incorporates knowledge from ethology (the naturalistic study of whole patterns of behavior), ecology (the study of the relationships of organisms to their environment), and genetics in order to derive general principles concerning the biological properties of entire societies. What is truly new about sociobiology is the way it has extracted the most important facts about social organization from their traditional matrix of ethology and psychology and reassembled them on a foundation of ecology and genetics studied at the population level in order to show how social groups adapt to the environment by evolution. Only within the past few years have ecology and genetics themselves become sophisticated and strong enough to provide such a foundation.

Sociobiology is a subject based largely on comparisons of social species. Each living form can be viewed as an evolutionary experiment, a product of millions of years of interaction between genes and environment. By examining many such experiments closely, we have begun to construct and test the first general principles of genetic social evolution. It is now within our reach to apply this broad knowledge to the study of human beings.

Sociobiologists consider man as though seen through the front end of a telescope, at a greater than usual distance and temporarily diminished in size, in order to view him simultaneously with an array of other social experiments. They attempt to place humankind in its proper place in a catalog of the social species on Earth. They agree with Rousseau that "One needs to look near at hand in order to study men, but to study man one must look from afar."

This macroscopic view has certain advantages over the traditional anthropocentrism of the social sciences. In fact, no intellectual vice is more crippling than defiantly self-indulgent anthropocentrism. I am reminded of the clever way Robert Nozick makes this point when he constructs an argument in favor of vegetarianism. Human beings, he notes, justify the eating of meat on the grounds that the animals we kill are too far below us in sensitivity and intelligence to bear comparison. It follows that if representatives of a truly superior extraterrestrial species were to visit Earth and apply the same criterion, they could proceed to eat us in good conscience. By the same token, scientists among these aliens might find human beings uninteresting, our intelligence weak, our passions unsurprising, our social organization of a kind already frequently encountered on other planets. To our chagrin they might then focus on the ants, because these little creatures, with their haplodiploid form of sex determination and bizarre female caste systems, are the truly novel productions of the Earth with reference to the Galaxy. We can imagine the log declaring, "A scientific breakthrough has occurred; we have finally discovered haplodiploid social organisms in the one- to ten-millimeter range." Then the visitors might inflict the ultimate indignity: in order to be sure they had not underestimated us, they would simulate human beings in the laboratory. Like chemists testing the structural characterization of a problematic organic compound by assembling it from simpler components, the alien biologists would need to synthesize a hominoid or two.

This scenario from science fiction has implications for the definition of man. The impressive recent advances by computer scientists in the design of artificial intelligence suggests the following test of humanity: that which behaves like man *is* man. Human behavior is something that can be defined with fair precision, because the evolutionary pathways open to it have not all been equally negotiable. Evolution has not made culture all-powerful. It is a misconception among many of the more traditional Marxists, some learning theorists, and a still surprising proportion of anthropologists and sociologists that social behavior can be shaped into virtually any form. Ultra-environmentalists start with the premise that man is the creation of his own culture: "culture makes man," the formula might go, "makes culture makes man." Theirs is only a half truth. Each person is molded by an interaction of his environment, especially his cultural environment, with the genes that affect social behavior. Although the hundreds of the world's cultures seem enormously variable to those of us who stand in their midst, all versions of human social behavior together form only a tiny fraction of the realized organizations of social species on this planet and a still smaller traction of those that can be readily imagined with the aid of sociobiological theory.

The question of interest is no longer whether human social behavior is genetically determined; it is to what extent. The accumulated evidence for a large hereditary component is more detailed and compelling than most persons, including even geneticists, realize. I will go further: it already is decisive.

That being said, let me provide an exact definition of a genetically determined trait. It is a trait that differs from other traits at least in part as a result of the presence of one or more distinctive genes. The important point is that the objective estimate of genetic influence requires comparison of two or more states of the same feature. To say that blue eyes are inherited is not meaningful without further qualification, because blue eyes are the product of an interaction between genes and the largely physiological environment that brought final coloration to the irises. But to say that the *difference* between blue and brown eyes is based wholly or partly on differences in genes is a meaningful statement because it can be tested and translated into the laws of genetics. Additional information is then sought: What are the eye colors of the parents, siblings, children, and more distant relatives? These data are compared to the very simplest model of Mendelian heredity, which, based on our understanding of cell multiplication and sexual reproduction, entails the action of only two genes. If the data fit, the differences are interpreted as being based on two genes. If not, increasingly complicated schemes are applied. Progressively larger numbers of genes and more complicated modes of interaction are assumed until a reasonably close fit can be made. In the example just cited, the main differences between blue and brown eyes are in fact based on two genes, although complicated modifications exist that make them less than an ideal textbook example. In the case of the most complex traits, hundreds of genes are sometimes involved, and their degree of influence can ordinarily be measured only crudely and with the aid of sophisticated mathematical techniques. Nevertheless, when the analysis is properly performed it leaves little doubt as to the presence and approximate magnitude of the genetic influence.

Human social behavior can be evaluated in essentially the same way, first by comparison with the behavior of other species and then, with far greater difficulty and ambiguity, by studies of variation among and within human populations. The picture of genetic determinism emerges most sharply when we compare selected major categories of animals with the human species. Certain general human traits are shared with a majority of the great apes and monkeys of Africa and Asia, which on grounds of anatomy and biochemistry are our closest living evolutionary relatives:

- Our intimate social groupings contain on the order of ten to one hundred adults, never just two, as in most birds and marmosets, or up to thousands, as in many kinds of fishes and insects.
- Males are larger than females. This is a characteristic of considerable significance within the Old World monkeys and apes and many other kinds of mammals. The average number of females consorting with successful males

closely corresponds to the size gap between males and females when many species are considered together. The rule makes sense: the greater the competition among males for females, the greater the advantage of large size and the less influential are any disadvantages accruing to bigness. Men are not very much larger than women; we are similar to chimpanzees in this regard. When the sexual size difference in human beings is plotted on the curve based on other kinds of mammals, the predicted average number of females per successful male turns out to be greater than one but less than three. The prediction is close to reality; we know we are a mildly polygynous species.
- The young are molded by a long period of social training, first by closest associations with the mother, then to an increasing degree with other children of the same age and sex.
- Social play is a strongly developed activity featuring role practice, mock aggression, sex practice, and exploration.

These and other properties together identify the taxonomic group consisting of Old World monkeys, the great apes, and human beings. It is inconceivable that human beings could be socialized into the radically different repertories of other groups such as fishes, birds, antelopes, or rodents. Human beings might self-consciously *imitate* such arrangements, but it would be a fiction played out on a stage, would run counter to deep emotional responses and have no chance of persisting through as much as a single generation. To adopt with serious intent, even in broad outline, the social system of a nonprimate species would be insanity in the literal sense. Personalities would quickly dissolve, relationships disintegrate, and reproduction cease.

At the next, finer level of classification, our species is distinct from the Old World monkeys and apes in ways that can be explained only as a result of a unique set of human genes. Of course, that is a point quickly conceded by even the most ardent environmentalists. They are willing to agree with the great geneticist Theodosius Dobzhansky that "in a sense, human genes have surrendered their primacy in human evolution to an entirely new, non-biological or superorganic agent, culture. However, it should not be forgotten that this agent is entirely dependent on the human genotype." But the matter is much deeper and more interesting than that. There are social traits occurring through all cultures which upon close examination are as diagnostic of mankind as are distinguishing characteristics of other animal species—as true to the human type, say, as wing tessellation is to a fritillary butterfly or a complicated spring melody to a wood thrush. In 1945 the American anthropologist George P. Murdock listed the following characteristics that have been recorded in every culture known to history and ethnography:

> Age-grading, athletic sports, bodily adornment, calendar, cleanliness training, community organization, cooking, cooperative labor, cosmology, courtship, dancing, decorative art, divination, division of labor, dream interpretation, education, eschatology, ethics, ethnobotany, etiquette, faith healing, family feasting, fire making, folklore, food taboos, funeral rites, games, gestures, gift giving, government, greetings, hair styles, hospitality, housing, hygiene, incest taboos, inheritance rules, joking, kin groups, kinship nomenclature, language, law, luck superstitions, magic, marriage, mealtimes, medicine, obstetrics, penal sanctions, personal names, population policy, postnatal care, pregnancy usages, property rights, propitiation of supernatural beings, puberty customs, religious ritual, residence rules, sexual restrictions, soul concepts, status differentiation, surgery, tool making, trade, visiting, weaving, and weather control.

Few of these unifying properties can be interpreted as the inevitable outcome of either advanced social life or high intelligence. It is easy to imagine nonhuman societies whose members are even more intelligent and complexly organized than ourselves, yet lack a majority of the qualities just listed. Consider the possibilities inherent in the insect societies. The sterile workers are already more cooperative and altruistic than people and they have a more pronounced tendency toward caste systems and division of labor. If ants were to be endowed in addition with rationalizing brains equal to our own, they could be our peers. Their societies would display the following peculiarities:

> Age-grading, antennal rites, body licking, calendar, cannibalism, caste determination, caste laws, colony-foundation rules, colony organization, cleanliness training, communal nurseries, cooperative labor, cosmology, courtship, division of labor, drone control, education, eschatology, ethics, etiquette, euthanasia, fire making, food taboos, gift giving, government, greetings, grooming rituals, hospitality, housing, hygiene, incest taboos, language, larval care, law, medicine, metamorphosis rites, mutual regurgitation, nursing castes, nuptial flights, nutrient eggs, population policy, queen obeisance, residence rules, sex determination, soldier castes, sisterhoods, status differentiation, sterile workers, surgery, symbiont care, tool making, trade, visiting, weather control,

and still other activities so alien as to make mere description by our language difficult. If in addition they were programmed to eliminate strife between colonies and to conserve the natural environment they would have greater staying power than people, and in a broad sense theirs would be the higher morality.

Civilization is not intrinsically limited to hominoids. Only by accident was it linked to the anatomy of bare-skinned, bipedal mammals and the peculiar qualities of human nature.

Freud said that God has been guilty of a shoddy and uneven piece of work. That is true to a degree greater than he intended: human nature is just one hodgepodge out of many conceivable. Yet if even a small fraction of the diagnostic human traits were stripped away, the result would probably be a disabling chaos. Human beings could not bear to simulate the behavior of even our closest relative among the Old World primates. If by perverse mutual agreement a human group attempted to imitate in detail the distinctive social arrangements of chimpanzees or gorillas, their effort would soon collapse and they would revert to fully human behavior.

It is also interesting to speculate that if people were somehow raised from birth in an environment devoid of most cultural influence, they would construct basic elements of human social life ab initio. In short time new elements of language would be invented and their culture enriched. Robin Fox, an anthropologist and pioneer in human sociobiology, has expressed this hypothesis in its strongest possible terms. Suppose, he conjectured, that we performed the cruel experiment linked in legend to the Pharaoh Psammetichus and King James IV of Scotland, who were said to have reared children by remote control, in total social isolation from their elders. Would the children learn to speak to one another?

> I do not doubt that they *could* speak and that, theoretically, given time, they or their offspring would invent and develop a language despite their never having been taught one. Furthermore, this language, although totally different from any known to us, would be analyzable to linguists on the same basis as other languages and translatable into all known languages. But I would push this further. If our new Adam and Eve could survive and breed—still in total isolation from any cultural influences—then eventually they would produce a society which would have laws about property, rules about incest and marriage, customs of taboo and avoidance, methods of settling disputes with a minimum of bloodshed, beliefs about the supernatural and practices relating to it, a system of social status and methods of indicating it, initiation ceremonies for young men, courtship practices including the adornment of females, systems of symbolic body adornment generally, certain activities and associations set aside for men from which women were excluded, gambling of some kind, a tool- and weapon-making industry, myths and legends, dancing, adultery, and various doses of homicide, suicide, homosexuality, schizophrenia, psychosis and neuroses, and various practitioners to take advantage of or cure these, depending on how they are viewed.

Not only are the basic features of human social behavior stubbornly idiosyncratic, but to the limited extent that they can be compared with those of animals they resemble most of all the repertories of other mammals and especially other primates. A few of the signals used to organize the behavior can be logically derived from the ancestral modes still shown by the Old World monkeys and great apes. The grimace of fear, the smile, and even laughter have parallels in the facial expressions of chimpanzees. This broad similarity is precisely the pattern to be expected if the human species descended from Old World primate ancestors, a demonstrable fact, and if the development of human social behavior retains even a small degree of genetic constraint, the broader hypothesis now under consideration.

STEPHEN JAY GOULD*

SOCIOBIOLOGY AND THE THEORY OF NATURAL SELECTION

NATURAL SELECTION AS STORYTELLING

Ludwig von Bertalanffy, a founder of general systems theory and a holdout against the neo-Darwinian tide, often argued that natural selection must fail as a comprehensive theory because it explains *too* much—a paradoxical, but perceptive statement. He wrote (1969, 24 and 11):

> If selection is taken as an axiomatic and a *priori* principle, it is always possible to imagine auxiliary hypotheses—unproved and by nature unprovable—to make it work in any special case.... Some adaptive value... can always be construed or imagined.
>
> I think the fact that a theory so vague, so insufficiently verifiable and so far from the criteria otherwise applied in "hard" science, has become a dogma, can only be explained on sociological grounds. Society and science have been so steeped in the ideas of mechanism, utilitarianism, and the economic concept of free competition, that instead of God, Selection was enthroned as ultimate reality.

Similarly, the arguments of Christian fundamentalism used to frustrate me until I realized that there are, in principle, no counter cases and that, on this ground alone, literal bibliolatry is bankrupt. The theory of natural selection is, fortunately, in much better straits. It could be invalidated as a general cause of evolutionary change. (If, for example, Lamarckian inheritance were true and general, then adaptation would arise so rapidly in the Lamarckian mode that nat-

*From Stephen Jay Gould, "Sociobiology and the Theory of Natural Selection," in *Sociobiology: Beyond Nature/Nurture?* ed. G. W. Barlow and James Silverberg (Boulder, CO: Westview Press, 1980), pp. 257–69. Reprinted by permission of Rhonda Roland Shearer.

ural selection would be powerless to create and would operate only to eliminate.) Moreover, its action and efficacy have been demonstrated experimentally by 60 years of manipulation within *Drosophila* bottles—not to mention several thousand years of success by plant and animal breeders.

Yet in one area, unfortunately a very large part of evolutionary theory and practice, natural selection has operated like the fundamentalist's God—he who maketh *all* things. Rudyard Kipling asked how the leopard got its spots, the rhino its wrinkled skin. He called his answers "just-so stories." When evolutionists try to explain form and behavior, they also tell just-so stories—and the agent is natural selection. Virtuosity in invention replaces testability as the criterion for acceptance. This is the procedure that inspired von Bertalanffy's complaint. It is also the practice that has given evolutionary biology a bad name among many experimental scientists in other disciplines. We should heed their disquiet, not dismiss it with a claim that they understand neither natural selection nor the special procedures of historical science.

This style of storytelling might yield acceptable answers if we could be sure of two things: (1) that all bits of morphology and behavior arise as direct results of natural selection, and (2) that only one selective explanation exists for each bit. But, as Darwin insisted vociferously, and contrary to the mythology about him, there is much more to evolution than natural selection. (Darwin was a consistent pluralist who viewed natural selection as the most important agent of evolutionary change, but who accepted a range of other agents and specified the conditions of their presumed effectiveness. In chapter 7 of the *Origin* (6th ed.), for example, he attributed the cryptic coloration of a flat fish's upper surface to natural selection and the migration of its eyes to inheritance of acquired characters. He continually insisted that he wrote his 2-volume *Variation of Animals and Plants Under Domestication* (1868), with its Lamarckian hypothesis of pangenesis, primarily to illustrate the effect of evolutionary factors other than natural selection. In a letter to *Nature* in 1880, he used the sharpest and most waspish language of his life to castigate Sir Wyville Thomson for caricaturing his theory by ascribing all evolutionary change to natural selection.)

Since God can be bent to support all theories, and since Darwin ranks closest to deification among evolutionary biologists, panselectionists of the modern synthesis tended to remake Darwin in their image. But we now reject this rigid version of natural selection and grant a major role to other evolutionary agents—genetic drift, fixation of neutral mutations, for example. We must also recognize that many features arise indirectly as developmental consequences of other features subject to natural selection—see classic (Huxley 1932) and modern (Gould 1966 and 1975; Cock 1966) work on allometry and the developmental consequences of size increase. Moreover, and perhaps most importantly, there are a multitude of potential selective explanations for each feature. There is no such thing in nature as a self-evident and unambiguous story.

When we examine the history of favored stories for any particular adaptation, we do not trace a tale of increasing truth as one story replaces the last, but rather a chronicle of shifting fads and fashions. When Newtonian mechanical explanations were riding high, G. C. Simpson wrote (1961, 1686):

> The problem of the pelycosaur dorsal fin . . . seems essentially solved by Romer's demonstration that the regression relationship of fin area to body volume is appropriate to the functioning of the fin as a temperature regulating mechanism.

Simpson's firmness seems almost amusing since now—a mere 15 years later with behavioral stories in vogue—most paleontologists feel equally sure that the sail was primarily a device for sexual display. (Yes, I know the litany: It might have performed both functions. But this too is a story.)

On the other side of the same shift in fashion, a recent article on functional endothermy in some large beetles had this to say about the why of it all (Bartholomew and Casey 1977, 883):

> It is possible that the increased power and speed of terrestrial locomotion associated with a modest elevation of body temperatures may offer reproductive advantages by increasing the effectiveness of intraspecific aggressive behavior, particularly between males.

This conjecture reflects no evidence drawn from the beetles themselves, only the current fashion in selective stories. We may be confident that the same data, collected 15 years ago, would have inspired a speculation about improved design and mechanical advantage.

SOCIOBIOLOGICAL STORIES

Most work in sociobiology has been done in the mode of adaptive storytelling based upon the optimizing character and pervasive power of natural selection. As such, its weaknesses of methodology are those that have plagued so much of evolutionary theory for more than a century. Sociobiologists have anchored their stories in the basic Darwinian notion of selection as individual reproductive success. Though previously underemphasized by students of behavior, this insistence on selection as individual success is fundamental to Darwinism. It arises directly from Darwin's construction of natural selection as a conscious analog to the laissez-faire economics of Adam Smith with its central notion that order and harmony arise from the natural interaction of individuals pursuing their own advantages (see Schweber 1977).

Sociobiologists have broadened their range of selective stories by invoking concepts of inclusive fitness and kin selection to solve (successfully I think) the vexatious problem of altruism—previously the greatest stumbling block to a Darwinian theory of social behavior. Altruistic acts are the cement of stable societies. Until we could explain apparent acts of self-sacrifice as potentially beneficial to the genetic fitness of sacrificers themselves—propagation of genes through enhanced survival of kin, for example—the prevalence of altruism blocked any Darwinian theory of social behavior.

Thus, kin selection has broadened the range of permissible stories, but it has not alleviated any methodological difficulties in the process of storytelling itself. Von Bertalanffy's objections still apply, if anything with greater force because behavior is generally more plastic and more difficult to specify and homologize than morphology. Sociobiologists are still telling speculative stories, still hitching without evidence to one potential star among many, still using mere consistency with natural selection as a criterion of acceptance.

David Barash (1976), for example, tells the following story about mountain bluebirds. (It is, by the way, a perfectly plausible story that may well be true. I only wish to criticize its assertion without evidence or test using consistency with natural selection as the sole criterion for useful speculation.) Barash reasoned that a male bird might be more sensitive to intrusion of other males before eggs are laid than after (when he can be certain that his genes are inside). So Barash studied two nests, making three observations at 10-day intervals, the first before the eggs were laid, the last two after. For each period of observation, he mounted a stuffed male near the nest while the male occupant was out foraging. When the male returned he counted aggressive encounters with both model and female. At time one, males in both nests were aggressive toward the model and less, but still substantially, aggressive toward the female as well. At time two, after eggs had been laid, males were less aggressive to models and scarcely aggressive to females at all. At time three, males were still less aggressive toward models, and not aggressive at all toward females. Barash concludes that he has established consistency with natural selection and need do no more (1976, 1099–1100):

> These results are consistent with the expectations of evolutionary theory. Thus aggression toward an intruding male (the model) would clearly be especially advantageous early in the breeding season, when territories and nests are nor-

> mally defended. . . . The initial, aggressive response to the mated female is also adaptive in that, given a situation suggesting a high probability of adultery (i.e., the presence of the model near the female) and assuming that replacement females are available, obtaining a new mate would enhance the fitness of male. . . . The decline in male-female aggressiveness during incubation and fledgling stages could be attributed to the impossibility of being cuckolded after the eggs have been laid. . . . The results are consistent with an evolutionary interpretation. In addition, the term "adultery" is unblushingly employed in this letter without quotation marks, as I believe it reflects a true analogy to the human concept, in the sense of Lorenz. It may also be prophesied that continued application of a similar evolutionary approach will eventually shed considerable light on various human foibles as well.

Consistent, yes. But what about the obvious alternative, dismissed without test in a line by Barash: male returns at times two and three, approaches the model a few times, encounters no reaction, mutters to himself the avian equivalent of "it's that damned stuffed bird again," and ceases to bother. And why not the evident test: expose a male to the model for the *first* time *after* the eggs are laid.

We have been deluged in recent years with sociobiological stories. Some, like Barash's are plausible, if unsupported. For many others, I can only confess my intuition of extreme unlikeliness, to say the least—for adaptive and genetic arguments about why fellatio and cunnilingus are more common among the upper classes (Weinrich 1977), or why male panhandlers are more successful with females and people who are eating than with males and people who are not eating (Lockard et al. 1976).

Not all of sociobiology proceeds in the mode of storytelling for individual cases. It rests on firmer methodological ground when it seeks broad correlations across taxonomic lines, as between reproductive strategy and distribution of resources, for example (Wilson 1975), or when it can make testable, quantitative predictions as in Trivers and Hare's work on haplodiploidy and eusociality in Hymenoptera (Trivers and Hare 1976). Here sociobiology has had and will continue to have success. And here I wish it well. For it represents an extension of basic Darwinism to a realm where it should apply.

SPECIAL PROBLEMS FOR HUMAN SOCIOBIOLOGY

Sociobiological explanations of human behavior encounter two major difficulties, suggesting that a Darwinian model may be generally inapplicable in this case.

Limited Evidence and Political Clout

We have little direct evidence about the genetics of behavior in humans; and we do not know how to obtain it for the specific behaviors that figure most prominently in sociobiological speculation—aggression, conformity, etc. With our long generations, it is difficult to amass much data on heritability. More important, we cannot (ethically, that is) perform the kind of breeding experiments, in standardized environments, that would yield the required information. Thus, in dealing with humans, sociobiologists rely even more heavily than usual upon speculative storytelling.

At this point, the political debate engendered by sociobiology comes appropriately to the fore. For these speculative stories about human behavior have broad implications and proscriptions for social policy—and this is true quite apart from the intent or personal politics of the storyteller. Intent and usage are different things; the latter marks political and social influence, the former is gossip or, at best, sociology.

The common political character and effect of these stories lies in the direction historically taken by innatist arguments about human behavior and capabilities—a defense of existing social arrangements as part of our biology.

In raising this point, I do not act to suppress truth for fear of its political consequences. Truth, as we understand it, must always be our primary criterion. We live, because we must, with all manner of unpleasant biological truth—death being the most pervasive and ineluctable. I complain because sociobiological stories are not truth but unsupported speculations with political clout (again, I must emphasize, quite apart from the intent of the storyteller). All science is embedded in cultural contexts, and the lower the ratio of data to social importance, the more science reflects the context.

In stating that there is politics in sociobiology, I do not criticize the scientists involved in it by claiming that an unconscious politics has intruded into a supposedly objective enterprise. For they are behaving like all good scientists—as human beings in a cultural context. I only ask for a more explicit recognition of the context and, specifically, for more attention to the evident impact of speculative sociobiological stories. For example, when the *New York Times* runs a week-long front page series on women and their rising achievements and expectations, spends the first four days documenting progress toward social equality, devotes the last day to potential limits upon this progress, and advances sociobiological stories as the only argument for potential limits—then we know that these are stories with consequences:

> Sociologists believe that women will continue for some years to achieve greater parity with men, both in the work place and in the home. But an uneasy sense of frustration and pessimism is growing among some advocates of full female equality in the face of mounting conservative opposition. Moreover, even some staunch feminists are reluctantly reaching the conclusion that women's aspirations may ultimately be limited by inherent biological differences that will forever leave men the dominant sex (*New York Times*, Nov. 30, 1977).

The article then quotes two social scientists, each with a story.

> If you define dominance as who occupies formal roles of responsibility, then there is no society where males are not dominant. When something is so universal, the probability is—as reluctant as I am to say it—that there is some quality of the organism that leads to this condition.
>
> It may mean that there never will be full parity in jobs, that women will always predominate in the caring tasks like teaching and social work and in the life sciences, while men will prevail in those requiring more aggression—business and politics, for example—and in the 'dead' sciences like physics.

Adaptation in Humans Need Not Be Genetic and Darwinian

The standard foundation of Darwinian just-so stories does not apply to humans. That foundation is the implication: if adaptive, then genetic—for the inference of adaptation is usually the only basis of a genetic story, and Darwinism is a theory of genetic change and variation in populations.

Much of human behavior is clearly adaptive, but the problem for sociobiology is that humans have so far surpassed all other species in developing an alternative, non-genetic system to support and transmit adaptive behavior—cultural evolution. As adaptive behavior does not require genetic input and Darwinian selection for its origin and maintenance in humans; it may arise by trial and error in a few individuals that do not differ genetically from their groupmates, spread by learning and imitation, and stabilize across generations by value, custom, and tradition. Moreover, cultural transmission is far more powerful in potential speed and spread than natural selection—for cultural evolution operates in the "Lamarckian" mode by inheritance through custom, writing, and technology of characteristics acquired by human activity in each generation.

Thus, the existence of adaptive behavior in humans says nothing about the probability of a genetic basis for it, or about the operation of natural selection. Take, for example, Trivers' (1971) concept of "reciprocal altruism." The phenomenon exists, to be sure, and it is clearly adaptive. In honest moments, we all acknowledge that many of our "altruistic" acts are performed in the hope and expectation of future reward. Can anyone imagine a stable society without bonds of reciprocal obligation? But structural necessities do not imply direct genetic coding. (All human behaviors are, of course, part of the potential range permitted by our genotype—but sociobiological speculations posit direct natural selection for specific behavioral traits.) As Benjamin Franklin said: "We must all hang together, or assuredly we shall all hang separately."

FAILURE OF THE RESEARCH PROGRAM FOR HUMAN SOCIOBIOLOGY

The grandest goal—I do not say the only goal—of human sociobiology must fail in the face of these difficulties. That goal is no less than the reduction of the behavioral (indeed most of the social) sciences to Darwinian theory. Wilson (1975) presents a vision of the human sciences shrinking in their independent domain, absorbed on one side by neurobiology and on the other by sociobiology.

But this vision cannot be fulfilled, for the reasons cited above. Although we can identify adaptive behavior in humans, we cannot tell thereby if it is genetically based (while much of it must arise by fairly pure cultural evolution). Yet the reduction of the human sciences to Darwinism requires the genetic argument, for Darwinism is a theory about genetic change in populations. All else is analogy and metaphor.

My crystal ball shows the human sociobiologists retreating to a fall-back position—indeed it is happening already. They will argue that this fallback is as powerful as their original position, though it actually represents the unravelling of their fondest hopes. They will argue: yes, indeed, we cannot tell whether an adaptive behavior is genetically coded or not. But it doesn't matter. The same adaptive constraints apply whether the behavior evolved by cultural or Darwinism routes, and biologists have identified and explicated the adaptive constraints. (Steve Emlen reports, for example, that some Indian peoples gather food in accordance with predictions of optimal foraging strategy, a theory developed by ecologists. This is an exciting and promising result within an anthropological domain—for it establishes a fruitful path of analogical illumination between biological theory and non-genetic cultural adaptation But it prevents the assimilation of one discipline by the other and frustrates any hope of incorporating the human sciences under the Darwinian paradigm.)

But it does matter. It makes all the difference in the world whether human behaviors develop and stabilize by cultural evolution or by direct Darwinian selection for genes influencing specific adaptive actions. Cultural and Darwinian evolution differ profoundly in the three major areas that embody what evolution, at least as a quantitative science, is all about:

1. Rate. Cultural evolution, as a "Lamarckian" process, can proceed orders of magnitude more rapidly than Darwinian evolution. Natural selection continues its work within *Homo sapiens*, probably at characteristic rates for change in large, fairly stable populations, but the power of cultural evolution has dwarfed its influence (alteration in frequency of the sickling gene vs. changes in modes of communication and transportation). Consider what we have done with ourselves in the past 3,000 years, all without the slightest evidence for any biological change in the size or power of the human brain.
2. Modifiability. Complex traits of cultural evolution can be altered profoundly all at once (social revolution, for example). Darwinian change is much slower and more piecemeal.
3. Diffusibility. Since traits of cultural evolution can be transmitted by imitation and inculcation, evolutionary patterns include frequent and complex anastomosis among branches. Darwinian evolution in sexually reproducing animals is a process of continuous divergence and ramification with few opportunities for coming together (hybridization or parallel modification of the same genes in independent groups).

I believe that the future will bring mutual illumination between two vigorous, independent disciplines—Darwinian theory and cultural history. This is a good thing, joyously to be welcomed. But there will be no reduction of the human sciences to Darwinian theory and the research program of human sociobiology will fail. The name, of course, may survive. It is an irony of history that movements are judged successful if their label sticks, even though the emerging content of a discipline may lie closer to what opponents originally advocated. Modern geology, for example, is an even blend of Lyell's strict uniformitarianism and the claims of catastrophists (Rudwick 1972; Gould 1977). But we call the hybrid doctrine by Lyell's name and he has become the conventional hero of geology.

I welcome the coming failure of reductionistic hopes because it will lead us to recognize human complexity at its proper level. For consumption of *Time*'s millions, my colleague Bob Trivers maintained: "Sooner or later, political science, law, economics, psychology, psychiatry, and anthropology will all be branches of sociobiology" (*Time* Aug. 1, 1977, 54). It is one thing to conjecture, as I would allow, that common features among independently developed legal systems might reflect adaptive constraints and might be explicated usefully with some biological analogies. It is quite another to state, as Trivers did, that the mores of the entire legal profession will be subsumed, along with a motley group of other disciplines, as mere epiphenomena of Darwinian processes.

I read Trivers' statement the day after I had sung in a full production of Berlioz' *Requiem.* And I remembered the visceral reaction I had experienced upon hearing the 4 brass choirs, finally amalgamated with the 10 tympani in the massive din preceding the great *Tuba mirum*—the spine tingling and the involuntary tears that almost prevented me from singing. I tried to analyze it in the terms of Wilson's conjecture—reduction of behavior to neurobiology on the one hand and sociobiology on the other. And I realized that this conjecture might apply to my experience. My reaction had been physiological and, as a good mechanist, I do not doubt that its neurological foundation can be ascertained. I will also not be surprised to learn that the reaction has something to do with adaptation (emotional overwhelming to cement group coherence in the face of danger, to tell a story). But I also realized that these explanations, however "true," could never capture anything of importance about the meaning of that experience.

And I say this not to espouse mysticism or incomprehensibility, but merely to assert that the world of human behavior is too complex and multifarious to be unlocked by any simple key. I say this to maintain that this richness—if anything—is both our hope and our essence.

SUMMARY

Ever since Darwin proposed it, the theory of natural selection has been marred by an uncritical style of speculative application to the study of individual adaptations: one simply constructs a story to explain how a shape, function, or behavior might benefit its possessor. Virtuosity in invention replaces testability and mere consistency with evolutionary theory becomes the primary criterion of acceptance. Although this dubious procedure has been used throughout evolutionary biology, it has recently become the primary style of explanation in sociobiology.

Human sociobiology presents two major problems related to this tradition. First, evidence is so poor or lacking that speculative storytelling assumes even greater importance than usual. Secondly, the existence of behavioral adaptation does not imply the operation of Darwinian processes at all—for non-genetic cultural evolution, working in the Lamarckian mode, dwarfs by its rapidity the importance of slower Darwinian change. The sociobiological vision of a reduction of the human sciences to biology via Darwinism and natural selection will fail. Instead, I anticipate fruitful, mutual illumination by analogy between independent theories of the human and biological sciences.

LITERATURE CITED

Barash, D. "Male Response to Apparent Female Adultery in the Mountain Bluebird (*Sialia currocoides*): An Evolutionary Interpretation." *American Naturalist* 110 (1976): 1097–101.

Bartholomew, C. A., and T. M. Casey. "Endothermy During Terrestrial Activity in Large Beetles." *Science* 195 (1977): 882–83.

Bertalanify, L. von. "Chance or law." In *Beyond reductionism*, edited by A. Koestler. London: Hutchinson, 1969.

Cock, A. C. "Genetical Aspects of Metrical Growth and Form in Animals." *Quarterly Review of Biology* 41 (1966): 131–90.

Darwin, C. *The Variation of Animals and Plants Under Domestication*. London: John Murray, 1868.

———. "Sir Wyville Thomson and Natural Selection." *Nature* 23 (1880): 32.

Gould, S. J. "Allometry and Size in Ontogeny and Phylogeny." *Biological Reviews* 41 (1966): 587–640.

———. "Allometry in Primates, with Emphasis on Scaling and the Evolution of the Brain." In "Approaches to Primate Paleobiology." *Contributions to Primatology* 5 (1975): 244–92.

———. "Eternal Metaphors of Paleontology." In *Patterns of Evolution*, edited by A. Hallam, 1–26. Amsterdam: Elsevier, 1977.

Huxley, J. *Problems of Relative Growth*. London: MacVeagh, 1932.

Lockard, J. S., L. L. McDonald, D. A. Clifford, and R. Martinez. "Panhandling: Sharing of Resources." *Science* 191 (1976): 406–408.

Rudwick, M. J. S. *The Meaning of Fossils*. London: Macdonald, 1972.

Schweber, S. S. "The Origin of the *Origin* revisited." *Journal of the History of Biology* 10 (1977): 229–316.

Simpson, C. C. "Some Problems of Vertebrate Paleontology." *Science* 133 (1961): 1679–89.

Trivers, R. "The Evolution of Reciprocal Altruism." *Quarterly Review of Biology* 46 (1971): 35–57.

Trivers, R., and H. Hare. "Haplodiploidy and the Evolution of the Social Insects." *Science* 191 (1976): 249–63.

Weinrich, J. D. "Human Sociobiology: Pair-bonding and Resource Predictability (Effects of Social Class and Race)." *Behavioral Ecology and Sociobiology* 2 (1977): 91–118.

Wilson, E. O. *Sociobiology: The New Synthesis*. Cambridge: Harvard University Press, 1975.

EDITORIAL NOTE

In this article Gould refers to one of the very greatest triumphs of recent biological understanding, the solution to a puzzle that had troubled evolutionists from the time of Darwin, namely, the evolution of sterile worker castes in the social insects, especially the ants, the bees, and the wasps (the "hymenoptera"). The answer to this problem—an impetus for the whole sociobiological movement—came to the English biologist William Hamilton in the 1960s, when he realized that selection can work by proxy, as it were. As long as copies of one's genes are passed on, it matters not if one does it directly oneself or indirectly through bearers of similar genes, namely close relatives. Indeed, if relatives can do the job better, then selection may promote the evolution of "altruistic" features, where one is positively inclined to help. (Such an evolutionary process is now known generally as "kin selection.") In the particular case of the hymenoptera, Hamilton seized on their peculiar method of reproduction. They are "Haplodiploid." Females are like other sexual organisms, having father and mother, receiving (as is normal) a half set of chromosomes from each parent. Males have only mothers, receiving just one half set of chromosomes. As a result—unlike the normal situation, where the relationship is symmetrical—females are more closely related to sisters than to daughters. Hence, biologically it pays females to forego personal reproduction, even to the point of sterility, and raise fertile sisters.

This and other examples are superbly discussed in John Maynard Smith's "The Evolution of Behavior," *Scientific American* 239 (September 1978): 176–92.

KIM STERELNY

THE PECULIAR PRIMATE

I. THE GENIUS IN THE MIRROR

One of the most striking discoveries of recent cognitive psychology is that in solving the ordinary decision problems of our daily life, each of us seems to be a genius. Ordinary human decision making has a high cognitive load. To make good decisions, agents must be sensitive to complex, subtle features of their environment. The information-hungry nature of human action first became apparent in thinking about language. Language makes intensive demands on memory. The different parties to a conversation must remember who said what to whom. It also makes intensive demands on attention. In a conversation, you must do more than recall what has been said: you must monitor and act on the effects on your utterances and those of others. You need to be alert to the signs that conversation is going wrong. Moreover, you will often have to do this while also attending to your physical and social world, for often the point of talking is to co-ordinate joint action: linguistic acts interface with, and are smoothly integrated with the rest of our lives. Language is at the core of the cognitive revolution in psychology and of the most prominent attempt to synthesize psychology and evolutionary theory. No wonder: even without considering Chomsky's celebrated arguments for the unlearnablity of human language, we can see that our ability to use language seems almost miraculous: we must be immediately sensitive to a limitless array of features of our current circumstances.

This moral seems to hold of human decision making generally. Consider decision making in a forager's world. Foragers do not accumulate a surplus, so they must make good decisions. Yet a forager on a hunting expedition who sees an armadillo disappearing down its burrow faces tough choices. Should he try to dig it out, or try his luck further down the path? The optimal choice depends on subtle ecological, informational, and risk-assessment issues. The forager must consider the probability of catching the animal: is the burrow likely to end under an immovable obstacle? He must estimate the costs and risks of catching the animal. Those costs include opportunity costs. If it will take the rest of the day to dig the armadillo out, the forager has forgone the potential reward of a day's hunting. Finally, of course, he must factor in the benefits of catching the animal, remembering that in many cultures, large catches are shared but small catches are kept. The right armadillo choice requires an agent to make a clear-sighted assessment of the local ecology, his own technical skills and social location. The high information load on forager decision making is obvious to us, because we know we would fail hopelessly to make good decisions in the face of these challenges. But our lives are similarly demanding: imagine a forager trying to navigate through LAX or Heathrow.

So how do we acquire the information we need to face the challenge of ordinary life and the capacities to use it? By what evolutionary route have we acquired those capacities? There is a well-known response to these questions about the architecture of the human mind and its evolution: the massive modularity hypothesis. Our minds are ensembles of special purpose devices each of which is innately equipped to solve the information-hungry but repeated, predictable problems of human life. I shall argue against this hypothesis and propose another. Human minds are not adapted to a standard set of problems which our ancestors faced. Rather, they are adapted to the highly variable environments in which we evolved. We cope with high load problems by informationally engineering our own environments and those of the next generation. We have a good adaptive fit to our environment in part because we have altered the environment to fit us, rather than merely adapting to our environment. Humans have coevolved with their world. That is true of many organisms, but it's especially true of us, and in distinctive ways.

II. THE MASSIVE MODULARITY HYPOTHESIS

Evolutionary nativists defend a modular solution to the problem of information load on human decision making. On this view, human minds are ensembles of special purpose cognitive devices, modules. We have a language module, a module for interpreting the thoughts and intentions of others, a "naive physics" module for causal reasoning about sticks, stones, and similar inanimate objects, a natural history module for ecological decisions, a social exchange module for monitoring economic interactions with our peers, and so on. The human mind is not an immensely powerful general problem solving engine: it's an integrated array of devices each of which solves a particular type of problem with remarkable efficiency.

These modules evolved in response to the distinctive, independent, and reccurring problems our ancestors faced in their lives as Pleistocene foragers. At some stage language became crucial to human life: the alingual and barely lingual would have been under an ever increasing handicap. Hence there would have been selection for an innate language competence. Those lives also depended on co-operation, but on cautious co-operation, as free-riding would have been an ever-present temptation. Thus the need for a capacity to monitor social exchange. And so on. On this view, we cope with the information processing demands on human life because we come preequipped with the crucial information we need: information about human language, human minds, inanimate causal interactions, the biological world. Information is built into the relevant module. Our modules are innately equipped with the right information, and are innately designed to use information of that kind appropriately.

I think there is something right about this view. But I also think it's radically incomplete. Innate evolved modules play a role in explaining our capacity to solve high information load problems. However, this cannot be a general solution to the information load problem. Natural selection can build and equip a special purpose module only if the information an agent needs to know is stable over evolutionary time. That sometimes happens: the causal properties of sticks, stones, and bones do not change: hence a naive physics module could well be built into human heads. Likewise, the acquisition of language is crucial for every human and its basic structural properties are stable. So it would not surprise if there was an innate language module pre-programmed with this structural information. For once a system of linguistic practices has evolved, it is in everyone's interests to conform to those practices. There is no "temptation to defect" from the basic features of phonology, syntax, and morphology of your local community, and this stablizes the basic organizational features of language. So a modular solution to the informational load imposed by language is plausible.

But many features of human environments have not been stable. The social, physical, and cognitive environments in which we act have been extraordinarily varied. Yet we cope: that is a part of the explanation of our cosmopolitan distribution. Social worlds vary in size, family structure, economic basis, technological elaboration, the extent and kinds of social hierarchy, division of labor. They have varied enormously physically and biologically: the world has changed very dramatically in its physical and biological state over the last few hundred thousand years, as it has cycled through ice ages. Moreover, we are now spread over virtually the whole of the globe. Humans also vary cognitively: humans minds are not everywhere and everywhen the same. For human actions depend on what agents believe, what they want, and what they can do, and those conditions are not constant. If many salient features of human worlds vary across time and space, information about those features cannot be engineered into human minds by natural selection. Our ability to act aptly in such worlds cannot depend on innate modules.

So while our ability to make high load decisions depends on our having innately equipped, special purpose modules, this is only one aspect of intelligent decision making. In the rest of this talk I discuss the evolution of two other responses to the problem of high load decision making: the evolution of highly structured developmental environments which in turn make the acquisition of complex skills possible, and the evolution of cognitive technology. Moreover, I link these responses to a distinctive feature of human evolution. In his *The Third Chimpanzee*, Jared Diamond noted that a few million years ago, hominids were just large mammals of no special ecological significance. Our ancestors were woodland apes, just one element of a remarkably diverse East African mammal fauna. Something happened: these unremarkable animals become a world-wide, world-transforming swarm. How did just one ordinary lineage become an extraordinary one? There have been plenty of key innovations in evolutionary history but these have resulted in adaptive radiations, not a cosmopolitan ecological dominance by a single species. Something strange has happened. The massive modularity hypothesis throws no real light on this enigma. The evolutionary mechanism that builds modules is standard natural selection. Our minds are adapted to the demands imposed by our (ancestral) environments, just as the extraordinary spatial memory of Clark's Nutcracker is driven by its need to cache seeds for winter survival. On this perspective, we are just another unique species. Like every other species, we have a set of distinctive adaptations to our environment, but we are not the result of unusual evolutionary mechanisms. But we are unusual in a way Clark's Nutcracker is not.

III. LIVING IN A SELF-MADE WORLD

Natural selection is often conceived as a process by which lineages are shaped to fit the environments in which they live, as a key-cutter shapes a key to a lock. This picture is sometimes appropriate. Many Australian trees have adapted to arid conditions by changes which reduce water loss. But lineages can also respond to challenge by engineering their own environments to transform the ways they are affected by their physical, biological, and social environments. Termites, for example, create their own stable microclimate within their mounds, thus insulating themselves from changes in temperature and humidity. Some lineages partially construct their own niches. They are ecological engineers. While true of many lineages, this is true with a vengeance of us.

When plants and animals engineer their own environments, quite often they modify not just their own environment but also that of their immediate offspring. Their ecological engineering has downstream consequences. Human ecological engineering does have these downstream consequences; children typically live with their parents and benefit from their ecological engineering. So humans partially construct not just their own niche but also that of the next generation. Moreover, as Michael Tomasello has emphasized, with humans this engineering is cumulative. Modification increases over time, generation by generation. Downstream niche construction itself is not uniquely human, but *cumulative downstream niche construction* is our speciality.

So we live in worlds that we have made and remade. Our chimpanzee relatives do not. Chimpanzee material culture is quite varied. But it is also quite rudimentary: there is no evidence that any chimp tools exist in their current form as the result of a cycle of discovery and improvement. Cycles of discovery and improvement depend on reliable and high fidelity transmission between generations. If an innovative australopithecine discovers a more efficient way of flaking stone tools, without reliable social transmission the technique will disappear at the death of its discoverer. Imitation plays little role in chimp social learning, and they lack language. So chimp social learning is not adapted to a communal data base and a communal skill base that can be ratcheted up over the generations. But for the last couple hundred thousand years the human environment has been the result of a ratchet effect in operation: a cycle in which an innovation is made, becomes standard for the group, and it then becomes a basis for further innovation. So material culture—and surely knowledge too—is built by cumulative improvements. The niche construction of one generation becomes a basis for further niche construction.

We have progressively modified our own physical, social, and biological environment. Tools, clothes, shelters have changed the worlds we live in. But we have modified our learning environment too: reshaping both the information available to children and the access they have to that information.

IV. SOCIALITY, CO-OPERATION, AND INFORMATION POOLING

Humans do not just remake their worlds. They remake their worlds in groups and for groups. Humans have often lived in metapopulations of competing groups.

This helps explain a second unusual feature of human life. We are not just very adaptable, we are also extraordinarily co-operative. It is hundreds of thousands of years since any human survived by gathering resources unassisted. Intricately and stably co-operative life poses a problem for evolutionary biology. For though co-operation is often beneficial, it is also fragile. Co-operative social worlds can easily decay because there is a temptation to enjoy the benefits of co-operation without paying the costs: a temptation to share others' resources while hogging your own. How can this be prevented? Policing cheats individually is costly. Policing cheats as a team is itself an instance of co-operation. There is an expectation that more co-operative worlds will decay into less co-operative ones; an expectation not met in our case.

Co-operative groups would clearly outcompete Hobbsian groups, but evolutionary biologists have typically been skeptical of explaining co-operation this way, for the conditions that make group selection possible are quite onerous. They require a population to be structured into subpopulations. Those subpopulations must be in themselves relatively uniform but differ from one another in ways relevant to fitness. There must be mechanisms which prevent mixed groups of co-operators and cheats from forming, or mechanisms which limit the benefit of cheating in mixed groups. These conditions are rarely met. But they have often met in human evolution. For human cultural life tends to magnify behavioral differences between groups and to suppress individual differences within them. Cultural learning tends to have conformist effects within groups and differentiating effects between them. Copying the majority is often adaptive: it enables you to learn from others' unfortunate experiences. But it has conformist outcomes. Likewise, imitation allows an innovation made by an individual within a group to come to characterize the whole group, and only that group, and its descendants. Moreover, humans have various antifreerider mechanisms that tend to deter, punish, or exclude defectors. In short, group selection has probably been important for our lineage.

Group selection and downstream niche construction combine to transform the environment of human evolution. For they make the advantages of information pooling available at a generation. That pooling also makes the flow of information across generations much more reliable, as children have access to information controlled by the group as a whole. The cumulative improvement of a culture's knowledge base depends on the high fidelity transmission of skill. But ratchet of cumulative improvement also requires co-operation. Knowledge is fitness. I get no personal benefit from letting you learn my improved flint knapping techniques. So unless conflicts within the group are partially suppressed, innovators have no interest in passing on successful innovation except to their immediate descendants. But a ratchet that operated only by direct descent would be slow and uncertain. The ratchet is powerful when an improvement can spread through a whole group to become widely available as a foundation for further improvement. So it depends on the horizontal and oblique flows of information that co-operation makes possible: human co-operation and human capacities for high load decision making are intimately linked.

V. SELECTION IN AN UNCERTAIN WORLD

Where a single set of developmental resources builds one architecture in one environment, and a different architecture in another, the organism is developmentally plastic. Almost all organisms show some degree of adaptive flexibility. Many species of reptile are developmentally plastic with respect to their sex: a crocodile egg can develop into either a male or a female, depending on the temperature of the nest. If the environment is unpredictable, developmental plasticity is advantageous. Human environments have been exceptionally unpredictable. As with niche construction, so too with phenotypic plasticity: a general feature of animal evolution is especially important in hominin evolution. We evolved in increasingly variable environments. For one thing we evolved (as Richard Potts argues) in a time of increasing climatic variability. There are paleobiological signals both of the variability itself and its biological impact. For example, in Kenya about 400,000 there was a mammalian turnover pulse in which specialist grassland mammals disappeared but their more generalist relatives survived. Potts makes a good case for the idea that the hominin experience became increasingly variable in time as well as in space: there was nothing remotely like a single set of physical and ecological conditions in which human evolution was played out. Moreover we ourselves have been the cause of our unpredictable environments: human niche construction has altered our social, technological, and biological environments.

Our phenotypic plasticity is mediated largely through our minds, for it is an organ of plasticity par excellence. Consider automatized skills. Skills are slowly built, but once built they are enduring and automatic. I will be able to play chess till the day I die. Many of these skills are rather like nativist psychology's modules. Once they are up and running, they are fast, reliable, automatic, and domain specific. Chess players cannot help seeing a chess board and its pieces as a chess position. They are often adaptive: they equip agents for the specific features of their environment. The hard-won skills of natural history and bushcraft that enable a forager to move silently, see much that is invisible to others, and to find his or her way, are plainly critical to survival. But they are built into no-one's genes: those skills will be very different in an Australian aboriginal in the Pilbara, an Ache hunter-gatherer in a central American rainforest, or in an Inuit seal-hunter. Minds have contingent but stable features: features that persist once they have developed. The same initial set of developmental resources can differentiate into quite different final cognitive products. But the adaptive benefit of this plasticity in turn depends on group selection and downstream niche construction. These automated skills are acquired through the information pooling and cross-generational information flow that these evolutionary mechanisms make possible. They are the result of individual learning—lots of it—in a massively shaped, organized environment.

VI. ENGINEERING CHILDHOOD: THE EXAMPLE OF OTHER MINDS

Even though we are very complicated creatures, humans are very good at interpreting other humans: that is, we are very good at estimating what others will think and do, even in novel circumstances. If an arrangement goes wrong, and a friend fails to turn up to a meeting at a café, we are quite good at working out how to re-coordinate on the fly. We predict others' responses, even taking into account that fact that their response will depend on what they expect us to do. One explanation of this remarkable capacity is that we have an innate "folk psychology" module: a module equipped with a good model of human thought and decision. This is the massive modularity explanation of our capacity to understand others. There is an alternative: interpretation is an automated skill, a skill that is acquired very reliably because humans of one generation engineer the learning environment of the next generation. Moreover, the acquisition of this skill is further supported by the fact that we have perceptual systems tuned to facial expression; signs of affect in voice, posture, and movement; the behavioral signatures which distinguish between intentional and accidental action, and the like. These systems make the right aspects of behavior, voice, posture, and facial expression salient to us. They make learning easier because we are apt to notice the right things in other agents. These perceptual adaptations come to operate in a developmental environment that is the product of cumulative niche construction. We engineer the informational environment to scaffold the acquisition of interpretative skills.

Even so, mental states are unobservable causes of behavior. So the task of learning how to interpret others might seem especially difficult, depending as it does on an inference from effects to their hidden causes. In adults, the connection between psychological state and action can be very complex and indirect, and that may reinforce the suspicion that we must have an innate folk psychology. How could anyone learn that action depends on an agent's beliefs and goals, when those are so hidden. What people believe and want is not obvious in their actions. But the step from action to its hidden cause may itself be scaffolded. When children interact with their peers, the connections between desire, emotion, and action will often be very direct. As children mature, they learn to inhibit impulses, and their actions become much more sensitive to spatio-temporally displaced information and motivation. But when interacting with their peers, the inference from effect to cause will often be much less challenging. Children are less good at concealing overt signs of their emotion than adults, and less good at resisting the urge to act on those emotions. As three- and four-year-olds are making crucial developmental transitions, this lack of inhibition of their peers simplifies their epistemic environment.

As I see it, then, the acquisition of interpretive skills depends on perceptual preadaptation and individual exploration in a socially structured learning environment. In particular, the reliable development of interpretive capacities is supported by the following factors.

(i) Perceptual mechanisms make crucial clues of agents' intentions salient to us. Folk psychology is scaffolded by perceptual tuning.
(ii) Children live in an environment soaked with agents interpreting one another. They are exposed both to third party interpretation, and to others interpreting them. Much of this interpretation is linguistic but there are also contingent interactions in which one is treated as an agent: imitation games, joint attention, joint play.
(iii) Learning is scaffolded by particular cultural inventions: for example, narrative stories are full of simplified and explicit interpretative examples.
(iv) Parents make the interpretive task easier by offering models of their own and their children's actions: they often rehearse interpretations of both their own and their children's actions.
(v) Language scaffolds the acquisition of interpretative capacities by supplying a pre-made set of interpretative tools. Thus linguistic labels help make differences salient.
(vi) Interpretation is scaffolded by interacting with agents—your developing peers—who have not yet gained the abilities to mask their emotions, inhibit their desires, and suppress their beliefs. Such agents simplify the problem of inferring from action to its psychological root.

(vii) Belief and preference are often hidden, having no overt and distinctive behavioral signature. But many folk psychological concepts—those for sensations and emotions—do have a regular behavioral signature, and these scaffold the acquisition of less behaviorally overt concepts by making available easier examples of inner causes of outer actions.

Perhaps interpretation was once very difficult, or very inaccurate. But because these skills are now acquired in engineered learning environments, they have been ratcheted both to greater levels of precision, and to earlier and more uniform mastery. No doubt the cognitive capacities involved in understanding others would be very hard to acquire by one's own unaided efforts. But we do not have to acquire them that way. Our environments have been informationally engineered to circumvent the cognitive limits of individuals on their own. We do not have to appeal to innate and canalized development to explain the early and uniform development of fast, unreflective, powerful, and accurate cognitive mechanisms. We have a second model: automatized skills. Jared Diamond describes the enculturation of forager children in PNG as one long apprenticeship in natural history: those children live in a highly structured, information-soaked developmental environment. That environment results in the extraordinary impressive mature competence of the adults: a rich and very accurate understanding of their local natural history and of the techniques needed to make a living from it. Niche construction provides an alternative model of how a fast, automatic, and sophisticated cognitive specialization can develop without depending on specific innate structure.

VII. EPISTEMIC TECHNOLOGY

One aspect of niche construction is information engineering, and I have already spoken of how one generation engineers the learning environment of the next. But agents also informationally engineer their own environment, and that is one way we cope with high load decision making. We use epistemic technology, and our use of such technology is ancient. We alter our environment to ease memory burdens: think of marking a trail. We store information in the environment; we recode it, and we exploit our social organization through a division of intellectual labor. Our contemporary environment is full of purpose-built tools for easing burdens on memory. These include diaries, notebooks, and other "organizers." Filofaxes are new tools, but purpose-built aids to memory are certainly ancient. Pictorial representation is over 30,000 years old. Furthermore ecological tools have informational side-effects. A fish-trap can be used as a template for making more fish-traps. When enduring representational formats were not available, people re-coded information in public language to make it easier to recall. In songs, stories and rhyme, the organization of the information enables some elements to prime others. Such re-coding enables us to partially substitute *recognition* for *recall.*

We transform difficult learning problems into easier ones. We do not just provide information verbally: learning is scaffolded in many other ways. Skills are demonstrated in a form suited for learning. Completed and partially completed artifacts are used as teaching props. Practice is supervised and corrected. The decomposition of a skill into its components is made obvious; subtle elements will often be exaggerated, slowed down, or repeated. Moreover, skills are often taught in an optimal sequence, so that one forms a platform for the next. Engineered learning environments play their most obvious role in intergenerational information flow, but these techniques also mediate horizontal flows of information.

Moreover, as Dennett in particular argues, cognitive technology also has profound developmental effects. He distinguishes between the capacity to have beliefs about beliefs and the capacity to think about thinking. On his view, even if non-human primates have beliefs about beliefs, they cannot think about thinking. Agents in a culture with enduring public symbols inherit an ability to make those symbols themselves objects of perception and to manipulate them voluntarily. Imagine a group of friends making a sketch map in the sand to coordinate a hike. Those representations are voluntary and planned. Dennett suggests that we first learn to think about thoughts by thinking about these public representations. In drafting and altering a sketch map, we are using cognitive skills that are already available. They are just being switched to a new target. Moreover, manipulating such a public representation makes fewer demands on memory; no-one has to remember where on the map the camp site is represented. Rich metarepresentational capacities are developmentally scaffolded by an initial stage in which public representations are objects of thought and action.

Epistemic technology—building tools for thinking, and altering the informational character of your environment—makes possible much that would otherwise be impossible. Moreover, for the most part, the effectiveness of epistemic technology is not linked to the pace of environmental change. Learning from the previous generation works well, but only if the pace of change is not too fast. Turning memory tasks into perceptual ones; using templates, public representational media, and good notation systems all enhance your capacity to learn about your environment. And they do so independently of the pace at which that environment changes. But though epistemic technology plays a crucial role in explaining human intelligence, the use of epistemic technology is itself informationally demanding. The use of such technology is itself an aspect of the selective landscape that has transformed human cognitive capacities. Epistemic technology—storing information in the world, and improving the local epistemic environment—is not a way of making a dumb naked-brain smart by adding the right peripherals; it is not a way of making dumb brains part of smart systems. Epistemic technology is not a complete solution in itself to the problem of cognitive load. The use of epistemic technology itself must be supported by some mix of automatized skills and modules.

VIII. REPRISE

It is time to synthesize the disparate threads of this talk. Early in hominin evolution, it is very likely that biological inheritance was much as it now is in chimps. To a reasonable approximation, a chimp inherits only genes from its parents. Though social learning is important in chimp life, and there is some flow of information from mothers to offspring, that flow is diffuse and short-lived. There is no evidence of deep behavioral traditions in chimp life, nor of cumulative downstream niche construction. There is no sign that group selection is allowing co-operation to take off. There is certainly some limited forms of co-operation: males co-operate to defend territory against other chimp groups, to hunt, and to form coalitions against other males. Females too form coalitions. But there is little evidence of effective suppression of free-riding and defection. But over time in our lineage:

(i) Early humans began to fundamentally alter their environment and that of their immediate descendants. Even relatively simple weapons technology would have profound effects, as would the regular use of fire or of shelters.

(ii) Group selection became very important, and underwrote the evolution of a cooperation explosion, the effects of which include language, the division of labor, and resource sharing.

(iii) Co-operation itself accentuates niche construction: it becomes more powerful within and across generations. Information pooling puts in place one precondition for the preservation and improvement of innovation.

(iv) The geographic expansion of the hominin range; the transformation of hominin lifeways, and the intensification of climatic variability select for flexible response. Hominin environments became more variable at a time, and changed faster over time, for some of these changes were self-induced. These changes select for plasticity.

(v) As this transformation proceeds, elements of culture become elements of biology, as they become part of a developmental matrix which is transmitted from one generation to the next.

(vi) Once information transmission became reliable and of high fidelity, downstream niche construction becomes cumulative, and Tomasello's Ratchet begins to work. That Ratchet required both cognitive and social preconditions. But once these are met, the Ratchet will turn. As it turns, different human groups became more markedly differentiated. For their phenotypes come to reflect not just their current environmental differences but also the differences in their cultural lineages' learning history.

(vii) The conditions are then in place that support effective high load decision making. The developmental plasticity of human wetware conspires with the biocultural engineering of human learning environments to support the acquisition of automated skills and of cognitive technology. These in turn feedback: increasing both the variety of human environments and the pace at which they change.

Humans are strange primates, then, because evolutionary forces that act elsewhere in the biological world have come together in our history. We are groupish and cooperative, especially with our in-group, because group selection is relevant to our story. Our intelligence in part depends on our co-operativeness, because it involves adaptations to and for information pooling. But it also depends on our propensities to alter our own world and that of our children. That propensity intensifies selection for cognitive adaptability: for the capacity to cope with the new. For the new comes thick and fast. But it also gives us the capacity to cope with novelty, by enabling us to engineer the learning worlds of our children. This perspective still owes something to the massive modularity hypothesis. But the overall picture is very different. On the modularity hypothesis, there is a single (or perhaps one for each sex) architecture that characterizes the human mind. On the picture advertised here, that is not even a reasonable first approximation. Our Pleistocene forebears did not have contemporary minds in a Pleistocene world; and we do not have essentially Pleistocene minds in a contemporary world.

G. M. FOODS

*THE PRINCE OF WALES**

A ROYAL VIEW

Like millions of other people around the world, I've been fascinated to hear five eminent speakers share with us their thoughts, hopes, and fears about sustainable development based on their own experience. All five of those contributions have been immensely thoughtful and challenging. There have been clear differences of opinion and of emphasis between the speakers, but there have also been some important common themes, both implicit and explicit. One of those themes has been the suggestion that sustainable development is a matter of enlightened self-interest. Two of the speakers used this phrase, and I don't believe that the other three would dissent from it, and nor would I.

Self-interest is a powerful motivating force for all of us, and if we can somehow convince ourselves that sustainable development is in all our interests, then we will have taken a valuable first step toward achieving it. But self-interest comes in many competing guises—not all of which I fear are likely to lead in the right direction for very long, nor to embrace the manifold needs of future generations. I am convinced we will need to dig rather deeper to find the inspiration, sense of urgency, and moral purpose required to confront the hard choices which face us on the long road to sustainable development. So, although it seems to have become deeply unfashionable to talk about the spiritual dimension of our existence, that is what I propose to do.

The idea that there is a sacred trust between mankind and our Creator, under which we accept a duty of stewardship for the earth, has been an important feature of most religious and spiritual thought throughout the ages. Even those whose beliefs have not included the existence of a Creator have, nevertheless, adopted a similar position on moral and ethical grounds. It is only recently that this guiding principle has become smothered by almost impenetrable layers of scientific ratio-

*BBC Reith Lectures 2000 (Home/Respect for the Earth).

nalism. I believe that if we are to achieve genuinely sustainable development, we will first have to rediscover, or reacknowledge, a sense of the sacred in our dealings with the natural world, and with each other. If literally nothing is held sacred anymore—because it is considered synonymous with superstition or in some other way "irrational"—what is there to prevent us treating our entire world as some "great laboratory of life" with potentially disastrous long-term consequences?

Fundamentally, an understanding of the sacred helps us to acknowledge that there are bounds of balance, order, and harmony in the natural world which set limits to our ambitions, and define the parameters of sustainable development. In some cases nature's limits are well understood at the rational, scientific level. As a simple example, we know that trying to graze too many sheep on a hillside will, sooner or later, be counterproductive for the sheep, the hillside, or both. More widely we understand that the overuse of insecticides or antibiotics leads to problems of resistance. And we are beginning to comprehend the full, awful consequences of pumping too much carbon dioxide into the earth's atmosphere. Yet the actions being taken to halt the damage known to be caused by exceeding nature's limits in these and other ways are insufficient to ensure a sustainable outcome. In other areas, such as the artificial and uncontained transfer of genes between species of plants and animals, the lack of hard, scientific evidence of harmful consequences is regarded in many quarters as sufficient reason to allow such developments to proceed.

The idea of taking a precautionary approach, in this and many other potentially damaging situations, receives overwhelming public support, but still faces a degree of official opposition, as if admitting the possibility of doubt was a sign of weakness or even of a wish to halt "progress." On the contrary, I believe it to be a sign of strength and of wisdom. It seems that when we do have scientific evidence that we are damaging our environment, we aren't doing enough to put things right, and when we don't have that evidence, we are prone to do nothing at all, regardless of the risks.

Part of the problem is the prevailing approach that seeks to reduce the natural world including ourselves to the level of nothing more than a mechanical process. For while the natural theologians of the eighteenth and nineteenth centuries like Thomas Morgan referred to the perfect unity, order, wisdom, and design of the natural world, scientists like Bertrand Russell rejected this idea as rubbish. "I think the universe," he wrote, "is all spots and jumps without unity and without continuity without coherence or orderliness. "Sir Julian Huxley wrote in "Creation a Modern Synthesis" that "modern science must rule out special creation or divine guidance." But why?

As Professor Alan Linton of Bristol University has written, "Evolution is a manmade theory to explain the origin and continuance of life on this planet without reference to a Creator." It is because of our inability or refusal to accept the existence of a guiding hand that nature has come to be regarded as a system that can be engineered for our own convenience or as a nuisance to be evaded and manipulated, and in which anything that happens can be fixed by technology and human ingenuity. Fritz Schumacher recognized the inherent dangers in this approach when he said that "there are two sciences—the science of manipulation and the science of understanding."

In this technology-driven age, it is all too easy for us to forget that mankind is a part of nature and not apart from it. And that this is why we should seek to work with the grain of nature in everything we do, for the natural world is, as the economist Herman Daly puts it, "the envelope that contains, sustains and provisions the economy, not the other way round." So which argument do you think will win—the living world as one or the world made up of random parts, the product of mere chance, thereby providing the justification for any kind of development? This, to my mind, lies at the heart of what we call sustainable development. We need, therefore, to rediscover a reference for the natural world, irrespective of its usefulness to ourselves—to become more aware in Philip Sherrard's words of "the relationship of interdependence, interpenetration, and reciprocity between God, Man, and Creation."

Above all, we should show greater respect for the genius of nature's designs, rigorously tested and refined over millions of years. This means being careful to use science to understand how nature works, not to change what nature is, as we do when genetic manipulation seeks to transform a process of biological evolution into something altogether different. The idea that the different parts of the natural world are connected through an intricate system of checks and balances which we disturb at our peril is all too easily dismissed as no longer relevant.

So, in an age when we're told that science has all the answers, what chance is there for working with the grain of nature? As an example of working with the grain of nature, I happen to believe that if a fraction of the money currently being invested in developing genetically manipulated crops were applied to understanding and improving traditional systems of agriculture, which have stood the all-important test of time, the results would be remarkable. There is already plenty of evidence of just what can be achieved through applying more knowledge and fewer chemicals to diverse cropping systems. These are genuinely sustainable methods, and they are far removed from the approaches based on monoculture, which lend themselves to large-scale commercial exploitation, and which Vandana Shiva condemned so persuasively and so convincingly in her lecture. Our most eminent scientists accept that there is still a vast amount that we don't know about our world and the life-forms that inhabit it. As Sir Martin Rees, the Astronomer Royal, points out, it is complexity that makes things hard to understand, not size. In a comment which only an astronomer could make, he describes a butterfly as a more daunting intellectual challenge than the cosmos!

Others, like Rachel Carson, have eloquently reminded us that we don't know how to make a single blade of grass. And St. Matthew, in his wisdom, emphasized that not even Solomon in all his glory was arrayed as the lilies of the field. Faced with such unknowns, it is hard not to feel a sense of humility, wonder, and awe about our place in the natural order. And to feel this at all stems from that inner heartfelt reason which sometimes despite ourselves is telling us that we are intimately bound up in the mysteries of life and that we don't have all the answers. Perhaps even that we don't have to have all the answers before knowing what we should do in certain circumstances. As Blaise Pascal wrote in the seventeenth century, "It is the heart that experiences God, not the reason."

So do you not feel that, buried deep within each and every one of us, there is an instinctive, heartfelt awareness that provides—if we will allow it to—the most reliable guide as to whether or not our actions are really in the long-term interests of our planet and all the life it supports? This awareness, this wisdom of the heart, may be no more than a faint memory of a distant harmony, rustling like a breeze through the leaves, yet sufficient to remind us that the earth is unique and that we have a duty to care for it. Wisdom, empathy, and compassion have no place in the empirical world, yet traditional wisdoms would ask "Without them are we truly human?" And it would be a good question. It was Socrates who, when asked for his definition of wisdom, gave as his conclusion, "knowing that you don't know."

In suggesting that we will need to listen rather more to the common sense emanating from our hearts if we are to achieve sustainable development, I'm not suggesting that information gained through scientific investigation is anything other than essential. Far from it. But I believe that we need to restore the balance between the heartfelt reason of instinctive wisdom and the rational insights of scientific analysis. Neither, I believe, is much use on its own. So it is only by employing both the intuitive and the rational halves of our own nature—our hearts and our minds—that we will live up to the sacred trust that has been placed in us by our creator—or our "sustainer" as ancient wisdom referred to the creator. As Gro Harlem Brundtland has reminded us, sustainable development is not just about the natural world, but about people, too. This applies whether we are looking at the vast numbers who lack sufficient food or access to clean water, but also those living in poverty and without work. While there is no doubt that globalization has brought advantages, it brings dangers, too. Without the humility and humanity expressed by Sir John Browne in his notion of the "connected economy"—an economy which acknowledges the social and environmental context within which it operates—there is the risk that the poorest and the weakest will not only see very little benefit but, worse, they may find that their livelihoods and cultures have been lost.

So if we are serious about sustainable development, then we must also remember that the lessons of history are particularly relevant when we start to look further ahead. Of course, in an age when it often seems that nothing can properly be regarded as important unless it can be described as "modern" it is highly dangerous to talk about the lessons of the past. And are those lessons ever taught or understood adequately in an age when to pass on a body of acquired knowledge of this kind is often considered prejudicial to "progress"? Of course our descendants will have scientific and technological expertise beyond our imagining, but will they have the insight or the self-control to use this wisely, having learned both from our successes and our failures?

They won't, I believe, unless there are increased efforts to develop an approach to education which balances the rational with the intuitive. Without this, truly sustainable development is doomed. It will merely become a hollow-sounding mantra that is repeated ad nauseam in order to make us all feel better. Surely, therefore, we need to look toward the creation of greater balance in the way we educate people so that the practical and intuitive wisdom of the past can be blended with the appropriate technology and knowledge of the present to produce the type of practitioner who is acutely aware of both the visible and invisible worlds that inform the entire cosmos. The future will need people who understand that sustainable development is not merely about a series of technical fixes, about redesigning humanity or reengineering nature in an extension of globalized, industrialization—but about a reconnection with nature and a profound understanding of the concepts of care that underpin long-term stewardship.

Only by rediscovering the essential unity and order of the living and spiritual world—as in the case of organic agriculture or integrated medicine or in the way we build—and by bridging the destructive chasm between cynical secularism and the timelessness of traditional religion will we avoid the disintegration of our overall environment. Above all, I don't want to see the day when we are rounded upon by our grandchildren and asked accusingly why we didn't listen more carefully to the wisdom of our hearts as well as to the rational analysis of our heads; why we didn't pay more attention to the preservation of biodiversity and traditional communities or think more clearly about our role as stewards of creation? Taking a cautious approach or achieving balance in life is never as much fun as the alternatives, but that is what sustainable development is all about.

RICHARD DAWKINS

AN OPEN LETTER TO PRINCE CHARLES

Sunday May 21, 2000

Your Royal Highness,

Your Reith lecture saddened me. I have deep sympathy for your aims, and admiration for your sincerity. But your hostility to science will not serve those aims; and your embracing of an ill-assorted jumble of mutually contradictory alternatives will lose you the respect that I think you deserve. I forget who it was who remarked, "Of course we must be open-minded, but not so open-minded that our brains drop out."

Let's look at some of the alternative philosophies which you seem to prefer over scientific reason. First, intuition, the heart's wisdom "rustling like a breeze through the leaves." Unfortunately, it depends whose intuition you choose. Where aims (if not methods) are concerned, your own intuitions coincide with mine. I wholeheartedly share your aim of long-term stewardship of our planet, with its diverse and complex biosphere.

But what about the instinctive wisdom in Saddam Hussein's black heart? What price the Wagnerian wind that rustled Hitler's twisted leaves? The Yorkshire Ripper heard religious voices in his head urging him to kill. How do we decide *which* intuitive inner voices to heed?

This, it is important to say, is not a dilemma that science can solve. My own passionate concern for world stewardship is as emotional as yours. But where I allow feelings to influence my aims, when it comes to deciding the best method of achieving them, I'd rather think than feel. And thinking, here, means scientific thinking. No more effective method exists. If it did, science would incorporate it.

Next, Sir, I think you may have an exaggerated idea of the naturalness of "traditional" or "organic" agriculture. Agriculture has always been unnatural. Our species began to depart from our natural hunter-gatherer lifestyle as recently as 10,000 years ago—too short to measure on the evolutionary timescale.

Wheat, be it ever so wholemeal and stoneground, is not a natural food for *Homo sapiens*. Nor is milk, except for children. Almost every morsel of our food is genetically modified—admittedly by artificial selection not artificial mutation, but the end result is the same. A wheat grain is a genetically modified grass seed, just as a pekinese is a genetically modified wolf. Playing God? We've been playing God for centuries!

The large, anonymous crowds in which we now teem began with the agricultural revolution, and without agriculture we could survive in only a tiny fraction of our current numbers. Our high population is an agricultural (and technological and medical) artifact. It is far more unnatural than the population-limiting methods condemned as unnatural by the Pope. Like it or not, we are stuck with agriculture, and agriculture—all agriculture—is unnatural. We sold that pass 10,000 years ago.

Does that mean there's nothing to choose between different kinds of agriculture when it comes to sustainable planetary welfare? Certainly not. Some are much more damaging than others, but it's no use appealing to "nature," or to "instinct" in order to decide which ones. You have to study the evidence, soberly and reasonably—scientifically. Slashing and burning (incidentally, no agricultural system is closer to being "traditional") destroys our ancient forests. Overgrazing (again, widely practiced by "traditional" cultures) causes soil erosion and turns fertile pasture into desert. Moving to our own modern tribe, monoculture, fed by powdered fertilizers and poisons, is bad for the future; indiscriminate use of antibiotics to promote livestock growth is worse.

Incidentally, one worrying aspect of the hysterical opposition to the possible risks from GM [genetically modified] crops is that it diverts attention from definite dangers which are already well understood but largely ignored. The evolution of antibiotic-resistant strains of bacteria is something that a Darwinian might have foreseen from the day antibiotics were discovered. Unfortunately, the warning voices have been rather quiet, and now they are drowned by the baying cacophony: "GM GM GM GM GM GM!"

Moreover, if, as I expect, the dire prophecies of GM doom fail to materialize, the feeling of letdown may spill over into complacency about real risks. Has it occurred to you that our present GM brouhaha may be a terrible case of crying wolf?

Even if agriculture could be natural, and even if we could develop some sort of instinctive rapport with the ways of nature, would nature be a good role model? Here, we must think carefully. There really is a sense in which ecosystems are balanced and harmonious, with some of their constituent species becoming mutually dependent. This is one reason the corporate thuggery that is destroying the rain-forests is so criminal.

On the other hand, we must beware of a very common misunderstanding of Darwinism. Tennyson was writing before Darwin, but he got it right. Nature really is red in tooth and claw. Much as we might like to believe otherwise, natural selection, working within each species, does not favor long-term stewardship. It favors short-term gain. Loggers, whalers, and other profiteers who squander the future for present greed are only doing what all wild creatures have done for three billion years.

No wonder T. H. Huxley, Darwin's bulldog, founded his ethics on a repudiation of Darwinism. Not a repudiation of Darwinism as science, of course, for you cannot repudiate truth. But the very fact that Darwinism is true makes it even more important for us to fight against the naturally selfish and exploitative tendencies of nature. We can do it. Probably no other species of animal or plant can. We can do it because our brains (admittedly given to us by natural selection for reasons of short-term Darwinian gain) are big enough to see into the future and plot long-term consequences. Natural selection is like a robot that can only climb uphill, even if this leaves it stuck on top of a measly hillock. There is no mechanism for going downhill, for crossing the valley to the lower slopes of the high mountain on the other side. There is no natural foresight, no mechanism for warning that present selfish gains are leading to species extinction—and indeed, 99 percent of all species that have ever lived are extinct.

The human brain, probably uniquely in the whole of evolutionary history, can see across the valley and can plot a course away from extinction and toward distant uplands. Long-term planning—and hence the very possibility of stewardship—is something utterly new on the planet, even alien. It exists only in human brains. The future is a new invention in evolution. It is precious. And fragile. We must use all our scientific artifice to protect it.

It may sound paradoxical, but if we want to sustain the planet into the future, the first thing we must do is stop taking advice from nature. Nature is a short-term Darwinian profiteer. Darwin himself said it: "What a book a devil's chaplain might write on the clumsy, wasteful, blundering, low, and horridly cruel works of nature."

Of course that's bleak, but there's no law saying the truth has to be cheerful; no point shooting the messenger—science—and no sense in preferring an alternative world view just because it feels more comfortable. In any case, science isn't all bleak. Nor, by the way, is science an arrogant know-all. Any scientist worthy of the name will warm to your quotation from Socrates: "Wisdom is knowing that you don't know." What else drives us to find out?

What saddens me most, Sir, is how much you will be missing if you turn your back on science. I have tried to write about the poetic wonder of science myself, but may I take the liberty of presenting you with a book by another author? It is *The Demon-Haunted World* by the lamented Carl Sagan. I'd call your attention especially to the subtitle: *Science as a Candle in the Dark.*

*LEE M. SILVER**

THE ENVIRONMENT'S BEST FRIEND: GM OR ORGANIC?

Pigs raised on farms are dirty, smelly animals. Shunned by Jews and Muslims for millennia, they get no respect from most other people as well.

It's also not just our senses, though, that pig farms assault: it's the environment. Pig manure is loaded with phosphorus. A large sow can secrete from 18 to 20 kg per year, which runs off with rainwater into streams and lakes, where it causes oxygen depletion, algal blooms, dead fish, and greenhouse gases.[1] Ecological degradation has reached crisis levels in heavily populated areas of northern Europe, China, Korea, and Japan.

THE COST OF DIETARY PROTEIN

The problem is that—unlike cows, goats, and sheep—farmed pigs cannot extract sufficient phosphorus from the corn and other grains they are fed. Grains actually contain plenty of phosphorus, but it is mostly locked away in a large chemical called *phytate*, which is inaccessible to digestion by animal enzymes. Ruminants release the phosphorus during a lengthy digestive process in four stomachs, with the help of symbiotic bacteria.

To survive, the pig's wild ancestors depended on a varied diet, including worms, insects, lizards, roots, and eggs. But pig farming is most efficient with a simpler all-grain diet, supplemented with essential minerals. Although this feeding strategy works well for the animals, the inaccessible phosphorus in the grain passes entirely into the manure, which farmers use as crop fertilizer or otherwise distribute onto the land.

Today, in most rich and poor countries alike, pigs provide more dietary pro-

---

*Reprinted with permission of the Food and Drug Law Institute.

tein more cheaply and to more people than any other animal. Worldwide, pork accounts for 40 percent of total meat consumption.[2] While northern Europe still maintains the highest pig-to-human ratio in the world (2 to 1 in Denmark), the rapidly developing countries of east Asia are catching up. During the decade of the 1990s alone, pork production doubled in Vietnam and grew by 70 percent in China.

Along the densely populated coastlines of both countries, pig density exceeds 100 animals per square kilometer, and the resulting pollution is "threatening fragile coastal, marine habitats including mangroves, coral reefs, and sea grasses."[3] As the spending power of people in developing Asian countries continues to rise, pig populations will almost certainly increase further.

Pig-caused ecological degradation is a complex problem, and no single solution is in the offing. But any approach that allows even a partial reduction in pollution should be subject to serious consideration by policy makers and the public.

A prototypical example of what directed genetic modification (GM) technology can deliver is the transgenic Enviropig, developed by Canadian biologists Cecil Forsberg and John Phillips (see fig. 9.1). Forsberg and Phillips used an understanding of mammalian gene regulation to construct a novel DNA molecule programmed for specific expression of the *E. coli* phosphorus-extraction gene (*phytase*) in pig saliva. They then inserted this DNA construct into the pig genome.[4]

The results obtained with the first generation of animals were dramatic: the newly christened *Enviropigs* no longer required any costly dietary supplements and the phosphorus content of their manure was reduced by up to 75 percent. Subtle genetic adjustments could yield even less-polluting pigs, and analogous

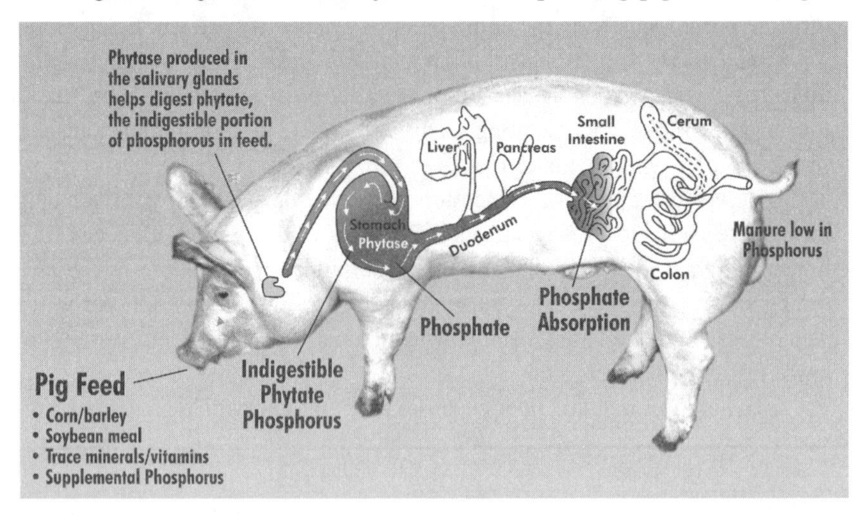

Figure 9.1. Courtesy of University of Guelph. Used by permission Macmillan Publishers Ltd: *Nature* 429: 119 © 2004.

genetic strategies can also be imagined for eliminating other animal-made pollutants, including the methane released in cow belches, which is responsible for 40 percent of total greenhouse gas emissions from New Zealand.[5]

ENZYMES IN NATURAL BACTERIA

An added advantage with the Enviropig is that the single extra enzyme in its saliva is also present naturally in billions of bacteria inhabiting the digestive tract of every normal human being. As bacteria continuously die and break apart, the naked enzyme and its gene both float free inside us without any apparent consequences, suggesting that the Enviropig will be as safe for human consumption as non-GM pigs. If the enzyme happened to escape into meat from modified pigs, it would be totally inactivated by cooking.

New varieties of animals with natural mutations undergo no safety testing at all.

Of course, empirical analysis is required to show that the modification does not make the meat any more harmful to human health than it would be otherwise. With modern technologies for analyzing genomes, transcriptosomes, proteomes, and metabolomes, any newly constructed transgenic animal can be analyzed in great molecular detail. "Isn't it ironic," Phillips and Forsberg commented, "that new varieties of animals with extreme but natural mutations undergo no safety testing at all?"[6]

ENVIRONMENTALLY FRIENDLY GM

Not all GM projects are aimed specifically at reducing the harmful effects of traditional agriculture on the environment. Other GM products approved to date, developed almost entirely in the private sphere, have aimed to reduce production costs on large-scale farms. But as molecular biology becomes ever more sophisticated, the universe of potential environmentally friendly GM applications will expand.

Scientists have begun research toward the goal of obtaining pigs modified to digest grasses and hay, much as cows and sheep do, reducing the land and energy-intensive use of corn and soy as pig feed. Elsewhere, trees grown on plantations for paper production could be made amenable to much more efficient processing. This would reduce both energy usage and millions of tons of the toxic chemical bleach in effluents from paper mills.[7]

The most significant GM applications will be ones that address an undeniable fact: every plot of land dedicated to agriculture is denied to wild ecosystems and species habitats. And that already amounts to 38 percent of the world's landmass. Genetic modifications that make crop production more efficient would give us opportunities to abandon farmland that, in many cases, should cede it back to forests and other forms of wilderness, as long as world population growth is ameliorated.

So why are many environmentally conscious people so opposed to all uses of GM technology? The answer comes from the philosophy of organic food promoters, whose fundamental principle is simply stated: *natural* is good; *synthetic* is bad.[8]

THE ROOTS OF ORGANIC FARMING

Before the eighteenth century, the material substance of living organisms was thought to be fundamentally different—in a vitalistic or spiritual sense—from that of nonliving things. Organisms and their products were *organic* by definition, while nonliving things were *mineral* or *inorganic*. But with the invention of chemistry, starting with Lavoisier's work in 1780, it became clear that all material substances are constructed from the same set of chemical elements.

As all scientists know today, the special properties of living organic matter emerge from the interactions of a large variety of complex, carbon-based molecules. Chemists now use the word organic to describe all complex, carbon-based molecules—whether or not they are actually products of any organism.

Through the nineteenth and twentieth centuries, increased scientific understanding, technological innovations, and social mobility changed the face of American agriculture. Large-scale farming became more industrialized and more efficient. In 1800, farmers made up 90 percent of the American labor force; by 1900, their proportion had decreased to 38 percent, and in 1990, it was only 2.6 percent.

However, not everyone was happy with these societal changes, and there were calls in the United States and Europe for a return to the preindustrial farming methods of earlier times. This movement first acquired the moniker organic in 1942, when J. I. Rodale began publication in America of *Organic Farming & Gardening,* a magazine still in circulation today.

According to Rodale and his acolytes, products created by—and processes carried out by—living things are fundamentally different from lab-based processes and lab-created products. The resurrection of this prescientific, vitalistic notion of organic essentialism did not make sense to scientists who understood that every biological process is fundamentally a chemical process. In fact, *all* food, by definition, is composed of organic chemicals. As a result, the US Department of Agriculture (USDA) refused to recognize organic food as distinguishable in any way from *nonorganic* food.

LEGISLATING MEANING

In 1990, lobbyists for organic farmers and environmental groups convinced the US Congress to pass the Organic Foods Production Act, which instructed the USDA to establish detailed regulations governing the certification of organic farms and foods.[9] After 12 years of work, the USDA gave organic farmers the certification standards that they wanted to prevent supposed imposters from using the word organic on their products.[10] Similar organic standards have been implemented by the European Commission and by the Codex Alimentarius Commission of the United Nations.[11]

In all instances, organic food is defined not by any material substance in the food itself, but instead by the so called natural methods employed by organic farmers. The USDA defines organic in essentially negative terms when it says, "the [organic] product must be produced and handled without the use of synthetic substances" and without the use of synthetic processes. The Codex Commission explains in a more positive light that "organic agriculture is a holistic production management system."

The physical attributes of organic products—and any effects they might have on the environment or health—are explicitly excluded from the definition. Nonetheless, the definitions implicitly assume that organic agriculture is *by its very nature* better for the environment than conventional farming.

The European Commission states as a fact that "organic farmers use a range of techniques that help sustain ecosystems and reduce pollution." Yet, according to self-imposed organic rules, precision genetic modification of any kind for any purpose is strictly forbidden, because it is a synthetic process. If conventional farmers begin to grow Enviropigs—or more sophisticated GM animals that reduce nearly all manure-based pollution—organic pig farmers will then blindly continue to cause much more pollution per animal, unless they are prevented from doing so by future EPA regulations.

Many organic advocates view genetic engineering as an unwarranted attack not just on the holistic integrity of organic farms, but on nature as a whole. On the other hand, spontaneous mutations caused by deep-space cosmic rays are always deemed acceptable since they occur "naturally." In reality, laboratory scientists can make subtle and precise changes to an organism's DNA, while high-energy cosmic rays can break chromosomes into pieces that reattach randomly and sometimes create genes that didn't previously exist.

Regardless, organic enthusiasts maintain their faith in the beneficence and superiority of nature over any form of modern biotechnology. Charles Margulis, a spokesman for Greenpeace USA, calls the Enviropig "a Frankenpig in disguise."[12]

CHEMICAL PESTICIDES AND ORGANIC FARMING

Although the market share held by organic products has yet to rise above single digits in any country, it is growing steadily in Europe, the United States, and Japan. Nearly all consumers assume that organic crops are, by definition, grown without chemical pesticides. However, this assumption is false.

Pyrethrin ($C_{21}H_{28}0_3$), for example, is one of several common toxic chemicals sprayed onto fruit trees by organic farmers—even on the day of harvesting. Another allowed chemical, rotenone ($C_{23}H_{22}0_6$), is a potent neurotoxin, long used to kill fish and recently linked to Parkinson's disease.13

How can organic farmers justify the use of these and other chemical pesticides? The answer comes from the delusion that substances produced by living organisms are not really chemicals, but rather organic constituents of nature. Since pyrethrin is produced naturally by chrysanthemums and rotenone comes from a native Indian vine, they are deemed organic and acceptable for use on organic farms.

However, the most potent toxins known to humankind are all-natural and organic. They include ricin, abrin, botulinum, and strychnine—highly evolved chemical weapons used by organisms for self-defense and territorial expansion. Indeed, every plant and microbe carries a variety of mostly uncharacterized, more or less toxic attack chemicals, and synthetic chemicals are no more likely to be toxic than natural ones.

LESS-ALLERGENIC GM FOOD

All currently used pesticides—both natural and synthetic—dissipate quickly and pose a miniscule risk to consumers. Nevertheless, faith in nature's beneficence can still be fatal to some children. About 5 percent express severe allergic reactions to certain types of natural food. Every year unintentional ingestion causes hundreds of thousands of cases of anaphylactic shock with hundreds of deaths.

The triggering agents are actually a tiny number of well-defined proteins that are resistant to digestive fluids. These proteins are found in such foods as peanuts, soybeans, tree nuts, eggs, milk, and shellfish. They linger in the intestines long enough to provoke an allergic immune response in susceptible people.

No society has been willing to ban the use of any allergenic ingredients in processed foods, even though this approach could save lives and reduce much human suffering. GM technology could offer a more palatable alternative: scientists could silence the specific genes that code for allergenic proteins. The subtly modified organisms would then be tested, in a direct comparison with unmodified organisms, for allergenicity as well as agronomic and nutritional attributes.

USDA-supported scientists have already created a less-allergenic soybean. Soy is an important crop used in the production of a variety of common foods, including baby formula, flour, cereals, and tofu. Eliot Herman and his colleagues embedded a transgene into the soy genome that takes advantage of the natural RNA interference system to turn off the soy gene responsible for 65 percent of allergic reactions.[14]

RNA interference can be made to work in a highly specific manner, targeting the regulation of just a single gene product. Not only was the modified soy less allergenic in direct tests, but the plants grew normally and retained a complex biochemical profile that was unaltered except for the absence of the major allergen. Further rounds of genetic surgery could eliminate additional allergenic soy proteins. Other scientists have reported promising results in their efforts to turn off allergy-causing genes in peanuts and shrimp.

Some day perhaps, conventional soy and peanut farmers will all switch production to low-allergenicity GM crop varieties. If that day arrives, organic food produced with *GM-free organic* soy or peanuts will be certifiably *more* dangerous to human health than comparable nonorganic products.

Unfortunately, conventional farmers have no incentive to plant reduced-allergy seeds when sales of their current crops are unrestricted, especially when the public has been led to believe that all genetic modifications create health risks. In the current social and economic climate, much of the critical research required to turn promising results into viable products is simply not pursued. Anti-GM advocates for organic food may be indirectly and unknowingly responsible for avoidable deaths in the future.

VEGETARIAN MEAT

Only three decades have passed since genetic modification technology was first deployed in a rather primitive form on simple bacteria. The power of the technology continues to explode with no end in sight, leading to speculation about how agriculture could *really* be transformed in the more distant future.

Chicken meat is actually cooked muscle, and muscle is a type of tissue with a particular protein, composition, and a particular structure. At some future date, as the power of biotechnology continues to expand, our understanding of plant and animal genes could be combined with the tools of genetic modification to create a novel plant that grows appendages indistinguishable in molecular composition and structure from chicken muscles.

Vegetative chickens—or perhaps muscular vegetables—could be grown just like other crops. Eventually, there could be fields of chicken, beef, and pork plants. At harvest time, low-fat boneless meat would be picked like fruit from a tree.

The advantages of genetically engineered vegetative meat are numerous and diverse. Without farm animals, there could be no suffering from inhumane husbandry conditions and no pollution from manure. Since the sun's rays would be used directly by the plant to make meat, without an inefficient animal intermediate, far less energy, land, and other resources would be required to feed people.

Up to 20 percent of the earth's landmass currently used for grazing or growing animal feed might be ceded back to nature for the regrowth of dense forests. As a result, biodiversity would expand, the extinctions of many species might be halted, and a large sink for extracting greenhouse gases from the atmosphere might be created (see fig. 9.2).

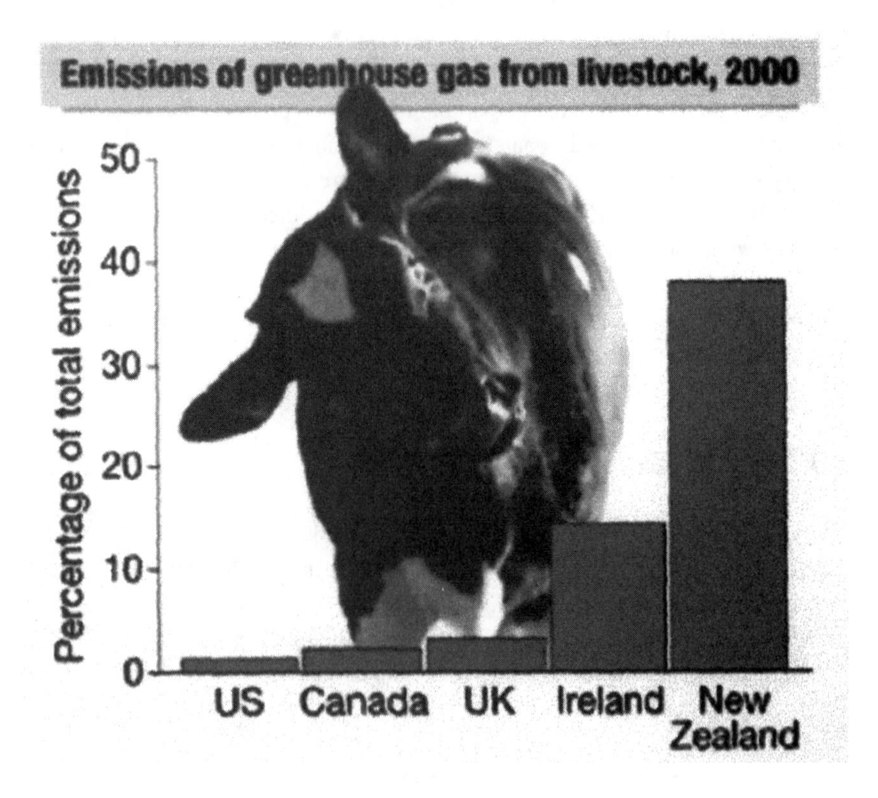

Figure 9.2. Forty percent of the greenhouse gas emissions in New Zealand come from livestock. (Used by permission Macmillan Publishers Ltd: *Nature* 429: 119 © 2004.)

Of course, this scenario is wild biotech speculation. But current-day organic advocates would reject any technology of this kind out of hand, even if it was proven to be beneficial to people, animals, and the biosphere as a whole. This categorical rejection of all GM technologies is based on a religious faith in the beneficence of nature and her processes under all circumstances, even when science and rationality indicate otherwise.

REFERENCES

1. OECD, "Agriculture, Trade and the Environment: The Pig Sector," *OECD Observer* (September, 2003).

2. Ibid.

3. FAO, *Livestock Policy Brief 02: Pollution from Industrialized Livestock Production* (Rome: UN Food and Agriculture Organization, 2006).

4. S. P. Golovan, R. G. Meidinger, A. Ajakaiye, et al., "Pigs Expressing Salivary Phytase Produce Low-Phosphorus Manure," *Nature Biotechnology* 19 (2001): 741–45.

5. C. Dennis, "Vaccine Targets Gut Reaction to Calm Livestock Wind," *Nature* 429 (2004): 119.

6. Personal communication.

7. V. L. Chiang, "From Rags to Riches," *Nature Biotechnology* 20 (2002): 557–58; G. Pilate, E. Guiney, and K. Holt, et al., "Field and Pulping Performances of Transgenic Trees with Altered Lignification," *Nature Biotechnology* 20 (2002): 607–12.

8. H. Verboog, M. Matze, E. Lammerts van Bueren, and T. Baars, "The Role of the Concept of the Natural (Naturalness) in Organic Farming," *Journal of Agricultural and Environmental Ethics* 16 (2003): 29–49.

9. M. Burros, "A Definition at Last, but What Does It All Mean?" *New York Times*, October 16, 2002.

10. USDA, *National Organic Program* (Washington, DC: US Department of Agriculture, 2000).

11. Codex Alimentarius Commission, *Guidelines for the Production, Processing, Labeling and Marketing of Organically Produced Foods* (Rome: U.N. Food and Agriculture Organization, 1999); EUROPA. *What Is Organic Farming?* (Belgium: European Commission, 2000).

12. C. Osgood, "Enviropigs May Be Essential to the Future of Hog Production," *Osgood File*, CBS News, August 1, 2003.

13. R. Betarbet, T. B. Sherer and G. MacKenzie, et al., "Chronic Systemic Pesticide Exposure Reproduces Features of Parkinson's Disease," *Nature Neuroscience* 3 (2000): 1301–1306; M. B. Isman, "Botanical Insecticides, Deterrents, and Repellents in Modern Agriculture and an Increasingly Regulated World," *Annual Review of Entomology* 51 (2006): 45–66.

14. E. M. Herman, R. M. Helm, R. Jung, and A. J. Kinney, "Genetic Modification Removes an Immunodominant Allergen from Soybean," *Plant Physiology* 132 (2003): 36–43.

DAVID CASTLE

ACCEPTANCE OF BIOTECHNOLOGY IN A RISK SOCIETY

INTRODUCTION

Biotechnology has been used in food production for thousands of years, but modern agricultural biotechnology elicits concerns one would not associate with the fermentation of beer, leavening of bread, and ripening of cheese. Proponents of modern agricultural biotechnology, particularly the genetic modification of food, often claim that biotechnology based on genomics and molecular genetics is on a continuum with its more humble forebears—we have just become a great deal more sophisticated in how we manipulate nature to make food. Opponents claim that the degree of human involvement in altering fundamental life processes in the laboratory makes these qualitatively different than using micro-organisms the brewery, bakery or cheesery.

The issue often turns on whether so-called "biotech foods" are less, or are no longer natural, whether "natural" and "synthetic" are proxies for "safe to eat" and "environmentally sustainable," and whether these in turn are proxies for "ethical" or "unethical food" choice. Some argue that these are questions to be answered by science, but others disagree that science arbitrates what is reasonable to think about new biotechnology. Science can show, for example, whether a biotechnology works as intended, but a decision to go ahead with it, particularly if there are new risks, is a decision that cannot be made on scientific grounds alone. The additional grounds can include non-scientific evidence, beliefs and attitudes, and can bring critical attention on the acceptability of the raison d'être of the innovation. Even in cases where scientific evidence suggests that benefits far outweigh the risks of a new biotechnology, it may be reasonable to forego a new biotechnology if the grounds for its existence are socially unacceptable.

WHY MAKE A GENETICALLY MODIFIED PIG?

In Canada alone, nearly 20 million hogs are slaughtered each year, which accounts for roughly 30 percent of livestock production in Canada (CPI 2005). Selective hog breeding has generated low-fat, fast growing and docile animals that are well-suited to intensive production techniques. Intensification of hog operations increases efficiency and profitability, but comes with more waste. Traditionally, waste from small hog farms is used to fertilize nearby fields and croplands because of its characteristically high nitrogen and phosphorous content. At industrial production levels an excess of manure can easily exceed environmentally safe applications as a fertilizer. When the amount of phosphorus applied to a field exceeds the soil's capacity to bind the nutrient, the excess leeches. Phosphorous can then find its way into water ecosystems, either through ground water leeching or through surface run-off.

The result is excessive nutrient loading or eutrophication of aquatic ecosystems which results in algal growth. Algal blooms are disastrous to freshwater systems because they deplete the oxygen content of the water, suffocating other aquatic life forms (Mallin 2000). In coastal marine ecosystems increased nutrient levels encourage toxic red algal species such as Pfisteria. The effects are often dramatic, as for example in the state of Santa Catarina in Brazil where nearly 85 percent of the waterways are contaminated with fecal organisms, and high nutrient levels are largely due to the nearly 30 million pigs produced annually in this region (Forsberg 2003). Similar environmental degradation is taking place in China, the world leader in hog production with nearly 465 million pigs produced annually. Concentrated farming operations surrounding Lake Taihu, the third largest lake in China, have resulted in eutrophication from excessive phosphorus (Forsberg 2003).

Researchers at the University of Guelph have developed a novel solution to the dual problem of swine nutrition and environmental pollution. The Enviropig is a common pig variety that has been genetically modified to secrete phytase in its saliva at levels at or above those previously supplemented to pig feed. Enviropigs do not require phosphate supplementation top-dressed in their feed, and they do not require enzyme supplements to break down plant phytate. When non-transgenic pigs were compared to transgenic pigs, the results indicate that the amounts of phosphorus in the manure of the transgenic pigs were 64 percent (gilts) and 67 percent (boars) lower than the amounts found in the non-transgenic manure, and there is a slight feed conversion advantage and hence reduced production cost (Golovan 2001). From an engineering standpoint, these are impressive results, and strictly speaking they could be part of a strategy to reduce nutrient loading of the environment. Were one to take two hog operations, one with conventional pigs and another with Enviropigs, with conventional waste management systems in place for each, the Enviropig operation would mean potentially two-thirds reduction in the release of a major pollutant.

LINES OF ARGUMENT

Suppose, for the sake of argument, that not only are Enviropigs safe to eat, but they offer the same nutrients and qualitative attributes as conventional pork products. They only differ with respect to phosphate metabolism, where they offer clear environmental benefits. Should everyone be convinced of the Enviropig's merits, or is recognizing the benefits of a feat of engineering a separate and perhaps partial consideration in making the decision to adopt the biotechnology? This is a difficult question to answer because it requires consideration of social and scientific dimensions of the problem. As it happens, much of the dispute about agricultural biotechnology and related terms is split into two main lines of argument. One line is amenable to the tools of science, and focuses on whether there are environmental and food safety benefits or risks that need to be considered. The questions here include: Are there novel allergens? Are there non-target effects on other species? Is this better for soil quality? Are these foods more nutritious than the others? People who believe that adopting biotechnology is primarily a matter of scientific risk assessment tend to exclude non-scientific evidence, attitudes, or belief from the discussion. The other line of argument addresses the social dimension of agri-food biotechnology, and includes questions like: Who benefits from this production technique? What does my religion tell me to eat? Will this technology disrupt or enhance rural communities? Is my qualitative experience of food changing? People interested in these issues tend to take seriously science's role in evaluating risks of new biotechnology, but couch that analysis in a more inclusive framework of evidence, attitudes, and beliefs used to make a decision.

There may be empirical answers, on a case-by-case basis, about the relative safety and sustainability of conventional and biotech foodstuffs, but anyone involved long enough in the debate about genetically modified foods knows that selecting and assembling the available evidence for the scientific support of new technology can be selective. Pro-biotechnology science progressivists like Norman Borlaug (and his acolytes) will show population growth curves and project the food shortfall for a population of 9 billion in 2050 unless biotech crops are deployed world-wide. Advocates of traditional, local forms of organic farming like the anti-globalization advocate, Vandana Shiva (and her acolytes), will point to the depletion of biodiversity in rural communities, soil erosion, and negative impact on nutrient intake. In each case, the data and techniques used to substantiate a case for or against biotech food are carefully chosen to advance a position founded on a blend of scientific and social considerations.

World views clash over agricultural biotechnology because of the complex and potent mixture of attitudes about the limits of science and future of society just described. The Enviropig provides an interesting and instructive case of how divergent viewpoints can be. An engineering success, the modified pig delivers such significant phosphate reduction that it has been conjectured that this biotechnology should be supported by even those normally opposed to genetically modified foods. This position, argued by Lee Silver, presumes that the pig will be shown to be safe to eat, and so it would be irrational on scientific grounds to forego the benefits of the technology (Silver 2006a). Silver's argument nicely illustrates the view that a scientific claim about the positive effects of biotechnology can, and ought to, outweigh what he considers to be non-scientific convictions of biotechnology opponents.

Silver argues that the genetic modification of agricultural organisms can lead to environmental benefits. If environmentally beneficial biotechnologies are used to improve upon farming practices known to be harmful, Silver believes that the biotechnology ought to be adopted. "Pig-caused ecological degradation is a complex problem, and no single solution is in the offing. But any approach that allows even a partial reduction in pollution should be subject to serious consideration by policy makers and the public (Silver 2006a)." While public and policy makers ought to consider this solution, and have been consulted (Castle 2005), Lee is mostly interested in convincing the organic movement that they should consider this technology seriously. The problem, as Silver sees it, is that there will be opponents to this technology who will not be persuaded to adopt the technology because of scientific evidence indicating an environmental benefit.

Opposition to the genetically modified pig, claims Silver, will arise with promoters of organic food whose fundamental principle is that "*natural* is good; *synthetic* is bad (Silver 2006a)." Silver attributes this position to J. I. Rodale, author of several books on the merits of composting and its relation to the preservation and promotion of soil quality. Silver thinks that the organic movement can be identified with vitalism, a conceptual viewpoint from which advocates for organic farming practices would distinguish organic agricultural processes and products from "inorganic" or "synthetic" food. It mattered to Rodale that inputs into agriculture came from natural systems as opposed to synthesis by humans. Silver attributes this view to current organic farmers, for whom the "fundamental distinction between organic and nonorganic food lies within the distinction between natural process or substances and *unnatural*, *synthetic*, or *artificial* processes . . ." Silver 2006b). This conception of organic has its roots in the work of Rudolph Steiner, who Silver describes as having "a scientific-sounding argument for the superiority of what we now call organic farming and organic food" (Silver 2006b).

> It is not just observant Catholics and other fundamentalists who are swayed, in at least some vague way, by notions of natural law. In western societies, nearly everyone perceives *natural* as a synonym for goodness, and *unnatural* as bad or wrong. Although left-leaning secularists typically imagine the natural in terms of soft feminine ecosystems rather than stern commandments from a masculine God, they respond to the same kind of language as those on the right. (Silver 2006b)

From these considerations, Silver concludes that the organic movement implicitly assumes "that organic agriculture is *by its very nature* better for the environment than conventional farming." The specter of a "Frankenpig" is, on this view, the apotheosis of the organic movement because it violates the laws of nature. Consequently, it can neither be safe, environmentally sound, nor an ethical food choice. Silver concludes that "current-day organic advocates would reject any technology of this kind out of hand, even if it was proven beneficial to people, animals, and the biosphere as a whole. This categorical rejection of all GM technologies is based on a religious faith in the beneficence of nature and her processes under all circumstances, even when science and rationality indicate otherwise" (Silver 2006a).

Pinning a single viewpoint to the organic movement is rather like saying there is a definitive version of feminism or Christianity. No doubt some take mystical conceptions of Gaia as literal truth and foolishly reject science out-right, but other conceptions of organic are more complementary with science. Rodale, and the foundation that bears his name, do not explicitly advocate for organic farming on the basis of a distinction, rooted in hostility to science, between natural and synthetic. Their interest in comparing the input costs and agricultural benefits of biodynamic and petro-chemical derived fertilizers is motivated in part by a presumption about the benefits of natural processes, but the methodology is more scientific than Silver admits. Consequently, Silver misses the mark when he attempts to turn tables by arguing that some natural substances (rotenone), processes (cosmic radiation), and organic foods (organic peanuts) are more dangerous than chemical pesticides or genetically modified crops. Indeed, the USDA regulated conception of "organic" makes no appeal to reified conceptions of nature, and organizations hoping to establish their own organic standards for animals often take scientific, not pop conceptions of ecology, as a cornerstone of their definition of organic (IFOAM 2006).

To summarize this discussion, several points can be made. First, the opposition between biotechnology and organic as worldviews is apparent in Silver's approach, and one might even go so far as to say that Silver's account perpetuates the rhetoric in both lines of argument. Second, the use of terms as proxies for other concepts is problematic for both advocates and opponents of biotechnology. "Science-based" or "rational" may not mean "safe" or "ethical," just as much as "natural" or "organic" need not mean "safe" or "ethical." More importantly, however, the terms of the debate ("natural" and "synthetic," "safe to eat" and "environmentally sustainable," "ethical" or "unethical food") prevent a more substantive and meaningful discussion about managing new risks from arising. Third, science-supported claims about the benefits of a biotechnology are not definitive and exclusionary reasons for adopting a new biotechnology, and neither is optimism about future technological developments in which "the universe of potential environmentally-friendly GM applications will expand (Silver 2006a)." There may be definitive and exclusionary scientific reasons for opting out of new biotechnology use, such as concerns that the risk of unforeseeable adverse events would outweigh the biotechnology's benefits. Equally, there can be definitive and exclusionary non-scientific reasons for opting out of new biotechnology use, usually when other social considerations are considered more important than the alleged benefits of the new technology.

ACCEPTING RISK

The motivation for developing and using a biotechnology like the modified pig is to address the harmful aspects of an antecedent technology—intensive hog farming. Enviropig technology introduces new risks, associated with the food safety, along with the intended environmental benefits. Silver's approach is to focus on the scientific evidence about the environmental benefits of genetically modified animals, the food safety issues, the naturalness of biotechnology, and so forth, because he is interested in showing that scientific rationality can defeat anti-biotechnology positions anchored in what he considers religious mysticism. This approach might win the battle, but will not win the war. It actually misses the war altogether because the issue is not about the merits of scientific versus non-scientific thought. The issue is about whether it is sensible to adopt a new biotechnology that introduces new risks, however small they may turn out to be, to correct the ills of an antecedent technology. It might be reasonable to object to a fresh overlay of new risks atop of an already objectionable set of existing risks.

Positivist conceptions of science, for which Silver has some sympathy, hold that science, and scientific reasoning, is the only source of truth. On this view, non-scientific reasoning, such as metaphysics or religion, is the source of social problems that science must clean up. Contrary to the claims of the positivists, social constructivists point out that science and society are in constant interaction, which is why historical contingencies can play a role in the development, retention, and change in scientific theories. Although it is still a point of contention, it can be argued that social constructivism does not imply that the content of science is contingent, just the processes by which science is generated. This view of science provides sufficient bedrock for Rivet's view that scientific research and technological developments may help to solve social problems, but the activity of research and development also generates social problems that cannot be separated from the R&D process itself (Rivets 1970). What is particularly interesting is that some of the effects on society caused by new science and technology are only discoverable through further scientific investigation. The discovery of the damage caused by chlorofluorocarbons on the ozone layer, and the impact that has on rates of skin cancer, is a case in point.

The use of science to evaluate the social effects of science and technology has been called "reflexive modernity" by Ulrich Beck (Beck 1992). In a risk society, "modern society becomes reflexive, that is becomes both an issue and a problem for itself" (Beck 1998). This seemingly abstract idea is really very concrete in cases like the BSE crises in Europe:

> Risk society begins where nature ends. As Giddiness has pointed out, this is where we switch the focus of our anxieties from what nature can do to us to what we have done to nature. The BSE crisis is not simply a matter of fate but a matter of decisions and options, science and politics, industries, markets and capital. This is not an outside risk but a risk generated right inside each person's life and inside a variety of institutions. A central paradox of risk society is that these internal risks are generated by the processes of modernization which try to control them. (Beck 1998)

A consequence of living in reflexive modernity with risks like BSE is that the constant engagement with science and technology, and the need to weigh the benefits and risks associated with new initiatives, creates a society accustomed to using the language of science and technology in the evaluation of social, as well as personal, risk.

If the activity of science generates social problems, then citizens affected by scientific research and technological development should be part of the "extended peer group" that assesses what science and technology projects are permitted in society (Funtowicz and Ravetz 1993). Failure to do so is not only undemocratic (Castle and Culver 2006), it also erodes trust in public institutions which could undermine their authority to make decisions (De Marchi and Ravetz 1999). Science no longer arbitrates the relationship between it and society. Instead, authority for making decisions about the science and technology introductions is devolved to publics capable of engaging the problem of reflexive modernity from within the set of norms that guide their lives.

The Enviropig presents a solution to a problem generated by antecedent applications of science, but the solution comes with new risks associated with food safety. These risks are now being investigated by the scientists who developed the Enviropig in partnership with the Canadian government. The principle concern is that the Enviropig might produce, in addition to the intended products of the inserted transgene, novel, allergenic proteins. There are ways to evaluate potential allergenicity by screening for high molecular weight and thermo-stable proteins, but the only real test is to wait and see after consuming the product. In this scenario, it is not unreasonable for people to think twice about using the technology. As it is, conventional industrialized hog farms involve some objectionable practices and have serious consequences for the environment. The Enviropig is a technology which would perpetuate industrialized farming, or perhaps permits its expansion where phosphorous rich manure is the limiting factor. So if one has reason to object to the conventional practice, one should object to the adoption of a technology that extends the practice. Since the new technology also introduces new risks that cannot be eliminated with certainty, it is not unreasonable for people to wish to forego the technology.

CONCLUSION

One way that opposition to new agricultural biotechnology can arise is response to innovations that are developed and used to mitigate or prevent the harmful consequences of antecedent technologies. Asking people if they want to adopt new technology to partially fix a problematic old technology, and accept new risks in the process, is democratic and autonomy-enhancing. One has to be prepared for a possible negative response. Of course, it is possible that in saying "no" the real objective is to bring about sweeping reform in agriculture. In crop biotechnology, for example, genetic modification of plants to make them herbicide and pesticide resistant is challenged because these modifications perpetuate the harmful effects of crop monocultures. Stop crop monoculture, and genetic modification of plants is no longer necessary. It is conceivable that some might try to block the use of the Enviropig to press for reforms to industrialized hog farming. It is true that preemptive reasoning of this kind can be used to stave off new biotechnology to gain time and refocus the debate on bringing about reform to convention practices. It is difficult to imagine, however, that there will be significant reform of intensive, industrialized farming practices in the near future. A technology like the Enviropig is thus all the more attractive because it offers a proximate solution to the broad problem. But if all science does is speak to this proximate issue, questions about impact in the broader context of risks and benefits go unanswered. Surely no one wishes that decisions about important matters like how our food is produced are made using scientific instrumentalism to decide what can and what cannot count as a good reasons for making decisions about what risks to accept.

REFERENCES

Beck, U. *Risk Society: Towards a New Modernity*. London: Sage, 1992.

———. "Politics of Risk Society." In *The Politics of Risk Society*, edited by J. Franklin, 9–22. Cambridge: Polity, 1998.

Castle, D. "The Balance between Expertise and Authority in Citizen Engagement about New Biotechnology." *Techne* 9 (2006): 1–13.

Castle, D., K. Finlay, and S. Clark. "Proactive Consumer Consultation: The Effect of Information Provision on Response to Transgenic Animals." *Journal of Public Affairs* 5 (2005): 200–16.

Castle, D., and K. Culver. "Public Engagement, Public Consultation, Innovation and the Market." *The Integrated Assessment Journal* 6 (2006): 137–52.

Canada Pork International. "Canadian Hog Production: Numbers of Hogs Slaughter in Canada 1998–2001." November 15, 2005. http://www.canadapork.com/english/pages/frmsts/page03.html

De Marchi, B., and J. Ravetz. "Risk Management and Governance: A Post-normal Science Approach." *Futures* 31 (1999): 743–57.

Forsberg, C. W., et al. "Genetic Opportunities to Enhance Sustainability of Pork Production in Developing Countries: A Model for Food Animals." Food and Agriculture Organization of the United Nations, 2003.

Funtowicz, S., and J. Ravetz. "Science for the Post-normal Age." *Futures* (September, 1993): 739–55.

Golovan, S., et al. "Pigs Expressing Salivary Phytase Produce Low Phosphorous Manure." *Nature Biotechnology* 19 (2001): 741–45.

International Federation of Organic Agriculture Movements (IFOAM). "The Principles of Organic Aquaculture." http://www.ifoam.org/about_ifoam/principles/index.html. (Accessed August 23, 2006.)

Mallin, M. A. "Impacts of Industrial Animal Production on Rivers and Estuaries." *American Scientist* (January–February, 2000): 26–37.

Ravetz, J. *Scientific Knowledge and Its Social Problems*. Oxford: Clarendon Press, 1970.
Silver, L. "The Environment's Best Friend: GM or Organic?" *Update: New York Academy of Sciences Magazine* (May/June 2006a): 14–17.
———. *Challenging Nature: The Clash of Science and Spirituality at the New Frontiers of Life*. New York: Harper Collins, 2006b.

BIOLOGICAL METAPHORS

JAMES LOVELOCK*

WHAT IS GAIA?

Most of us sense that the Earth is more than a sphere of rock with a thin layer of air, ocean, and life covering the surface. We feel that we belong here as if this planet were indeed our home. Long ago the Greeks, thinking this way, gave to the Earth the name Gaia or, for short, Ge. In those days, science and theology were one and science, although less precise, had soul. As time passed this warm relationship faded and was replaced by the frigidity of the schoolmen. The life sciences, no longer concerned with life, fell to classifying dead things and even to vivisection. Ge was stolen from theology to become no more the root from which the disciplines of geography and geology were named. Now at last there are signs of a change. Science becomes holistic again and rediscovers soul, and theology, moved by ecumenical forces, begins to realize that Gaia is not to be subdivided for academic convenience and that Ge is much more than just a prefix.

The new understanding has come from going forth and looking back to see the Earth from space. The vision of that splendid white flecked blue sphere stirred us all, no matter that by now it is almost a visual cliché. It even opens the mind's eye, just as a voyage away from home enlarges the perspective of our love for those who remain there.

The first impact of those voyages was the sense of wonder given to the astronauts and to us as we shared their experience vicariously through television, but at the same time the Earth was viewed from outside by the more objective gaze of scientific instruments. These devices were quite impervious to human emotion yet they also sent back the information that let us see the Earth as a strange and beautiful anomaly. They showed our planet is made of the same elements and in much the same proportions as are Mars and Venus, but they also revealed our sibling planets to be bare and barren and as different from the Earth as a robin from a rock.

We now see that the air, the ocean and the soil are much more than a mere

*Source: http://www.ozi.com/ourplanet/lovelock2.html

environment for life; they are a part of life itself. Thus the air is to life just as is the fur to a cat or the nest for a bird. Not living but something made by living things to protect against an otherwise hostile world. For life on Earth the air is our protection against the cold depths and fierce radiations of space.

There is nothing unusual in the idea of life on Earth interacting with the air, sea and rocks, but it took a view from outside to glimpse the possibility that this combination might consist of a single giant living system and one with the capacity to keep the Earth always at a state most favorable for the life upon it.

An entity comprising a whole planet and with a powerful capacity to regulate the climate needs a name to match. It was the novelist William Golding who proposed the name Gaia. Gladly we accepted his suggestion and Gaia is also the name of the hypothesis of science which postulates that the climate and the composition of the Earth always are close to an optimum for whatever life inhabits it.

The evidence gathered in support of Gaia is now considerable but as is often the way of science, this is less important than is its use as a kind of looking glass for seeing the world differently, and which makes us ask new questions about the nature of Earth.

If we are "all creatures great and small," from bacteria to whales, part of Gaia then we are all of us potentially important to her well being. We knew in our hearts that the destruction of a whole range of other species was wrong but now we know why. No longer can we merely regret the passing of one of the great whales, or the blue butterfly, nor even the smallpox virus. When we eliminate one of these from Earth, we may have destroyed a part of ourselves, for we also are a part of Gaia.

There are many possibilities for comfort as there are for dismay in contemplating the consequences of our membership in this great commonwealth of living things. It may be that one role we play is as the senses and nervous system for Gaia. Through our eyes she has for the first time seen her very fair face and in our minds become aware of herself. We do indeed belong here. The earth is more than just a home, it's a living system and we are part of it.

*JAMES W. KIRCHNER**

THE GAIA HYPOTHESIS: FACT, THEORY, AND WISHFUL THINKING

1. INTRODUCTION

Several years ago I overheard a radio interview with Douglas Adams, the author of *The Hitchhiker's Guide to the Galaxy*, in which he was asked to comment on the Gaia hypothesis. His answer, as best I can reconstruct it now, was roughly this:

> Imagine a puddle, waking up in the morning, and examining its surroundings (a brief pause here, to let the audience grapple with this rather odd image). The puddle would say, "Well, this depression in the ground here, it's really quite comfortable, isn't it? It's just as wide as I am, it's just as deep as I am, it's the same shape as I am . . . in fact, it conforms exactly to me, in every detail. This depression in the ground, it must have been made just for me!"

Adams' fanciful image illustrates a central problem in some of the current incarnations of the Gaia hypothesis, and in all other theories that find, in Earth's obvious suitability for our particular form of life, evidence that the environment must be conforming to life's needs (e.g., Henderson 1913; Redfield 1958). The problem is this: given that organisms must adapt to the constraints of their environment—or else they don't survive—the particular forms of life that we observe will always be those that are reasonably well matched to their environmental conditions. Those that are not well matched to their environment will not thrive and will not be noticed.

Thus the environment and its life forms will always appear well suited to each other, whether or not the environment is in any sense adjusted to life's

*From James W. Kirchner, The Gaia Hypothesis: Fact, Theory, and Wishful Thinking," *Climatic Change* 52 (2002): 391–408.

requirements. It seems inevitable that sentient life should view its world as an Eden, because if there were any evolutionary lineages for which that world were a Hell, they would not persist long enough to develop intelligent life forms. To me, the Earth seems to be remarkably well suited to human needs. But I also understand that evolution has made it virtually inevitable that I should believe this, since any would-be ancestors of mine for whom the Earth were too hostile would have been removed from the gene pool long before their traits could have been passed on to me. Perhaps this helps to explain how the beneficence of nature has become a theme in human thought. As winners of the evolutionary lottery, it is not surprising that we would view ourselves and the life forms around us—with whom we share the winner's circle—as the beneficiaries of an environment has been tailored to our needs.

What makes the Gaia hypothesis interesting is that it proposes that the benificence of Nature is neither an accident nor the work of a benevolent diety, but instead is the inevitable result of interactions between organisms and their environment. Simply put, if organisms have a significant influence on their environment, then . . . those species of organisms that retain or alter conditions optimizing their fitness (i.e., proportion of offspring left to the subsequent generation) leave more of the same. In this way conditions are retained or altered to their benefit (Lovelock and Margulis 1974a). Life and the environment evolve together as a single system so that not only does the species that leaves the most progeny tend to inherit the environment, but also the environment that favors the most progeny is itself sustained (Lovelock 1986).

Thus the Gaia hypothesis validates our sense of wonder and reverence for the natural world, by proposing a scientific basis for our sense that the Earth is indeed tailored to our needs and those of the organisms that share the Earth with us.

Some have hailed Gaia as a profound discovery, while others have dismissed it as a "just-so story" that is more entertaining than informative. Here I will argue that Gaia, in its different guises, is a mixture of fact, theory, metaphor, and wishful thinking. It will be necessary to untangle these from each other, and put each in its proper place, in order to get a clearer view of the Earth system as it is, in all its intriguing complexity. It has been a dozen years since my critique of the Gaia hypothesis appeared in print (Kirchner 1989; Kirchner 1991). My intent was to clarify and focus the Gaia debate, which seemed to be at risk of becoming a shouting match between disciples and detractors. I pointed out that the singular term, "the Gaia hypothesis," was being applied to many different propositions, ranging from ideas that most modern Earth scientists would consider self-evident, to notions that most would consider outlandish. The weak forms of the Gaia hypothesis hold that life collectively has a significant effect on Earth's environment

("Influential Gaia"), and that therefore the evolution of life and the evolution of its environment are intertwined, with each affecting the other ("Coevolutionary Gaia"). I argued that abundant evidence supports these weak forms of Gaia, and that they are part of a venerable intellectual tradition (Spencer 1844; Huxley 1877; Hutchinson 1954; Harvey 1957; Holland 1964; Sillen 1966; Schneider and Londer 1984). By contrast, the strongest forms of Gaia depart from this tradition, claiming that the biosphere can be modeled as a single giant organism ("Geophysiological Gaia") or that life optimizes the physical and chemical environment to best meet the biosphere's needs ("Optimizing Gaia"). I argued that the strong forms of Gaia may be useful as metaphors but are unfalsifiable, and therefore misleading, as hypotheses (Kirchner 1989). Somewhere between the strongest and the weakest forms of Gaia is "Homeostatic Gaia," which holds that atmosphere-biosphere interactions are dominated by negative feedback, and that this feedback helps to stabilize the global environment. I argued that if one defines it carefully enough, Homeostatic Gaia may be testable, but I also pointed out that there was abundant evidence that the biota can also profoundly destabilize the environment.

How has the Gaia debate progressed? One can discern several encouraging trends. The view that organisms have no effect on their environment has not been taken seriously for several decades now. The most extreme forms of the Gaia hypothesis have generally been abandoned, particularly those that impute a sense of purpose to the global biosphere, and Gaia's proponents have instead searched for mechanisms by which Gaian regulation might evolve by natural selection. Thus the debate has become focused on a more promising (though somewhat ill-defined) middle ground between the mundane versions of Gaia and the extravagant ones. This is progress, and I wish that I could claim some credit for it, but Gaia enthusiasts have rarely come to grips with my critique (which extends well beyond just the taxonomy outlined above), and Gaia skeptics have generally ignored the debate entirely. Thus the Gaia debate has not been much of a dialogue, with Gaia's proponents repeatedly putting forward their case in print, and Gaia's detractors casually dismissing it rather than taking the time to respond. Therefore, my purpose here is to assess the current, more focused Gaia hypothesis, and to outline the difficulties that it poses.

Gaia contains elements of fact, theory, and wishful thinking. One part of Gaia that is clearly fact is the recognition that Earth's organisms have a significant effect on the physical and chemical environment. Biogeochemists have devoted decades of painstaking work to tracing the details of these interactions. Many important chemical constituents of the atmosphere and oceans are either biogenic or biologically controlled, and many important fluxes at the Earth's surface are biologically mediated (see Kirchner 1989, and references therein). This was well understood among biogeochemists well before Gaia, although Gaia's proponents have helped to educate a much wider audience about the pervasive influence of organisms on their environment. Gaia has also inspired original research exploring several biologically mediated processes, including production of dimethyl sulfide by phytoplankton (Charlson, et al. 1987) and microbial acceleration of mineral weathering (Schwartzmann and Volk 1989). This search for Gaian leverage points has proceeded in parallel with a much larger effort by the whole biogeochemical community to trace the mechanisms regulating global geochemical cycles.

Another well-established fact that is incorporated in Gaia is the notion that Earth's organisms and their environment form a coupled system; the biota affect their physical and chemical environment, which in turn shapes their further evolution. In this way, Earth's environment and life co-evolve through geologic time (Schneider and Londer 1984). Over the long term, organisms do not simply adapt to a fixed abiotic environment; nor is the environment sculpted to conform to fixed biotic needs (as early versions of Gaia appeared to argue; e.g., Lovelock and Margulis 1974b). A key theoretical element of Gaia is that, as with any complex coupled system, the atmosphere/biosphere system should be expected to exhibit "emergent" behaviors, that is, ones that could not be predicted from its components alone, considered in isolation from one another. So understanding Earth's history and global biogeochemistry requires systems thinking (in addition to lots of reductionist science as well). Gaia's proponents have helped to promote a systems-analytic approach to the global environment, in parallel with a much larger and broader-based effort in the biogeochemical community as a whole.

Up to this point, nothing that I have said about Gaia would seem controversial to modern Earth scientists. These themes in Gaia are consistent with a broad effort in the biogeochemical community to better understand the Earth system, but carried out in parallel, with different emphasis and different language.

At the same time, Gaia's proponents have consistently held that the Gaia hypothesis means something more than just the co-evolution of climate and life, and something more than just the idea that the Earth system can exhibit interesting system-level behaviors. Coupling between the biosphere and the physical environment can potentially give rise to either negative (stabilizing) feedback, or positive (destabilizing) feedback, and the consequences of this feedback can potentially be either beneficial or detrimental for any given group of organisms. But in the Gaia literature, mechanisms linking organisms to their environment are generally termed "Gaian" only if they create negative feedbacks, and only if they are beneficial to the organisms involved, or to the biota as a whole (Lenton 1998; Gillon 2000). Positive feedbacks, or those that seem detrimental, are typically referred to as "non-Gaian" or "anti-Gaian" mechanisms. Thus Gaia's proponents appear to view the

Gaia hypothesis as combining elements of what I have termed "Homeostatic Gaia" (i.e., biologically mediated feedbacks stabilize the global environment) and a qualified form of "Optimizing Gaia" (i.e., biological modifications of the environment make it more suitable for life). As Hamilton (1995) has put it, "right or wrong, Gaia presents the claim for an evolution of a supreme 'balance of nature.'"

This is what makes Gaia interesting, but this is also what makes Gaia difficult. If Gaia meant only that organisms influence their environment, and that these interactions may give rise to interesting system-level behaviors, then Gaia would add little—apart from different language and different metaphors—to the general consensus of the biogeochemical community. By claiming that organisms stabilize the global environment and make it more suitable for life, Gaia's proponents advance a much more ambitious argument, but one that is less clearly consistent with the available data, and one that sometimes may be difficult to test against data at all.

2. GAIA AND HOMEOSTASIS

Do biological feedbacks stabilize, or destabilize, the global environment? That is, is the "Homeostatic Gaia" hypothesis correct? This is not just a matter for theoretical speculation; there is a large and growing body of information that provides an empirical basis for evaluating this question. Biogeochemists have documented and quantified many important atmosphere-biosphere linkages (particularly those associated with greenhouse gas emissions and global warming), with the result that one can estimate the sign, and sometimes the magnitude, of the resulting feedbacks. Such analyses are based on three types of evidence: biological responses to plot-scale experimental manipulations of temperature and/or CO_2 concentrations (e.g., Saleska, et al. 1999), computer simulations of changes in vegetation community structure (e.g., Woodward, et al. 1998), and correlations between temperature and atmospheric concentrations in long-term data sets (e.g., Tans, et al. 1990; Keeling, et al. 1996a; Petit, et al. 1999). Below, I briefly summarize some of the relevant biological feedbacks affecting global warming; more detailed explanations, with references to the primary literature, can be found in reviews by Lashof (1989), Lashof, et al. (1997) and Woodwell, et al. (1998):

- Increased atmospheric CO_2 concentrations stimulate increased photosynthesis, leading to carbon sequestration in biomass (negative feedback).
- Warmer temperatures increase soil respiration rates, releasing organic carbon stored in soils (positive feedback).

- Warmer temperatures increase fire frequency, leading to net replacement of older, larger trees with younger, smaller ones, resulting in net release of carbon from forest biomass (positive feedback).
- Warming may lead to drying, and thus sparser vegetation and increased desertification, in mid-latitudes, increasing planetary albedo and atmospheric dust concentrations (negative feedback).
- Conversely, higher atmospheric CO_2 concentrations may increase drought tolerance in plants, potentially leading to expansion of shrublands into deserts, thus reducing planetary albedo and atmospheric dust concentrations (positive feedback).
- Warming leads to replacement of tundra by boreal forest, decreasing planetary albedo (positive feedback).
- Warming of soils accelerates methane production more than methane consumption, leading to net methane release (positive feedback).
- Warming of soils accelerates N20 production rates (positive feedback).
- Warmer temperatures lead to release of CO_2 and methane from high-latitude peatlands (positive, potentially large, feedback).

This list of feedbacks is not comprehensive, but I think it is sufficient to cast considerable doubt on the notion that biologically mediated feedbacks are necessarily (or even typically) stabilizing. As Lashof, et al. (1997) conclude,

> While the processes involved are complex and there are both positive and negative feedback loops, it appears likely that the combined effect of the feedback mechanisms reviewed here will be to amplify climate change relative to current projections, perhaps substantially . . . The risk that biogeochemical feedbacks could substantially amplify global warming has not been adequately considered by the scientific or the policymaking communities.

Most of the work to date on biological climate feedbacks has focused on terrestrial ecosystems and soils; less is known about potential biological feedbacks in the oceans. One outgrowth of the Gaia hypothesis has been the suggestion that oceanic phytoplankton might serve as a planetary thermostat by producing dimethyl sulfide (DMS), a precursor for cloud condensation nuclei, in response to warming (Charlson, et al. 1987). Contrary to this hypothesis, paleoclimate data now indicate that to the extent that there is such a marine biological thermostat, it is hooked up backwards, making the planet colder when it is cold and warmer when it is warm (Legrand, et al. 1988; Kirchner 1990; Legrand, et al. 1991). It now appears that DMS production in the Southern Ocean may be controlled by atmospheric dust, which supplies iron, a limiting nutrient (Watson, et al. 2000). The Antarctic ice core record is consistent with this view, showing greater deposition of atmospheric dust during glacial periods, along with higher levels of DMS proxy compounds, lower concentrations of CO_2 and CH4, and lower temperatures (figure 10.1). Watson and Liss (1998) conclude, "It therefore seems very likely that, both with respect to CO_2 and DMS interactions, the marine biological changes which occurred across the last glacial-interglacial transition were both positive feedbacks."

This example illustrates how the Gaia hypothesis can motivate interesting and potentially important research, even if, as in this case, the hypothesis itself turns out to be incorrect. But it is one thing to view Gaia as just a stimulus for research, and quite another to view it as "the essential theoretical basis for the putative profession of planetary medicine" (Lovelock 1986). To the extent that the Gaia hypothesis posits that biological feedbacks are typically stabilizing, it is contradicted both by the ice core records and by the great majority of the climate feedback research summarized above. Given the available evidence that biological feedbacks are often destabilizing, it would be scientifically unsound—and unwise as a matter of public policy—to assume that biological feedbacks will limit the impact of anthropogenic climate change. As Woodwell, et al. (1998) have put it, "The biotic feedback issue, critical as it is, has been commonly assumed to reduce the accumulation of heat trapping gases in the atmosphere, not to amplify the trend. The assumption is serious in that the margin of safety in allowing a warming to proceed may be substantially less than has been widely believed."

3. GAIA AND ENVIRONMENTAL ENHANCEMENT

I now turn to the second major theme in the Gaia hypothesis, namely the claim that the biota alter the physical and chemical environment to their own benefit. We have seen that Gaian claims for biologically mediated homeostasis are inconsistent with much of the available empirical evidence. But by contrast, our day-to-day experience gives the strong impression that our environment is, in fact, very well suited to the organisms that live in it. Given that our environment is partly a biological by-product, it seems reasonable to speculate that biological feedbacks make our environment a better place to live. If we think about how thoroughly our world would be disrupted by the loss of any of the environmental services that its organisms provide, it seems natural to say that biological influences greatly enhance our environment. But it is one thing to say that we benefit from the environmental services that our ecosystem provides, and entirely another to say that our environment is in any sense tailored to our needs.

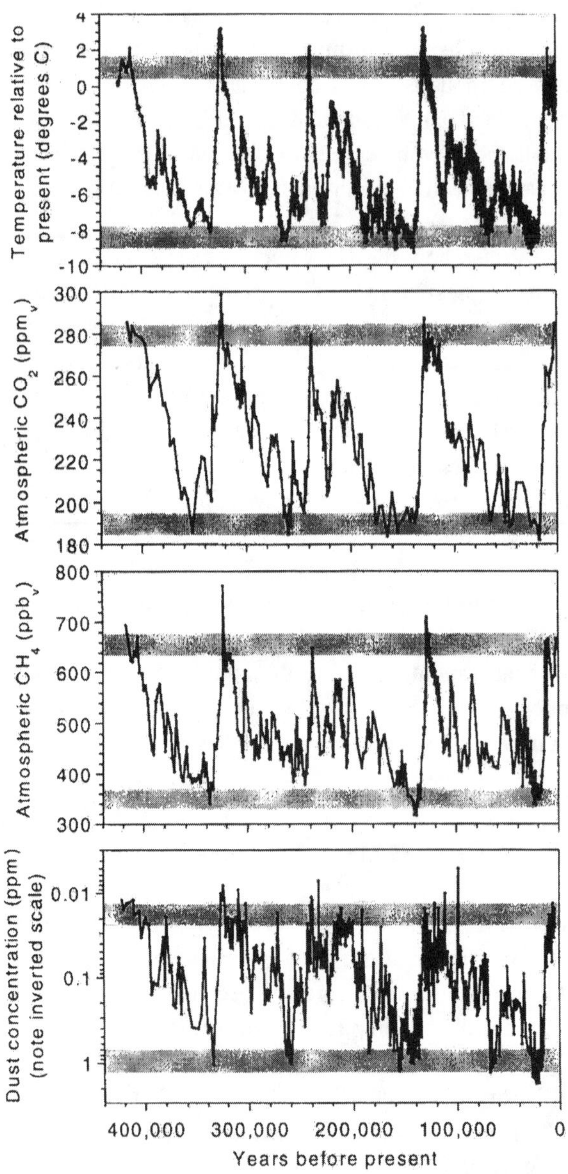

Figure 10.1. The Vostok ice core record of atmospheric temperature, CO_2, methane, and dust loading (data of Petit et al., 1999). Glacial maxima and interglacials to exhibit relatively narrow ranges of characteristic concentrations and temperatures, indicated by the shaded bands. Temperature, CO_2, and methane are strongly correlated. High dust concentrations (note the inverted logarithmic scale) are correlated with low CO_2, methane, and temperature.

Organisms can strongly influence their environments. Organisms are also naturally selected to do well in their environments—which means doing well under the conditions that they, and their co-occurring species, have created. Thus it is likely that those particular organisms will be better off under those particular conditions than they would be without the environmental services that they, and their co-occurring species, are responsible for. In any such system, it will be true that "biotic effects strongly enhance the biogeophysical and biogeochemical conditions for life" (Kleidon 2002), precisely because the particular life in question is that which has been naturally selected to thrive under the conditions that those particular biotic effects promote.

Let me illustrate this concept with a simple example. Rainforests are wet, in large part because of water recycled by transpiration from the rainforest vegetation itself. Rainforest vegetation is thus spared the heat and drought stress that it would encounter in an arid environment, but instead must cope with other environmental challenges that the rainforest presents, such as light and nutrient stress due to crowding, or parasitism by pathogens that flourish under damp conditions. Organisms that thrive in the rainforest will be highly evolved to cope with these threats to survival, and to depend on (and exploit, insofar as possible) the prevailing wetness of the rainforest environment. If the recycling of water by transpiration were somehow disrupted, the rainforest would become much drier and its vegetation would suffer. Thus the rainforest's health depends on an environmental service provided by its own organisms.

In cases like this, one might be tempted to say that rainforest vegetation influences its climate to its own benefit, or that the rainforest enhances the physical conditions required for life. These statements are semantically correct but mechanistically misleading, because they suggest that the environmental conditions have somehow been adjusted to the needs of the organisms. Instead, it is more mechanistically accurate to say that natural selection has made rainforest organisms dependent on rainforest conditions, which are partly of their own making.

The Gaia literature further compounds this semantic confusion by making claims such as, "vegetation almost always influences climate *for* its own benefit" (Lenton 1998, emphasis added), rather than *to* its own benefit. What a difference a word makes! Saying that vegetation influences climate *to* its own benefit implies that vegetation alters its environment in ways that are beneficial. But saying that vegetation influences climate *for* its own benefit advances the much stronger claim that vegetation modifies its environment in order to reap the benefits that will result from doing so. Although Gaia's proponents have tried hard to avoid teleology (Kerr 1988), it occasionally slips in. This is almost inevitable, given that Gaia's metaphorical models are based on thinking about how the biota could benefit from altering their environment. Thinking mechanistically instead—that is, not thinking about what would be good for the biota, but instead thinking about how the environment will affect natural selection on the biota (and how that, in turn, will further alter the environment)—leads to a more scientifically defensible conclusion: Because natural selection will favor organisms that can best exploit their environment, organisms will often be dependent on the environmental services that their ecosystem provides. Thus they will benefit from their environment, even though it has not been constructed to conform to their needs.

The same reasoning holds at global scale. The global environment seems very well suited to the needs of the organisms that inhabit it. Change any major parameter of the global environment (temperature, pH, redox potential, etc.) by very much, and Earth would become a much less hospitable place—at least for the organisms that are now dominant precisely because its conditions suit them. But inferring from this that the environment has been tailored to the needs of the biota is a bit like evaluating the million-dollar lottery by polling only those who have won the jackpot. The life forms that we observe today are descended from a very select subset of evolutionary lineages, namely those for which Earth's conditions have been favorable. The other lineages, for whom Earth's conditions are hostile, have either gone extinct or are found in refugia (such as anaerobic sediments) which protect them from the conditions that prevail elsewhere. As Holland (1984) has put it, "We live on an Earth that is the best of all possible worlds only for those who are well adapted to its current state."

4. GAIA AND NATURAL SELECTION

Nevertheless, it is striking to observe that Earth's environment has remained suitable for advanced forms of life for hundreds of millions of years. Before getting too wrapped up in this observation, we should remember that this is also a necessary prerequisite for our being here to observe it, and we are in no position to judge how improbable a circumstance that is. It is therefore possible that the Earth has gotten by on sheer good fortune alone. But although such a hypothesis is logically defensible it is scientifically unsatisfying, since if there is a mechanistic explanation for Earth's resilience through time, there is a lot to be learned by finding it. Furthermore, paleoclimate records do show patterns that suggest a planetary-scale feedback system. For example, ice core records show that through the last four glacial/interglacial cycles, atmospheric chemistry and temperature have oscillated between the same narrow upper and lower bounds (the shaded bands in figure 10.1). What mechanisms set these upper and lower limits, and what processes control how the climate flips from one to the other? The answers are not known at present, but the search for them is a scientific challenge of the highest order (Falkowski, et al. 2000).

As an answer to the mystery of how life has persisted, the Gaia hypothesis proposes that the Earth is self-regulated, in a state that is favorable for life, by biologically mediated feedbacks. It further proposes that these stabilizing, environment-enhancing feedbacks should arise naturally as the result of natural selection acting on individual organisms. The primary vehicle through which this theory has been promoted is the Daisyworld model. The Daisyworld model has been modified and elaborated in many different ways (see Lenton 1998, and references therein), but the original version (Watson and Lovelock 1983) presents the clearest picture of the underlying mechanisms, and will suffice for the present discussion.

Daisyworld was designed to explain how biologically mediated feedbacks—arising through natural selection alone—could hypothetically stabilize a planet's temperature in the face of an increase in solar luminosity, similar to that experienced by the Earth on billion-year timescales. Daisyworld is a planet on which the albedo (and thereby temperature) is determined by the black and white daisies growing on its surface. The black daisies and white daisies are assumed to have the same growth response to temperature, but the black daisies are assumed to be warmer than the white daisies. Thus the black daisies thrive in cooler ambient conditions, lower the albedo, and warm the surface, whereas the white daisies thrive in warmer ambient conditions, raise the albedo, and cool the surface. Schneider (2001) summarizes the climatic evolution of Daisyworld as follows: As the sun heats up over hundreds of millions of years, black daisies approach their optimum temperatures, become more fit, and thereby increase their numbers, causing the albedo (reflectivity) of the planet's surface to drop. This is a positive feedback, because while more sunlight is absorbed by the dark flowers the planet further warms. Black daisies increase until the temperature passes their fitness peak and moves into the fitness range for the white daisies. These then begin to multiply and replace the black daisies; this shift increases the planet's albedo, which serves as a negative feedback on further warming. The planet's overall temperature is stabilized for eons even though the sun inexorably increases its luminosity. But eventually, the white daisies are heated past their fitness range and can't resist further warming. The biota then collapses and temperatures rise rapidly to the level an inorganic rock would experience.

Schneider (2001) points out that on the real Earth, vegetation albedo has a relatively weak effect on climate because clouds and haze obscure most of the surface. Likewise, vegetation response to temperature is weaker on the real Earth than it is on Daisyworld, where a temperature shift of only 1 C can expand daisy cover from zero to 45 percent of the planet's surface (Kirchner 1989). And on the real Earth, vegetation would actually respond to temperature in the *opposite direction* from what Daisyworld predicts; all else equal, warmer temperatures would expand forests poleward, making the surface darker and thus amplifying the warming (Kirchner 1989).

But despite the fact that Daisyworld does not apply in any literal sense to the real Earth, it seems to embody a general principle that deserves further exploration. Daisyworld appears to demonstrate that whenever organisms are strongly coupled to their environment, Darwinian evolution will naturally generate a system that self-regulates near the biological optimum. What Daisyworld actually demonstrates is that it is *possible* for such a system to arise by natural selection, but only given a very specific assumption embedded in the model. Because this assumption may not be true on the real Earth, Daisyworld only demonstrates a theoretical possibility rather than a guiding principle of the natural world. To explain what this assumption is, and why it matters, requires a brief synopsis of evolutionary theory.

Heritable traits become more common over time if the individuals who have those traits have a reproductive advantage over those who lack them. They become more common by virtue of the simple fact that individuals who carry the advantageous trait will have more offspring on average than those who do not (that is what it means for the trait to be "advantageous"), which implies that in each generation a larger fraction of the population will carry the trait. If a trait were equally beneficial to those who carried it and those who did not, it would not become more widespread in the population, since carriers and non-carriers of the trait would pass on their genes to the next generation at equal rates. Thus traits that confer a *differential advantage* (for carriers over non-carriers) will become more common via natural selection, but traits that confer a *general benefit* (to carriers and non-carriers alike) will undergo genetic drift; natural selection cannot have any effect on them.

Here, then, is the crux of the matter. The environment is that which is shared among organisms. To the extent that a trait improves the environment for life in general—and thus benefits its carriers and non-carriers alike—natural selection will not have any effect on it. Thus claims that "life-enhancing effects would be favored by natural selection" (Kleidon 2002) are not generally valid.

Of course, it is always possible that a particular trait could confer both a differential advantage to the individual and a general benefit to the environment. Such a trait would be favored by natural selection, but only because of the differential advantage it confers to the individual, not because of any general benefit that is shared with others who do not carry the trait. Thus environmentally beneficial traits and environmentally detrimental ones will both be favored by natural selection, as long as each confers a differential reproductive advantage to the individuals that carry it. Even if an individual organism benefits from its own environmental good deeds, it will only gain a reproductive advantage (and be favored by natural selection) to the extent that it benefits more than its neighbors do.

Which brings us to Daisyworld. The Daisyworld model assumes that traits that benefit the environment also give an individual a reproductive advantage over its neighbors. Thus Daisyworld gives the impression that traits are favored by natural selection if they are environmentally beneficial, even though natural selection does not—indeed cannot—act on environmental benefits per se, precisely because they are shared between carriers and non-carriers alike. By assuming that individual reproductive success and environmental good deeds are linked, and thus that organisms will do well by doing good, Daisyworld to some extent assumes what it sets out to prove. Environmentally beneficial traits are not favored by natural selection in Daisyworld because they are environmentally beneficial; they are instead favored because they also, coincidentally, confer a reproductive advantage.

When Daisyworld is cool, the black daisies are *not* favored by natural selection because they warm the environment. Instead, they are favored by natural selection because they warm *themselves*, and therefore thrive better in cool temperatures than the white daisies do. To the extent that they also warm the *environment*, and therefore make it better for white daisies and worse for themselves, they diminish (not enhance) their evolutionary advantage. In competition with white daisies, black daisies would have a greater evolutionary advantage if they had no effect on their environment at all, because the planet would remain cooler and thus more favorable for black daisies. The black daisies could increase their evolutionary advantage still further if, while warming themselves, they *cooled* the environment, thus suppressing competition from the white daisies. If the black daisies only warmed the environment, and thus warmed themselves and the white daisies equally, they would never be favored by natural selection. Thus the Daisyworld model behaves the way it does specifically because it assumes that a given trait (here, daisy color) will affect the individual and its environment in the same way (Lenton 1998), with a bigger benefit to the individual than to the environment as a whole.

The numerous elaborations of the Daisyworld model all share this premise. It might be true for the particular case of variations in an organism's albedo, but a moment's reflection will show that it is not universally, or even generally, true in the natural world. For example, all organisms must consume resources, and by doing so they deplete their local environments of those resources. Likewise, all organisms must eliminate wastes, and by doing so they pollute their environments. Traits that enable organisms to better consume resources or eliminate wastes will benefit the individual, and thus will be favored by natural selection, even though they also degrade the environment. Examples of such traits abound. Trees are highly evolved to catch sunlight, and thus shade their neighbors. Plants in arid zones are highly evolved to intercept moisture before it reaches their competitors. Some tree species (such as eucalyptus and black walnut) even conduct a form of chemical warfare against potential competitors, by dropping leaves or fruits that make the surrounding soils toxic for other species.

These evolutionary strategies may offend our human sense of fair play, but they also serve to illustrate an important point about natural selection. Natural selection is a *mechanism*, not a *principle*. It does not seek a goal; it just passes traits from one generation to the next, with the reproductively successful ones becoming more common over time. Thus there is no direction to evolution beyond the fact that whatever works (in a reproductive sense) works, and will be passed on to the next generation. Natural selection will favor both environment-enhancing and environment-degrading traits, as long as those traits confer a reproductive advantage, that is, as long as those who carry them have greater reproductive success than those who don't.

Thus I would agree with Volk (1998, p. 239) that "[w]hat organisms do to help themselves survive may affect the planet in enormous ways that are not at all the reasons those survival skills were favored by evolution." But if the connection between environmental good deeds and individual reproductive advantage is only coincidental, we should not expect Gaian traits to evolve any more frequently than anti-Gaian ones.

It is still theoretically possible for evolution to systematically favor Gaian traits over anti-Gaian ones, but only if environments and their organisms are jointly subject to some form of natural selection, in competition with other environment/ organism assemblages. This would create a kind of metapopulation, and natural selection in metapopulations can produce evolutionary outcomes that would otherwise seem counterintuitive (e.g., Kirchner and Roy 1999). However, a mechanism needs to be demonstrated by which such natural selection could occur in this context.

Obviously, no biologically mediated feedback can be too environment-degrading, in the limiting sense that no simple feedback loop can lead to the extinction of the organisms responsible for it. Thus, for example, photosynthetic organisms cannot sequester organic carbon beyond the point at which they starve themselves for CO_2; in this way, the compensation point for photosynthesis (roughly 100–150 ppm CO_2) may set the limit to the glacial deep freeze. Likewise, organisms that consume plants and respire their carbon could not multiply beyond the point at which they either run out of food or poison themselves with waste CO_2. But these are far from the environment-enhancing feedbacks envisioned by the Gaia hypothesis.

One should remember that environment-enhancing feedbacks, if they occur, are intrinsically destabilizing (Kirchner 1989). Organisms that make their environment more suitable for themselves will grow, and thus affect their environment still more, and thus grow still further. This is positive feedback, not negative feedback. Negative feedback arises when a growing population makes its environment less suitable for itself, and thus limits its growth. Environment-enhancing feedbacks are destabilizing; environment-degrading feedbacks are stabilizing. The Gaian notion of environment-enhancing negative feedbacks is, from the standpoint of control theory, a contradiction in terms.

5. CONCLUSION

Organisms are not merely passengers, riding passively on spaceship Earth. But to the extent that the biota are piloting the craft, they are flying blind, and the various life forms are probably wrestling over the controls. Earth's surface environment has remained stable enough to allow life in some form to persist for billions of years, but it is not clear whether this has occurred because of biological feedbacks, or in spite of them.

Gaia's proponents have done a great service by championing the need to consider the Earth as a coupled system. We now need to figure out how that system works, and it is crucial that we get it right. As figure 10.2 shows, anthropogenic emissions have pushed greenhouse gas concentrations far beyond the limits that they had previously remained within, for over 400,000 years. This indicates that the composition of the atmosphere is not tightly regulated, by either biotic or abiotic feedbacks, on human timescales. Yet both CO_2 and methane are biologically active gases. And at least in the case of CO_2, the anthropogenic fluxes are a tiny fraction of the gross fluxes entering and leaving the biosphere; only a small adjustment to those biological fluxes would have been needed to keep CO_2 concentrations stable. There has been a modest increase in terrestrial photosynthetic uptake of CO_2 (Ciais, et al. 1995; Keeling, et al. 1996b; Myneni, et al. 1997), but not nearly enough to keep CO_2 concentrations within their natural limits. This is an empirical rebuttal to Gaian notions of homeostasis and optimization, as it indicates that atmospheric CO_2 is not tightly regulated at a biological set point.

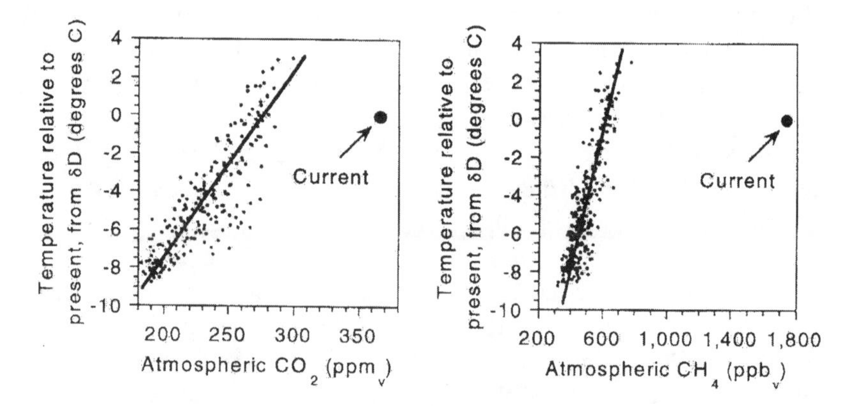

Figure 10.2. Carbon dioxide and methane concentrations over the last four glacial cycles, from figure 10.1, re-scaled to show the current concentrations that have resulted from anthropogenic emissions. The current chemistry of the atmosphere is unprecedented in recent Earth history.

This could be the signature of a biological thermostat that has broken down (Lovelock and Kump 1994), or it could be the signature of a complex system that combines both positive and negative feedbacks. The latter hypothesis is more consistent with what we know about the mechanisms underlying climate feedbacks, and it points to the urgent need to figure out their relative strengths and the timescales over which they operate. In particular, we need to understand why, despite greenhouse gas concentrations that are unprecedented in recent Earth history, global temperatures have not (yet) risen nearly as much as the correlations in the ice core records would indicate that they could. The correlations in the ice core data suggest that, for the current composition of the atmosphere, current temperatures are anomalously cool by many degrees (figure 10.3). This might indicate that glacial-interglacial temperature shifts were amplified by ice-albedo feedbacks that are now much less influential (since the ice caps are already nearly gone, relative to ice age conditions). Alternatively, it may indicate that we have not yet fully felt the effects of important positive feedbacks (northward-spreading boreal forests, respiring soils, and so forth) that will amplify the warming experienced to date. If this is the case, it is vitally important that we appreciate it. Believing that biological feedbacks are generally stabilizing and beneficial will not help us unravel this puzzle.

In the human enterprise of science, our most daunting task is to see things as they are, rather than as we wish they were. Gaia's vision of Earth as a harmonious whole, engineered by and for the organisms that live on it, is a deeply evocative

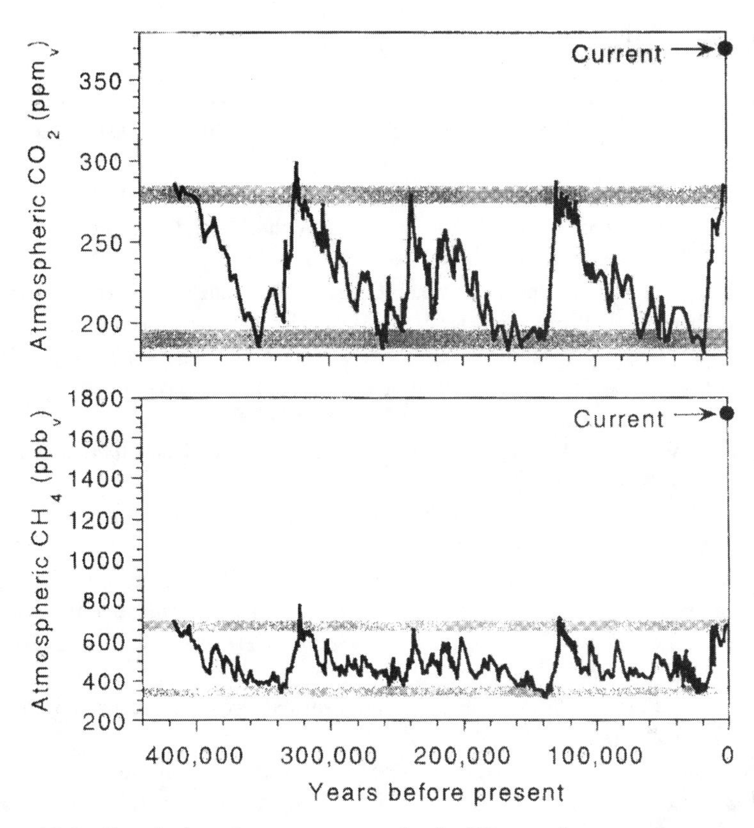

Figure 10.3. Correlations between atmospheric CO_2, methane, and temperature over the last 400,000 years, from figure 10.1, compared to current conditions. Current conditions lie far outside the envelope of the prehistoric data, and far below any extrapolation from them.

notion. It is emotionally very appealing to me. But I suspect that compared to the Gaian vision of global harmony, the actual Earth system—as it comes into clearer focus—will prove to be more complicated, more intriguing, and perhaps more challenging to our notions of the way things should be. Understanding the Earth system, in all of its fascinating complexity, is the most important scientific adventure of our time. We should get on with it, as free as possible from our preconceptions of the way the world ought to work.

REFERENCES

Charlson, R. J., J. E. Lovelock, M. O. Andreae, and S. G. Warren. "Oceanic Phytoplankton, Atmospheric Sulphur, Cloud Albedo and Climate." *Nature* 326 (1987): 655–61.

Ciais, P., P. P. Tans, M. Trolier, J. C. W. White, and R. J. Francey. "A Large Northern Hemisphere Terrestrial CO_2 Sink Indicated by the 13C/12C Ratio of Atmospheric CO_2." *Science* 269 (1995): 1098–102.

Falkowski, P., R. J. Scholes, E. Boyle, J. Canadell, D. Canfield, J. Elser, N. Gruber, K. Hibbard, P. Hogberg, S. Linder, F. T. Mackenzie, B. Moore, T. Pedersen, Y. Rosenthal, S. Seitzinger, V. Smetacek, and W. Steffen. "The Global Carbon Cycle: A Test of Our Knowledge of Earth as a System." *Science* 290 (2000): 291–96.

Gillon, J. "Feedback on Gaia." *Nature* 406 (2000): 685–86.

Hamilton, W. D. "Ecology in the Large: Gaia and Ghengis Khan." *Journal of Applied Ecology* 32 (1995): 451–53.

Harvey, H. W. *The Chemistry and Fertility of Sea Waters*. New York: Cambridge University Press, 1957.

Henderson, L. J. *The Fitness of the Environment.* New York: MacMillan, 1913.

Holland, H. D. "The Chemical Evolution of the Terrestrial and Cytherian Atmospheres." In *The Origin and Evolution of Atmospheres and Oceans*, edited by Brancazio, P. J. and Cameron, A. G. W. New York: Wiley, 1964.

———. *The Chemical Evolution of the Atmosphere and Oceans*. Princeton, NJ: Princeton University Press, 1984.

Hutchinson, G. E. "The Biogeochemistry of the Terrestrial Atmosphere." In *The Earth as a Planet*, edited by Kuiper, G. P, 371–43. Chicago: University of Chicago Press, 1954.

Huxley, T. H. *Physiography.* London: MacMillan and Co., 1877.

Keeling, C. D., J. F. S. Chin, and T. P. Whorf. "Increased Activity of Northern Vegetation Inferred from Atmospheric CO_2 Measurements." *Nature* 382 (1996a): 146–49.

Keeling, R. F., S. C. Piper, and M. Heimann. "Global and Hemispheric CO_2 Sinks Deduced from Changes in Atmospheric O2 Concentration." *Nature* 381 (1996b): 218–21.

Kerr, R. A. "No Longer Willful, Gaia Becomes Respectable." *Science* 240 (1988): 393–95.

Kirchner, J. W. "The Gaia Hypothesis: Can It Be Tested?" *Reviews of Geophysics* 27 (1989): 223–35.

———. "Gaia Metaphor Unfalsifiable." *Nature* 345 (1990): 470.

———. "The Gaia Hypotheses: Are They Testable? Are They Useful?" In *Scientists on Gaia*. edited by S. H. Schneider, and P. J. Boston. 38–46. Cambridge, MA: MIT Press, 1991.

Kirchner, J. W., and B. A. Roy. "The Evolutionary Advantages of Dying Young: Epidemiological Implications of Longevity in Metapopulations." *American Naturalist* 154 (1999): 140–59.

Kleidon, A. "Testing the Effect of Life on Earth's Functioning: How Gaian Is the Earth System?" *Climitic Change* 52 (2002).

Lashof, D. A. "The Dynamic Greenhouse: Feedback Processes That May Influence Future Concentrations of Atmospheric Trace Gases in Climatic Change." *Climatic Change* 14 (1989): 213–42.

Lashof, D. A., B. J. DeAngelo, S. R. Saleska, and J. Harte. "Terrestrial Ecosystem Feedbacks to Global Climate Change." *Ann Rev. Energy Environ.* 22 (1997): 75–118.

Legrand, M., C. Feniet-Saigne, E. S. Saltzman, C. Germain, N. I. Barkov, and V. Petrov. "Ice-Core Record of Oceanic Emissions of Dimethylsulphide during the Last Climate Cycle." *Nature* 350 (1991): 144–46.

Legrand, M. R., R. J. Delmas, and R. J. Charlson. "Climate Forcing Implications from Vostok Ice-Core Sulphate Data." *Nature* 334 (1988): 418–20.

Lenton, T. M. "Gaia and Natural Selection." *Nature* 394 (1998): 439–47.

Lovelock, J. E. "Geophysiology: A New Look at Earth Science." In *The Geophysiology of Amazonia: Vegetation and Climate Interactions*, edited by Dickinson, R. E., 11–23. New York: Wiley, 1986.

Lovelock, J. E., and L. R. Kump. "Failure of Climate Regulation in a Geophysiological Model." *Nature* 369 (1994): 732–34.

Lovelock, J. E., and L. Margulis. "Homeostatic Tendencies of the Earth's Atmosphere." *Origins Life* 5 (1974a): 93–103.

———. "Atmospheric Homeostasis by and for the Biosphere: The Gaia Hypothesis." *Tellus* 26 (1974b): 2–9.

Myneni, R. B., C. D. Keeling, C. J. Tucker, G. Asrar, and R. R. Nemani. "Increased Plant Growth in the Northern High Latitudes from 1981 to 1991." *Nature* 386 (1997): 698–702.

Petit, J. R., J. Jouzel, D. Raynaud, N. I. Barkov, J. M. Barnola, I. Basile, M. Bender, J. Chappellaz, M. Davisk, G. Delaygue, M. Delmotte, V. M. Kotlyakov, M. Legrand, V. Y. Lipenkov, C. Lorius, L. Pepin, C. Ritz, E. Saltzmank, and M. Stievenard. "Climate and Atmospheric History of the Past 420,000 Years from the Vostok Ice Core, Antarctica." *Nature* 399 (1999): 429–36.

Redfield, A. C. "The Biological Control of Chemical Factors in the Environment." *American Journal of Science* 46 (1958): 205–21.

Saleska, S. R., J. Harte, and M. S. Torn. "The Effect of Experimental Ecosystem Warming on CO_2 Fluxes in a Montane Meadow." *Global Change Biology* 5 (1999): 125–41.

Schneider, S. H. "A Goddess of Earth or the Imagination of a Man?" *Science* 291 (1999): 1906–07.

Schneider, S. H., and R. Londer. *The Coevolution of Climate and Life.* San Francisco: Sierra Club Books, 1984.

Schwartzmann, D. W., and T. Volk. "Biotic Enhancement of Weathering and the Habitability of Earth." *Nature* 340 (1989): 457–60.

Sillen, L. G. "Regulation of O2, N2, and CO_2 in the Atmosphere; Thoughts of a Laboratory Chemist." *Tellus* 18 (1966): 198–206.

Spencer, H. "Remarks upon the Theory of Reciprocal Dependence in the Animal and Vegetable Creations, as Regards Its Bearing upon Paleontology." *London Edinburgh Dublin Philosophical Magazine and Journal of Science* 24 (1844): 90–94.

Tans, P. P., I. Y. Fung, and T. Takahashi. "Observational Constraints on the Global Atmospheric CO_2 Budget." *Science* 247 (1990): 1431–38.

Volk, T. *Gaia's Body: Toward a Physiology of Earth.* New York: Copernicus, 1998.

Watson, A. J., D. C. E. Bakker, A. J. Ridgwell, P. W. Boyd, and C. S. Law. "Effect of Iron Supply on Southern Ocean CO_2 Uptake and Implications for Glacial Atmospheric CO_2." *Nature* 407 (2000): 730–33.

Watson, A. J., and P. S. Liss. "Marine Biological Controls on Climate via the Carbon and Sulphur Geochemical Cycles." *Philosophical Transactions of the Royal Society of London, Series B* 353 (1998): 41–51.

Watson, A. J., and J. E. Lovelock. "Biological Homeostasis of the Global Environment: The Parable of Daisyworld." *Tellus, Series B: Chemical and Physical Meteorology* 35 (1983): 284–89.

Woodward, F. I., M. R. Lomas, and R. A. Betts. "Vegetation-Climate Feedbacks in a Greenhouse World." *Philosophical Transactions of the Royal Society of London, Series B* 353 (1998): 29–39.

Woodwell, G. M., F. T. Mackenzie, R. A. Houghton, M. Apps, E. Gorham, and E. Davidson. "Biotic Feedbacks in the Warming of the Earth." *Climatic Change* 40 (1998): 495–518.

DEREK D. TURNER

ARE WE AT WAR WITH NATURE?

ABSTRACT

A number of people, from William James to Dave Foreman and Vandana Shiva, have suggested that humans are at war with nature. Moreover, the analogy with warfare figures in at least one important argument for strategic monkey-wrenching. In general, an analogy can be used for purposes of (1) justification; (2) persuasion; or (3) as a tool for generating novel hypotheses and recommendations. This paper argues that the analogy with warfare should not be used for justificatory or rhetorical purposes, but that it may nevertheless have a legitimate heuristic role to play in environmental philosophy.

1. INTRODUCTION

In his 1910 essay, "The Moral Equivalent of War," William James argues that instead of waging war against one another, we should conscript young people to serve in the ongoing war against nature:

> If now—and this is my idea—there were, instead of military conscription a conscription of the whole youthful population to form for a certain number of years a part of the army enlisted against *Nature*, the injustice would tend to be evened out, and numerous other goods to the commonwealth would follow. (1910/1977, p. 669)

The injustice to which James refers is *social* injustice. Why should working class people experience toil and pain while the young people of the leisure class take it easy?

> To coal and iron mines, to freight trains, to fishing fleets in December, to dish-washing, clothes-washing, and window-washing, to road-building and tunnel-making, to foundries and stoke-holes, and to the frames of sky-scrapers, would our gilded youths be drafted off, according to their choice, to get the childishness knocked out of them, and to come back into society with healthier sympathies and soberer ideas. They would have paid their blood-tax, done their own part in the immemorial human warfare against nature; they would tread the earth more proudly, the women would value them more highly, there would be better fathers and teachers of the following generation. (1910/1977, p. 669)

James argues that pacifists must propose ways of cultivating the traditional military virtues: courage, discipline, loyalty, fortitude, obedience, hardihood, and so on. He characterizes them as "absolute and permanent human goods" (1910/1977, p. 668). By conscripting the youth to wage a war against nature, we can continue to foster the martial virtues without the destructiveness of real war. Humans, he thinks, have been engaged in a war with nature for a very long time, but in a haphazard, undisciplined, and disorganized way. Like the general who is glad when the war ends on one front so that he can divert troops and equipment to another front, James seems to think that we could finally transform nature into an instrument of human flourishing if only we stopped fighting other nations and transferred all of our equipment and personnel to the environmental theater.

Nowadays the people who argue that our relationship to the environment resembles open warfare—or in areas where the wilderness has been conquered, a state of armed occupation—are writers and activists who wish to recruit others to the cause of environmental defense. Occasionally resorting to the language of the just war tradition, they argue that non-human nature is the victim of unjust aggression by humans. Derrick Jensen and George Draffan, in a popular book on deforestation entitled *Strangely Like War*, write that forests are "under attack" (2003, p. 4) and that the destruction of forests is an "atrocity" (2003, p. 142). In her preface to that book, Vandana Shiva agrees that our forests are becoming "victims of war" (Jensen and Draffan 2003, p. ix). If the forests are under attack, if they are innocent victims, it would seem to follow that someone ought to defend them. Thus Jensen and Draffan argue that "They will not leave the forests, and leave the forests alone, until either the forests are gone, or until those of us who love the land force them out of the forests" (2003, p. 141). Who "they" are is left somewhat vague. In his influential novel, *Ishmael*, Daniel Quinn's title character tells us that

> I no longer think of what we're doing as a blunder. We're not destroying the world because we're clumsy. We're destroying the world because we are, in a very literal and deliberate way, at war with it. (1992, p. 130)

Environmentalists such as Quinn and Shiva take the injustice of our war against non-human nature for granted.

The proposition that humans are at war with nature sometimes figures as a premise in the arguments used to justify monkeywrenching and ecosabotage. Dave Foreman, the founder of *Earth First!,* predicted in 1987 that "in half a decade, the saw, 'dozer, and drill will devastate most of what is unprotected. The battle for wilderness will be over" (1987, p.14). He describes monkeywrenchers as "eco-defenders," "Earth defenders," and "warriors." And he concludes with something resembling a call to arms:

> John Muir said that if it ever came to a war between the races, he would side with the bears. That day has arrived. (1987, p. 17)

Like James, Foreman claims to reject the use of violence. He conceives of ecosabotage as a form of nonviolent resistance on behalf of non-human nature. In his view, there is nothing violent about destroying the weapons—saws, 'dozers, and drills—that people are using in their war against the environment, so long as the saboteur takes care not to injure any people. There are serious problems here concerning the definitions of terms such as "violence" and "injury." It is by no means obvious that we cannot injure people by destroying their property. Nor is it obvious that arson is non-violent.

On the one hand, we have William James arguing that what is needed in order to make pacifism an attractive and realistic alternative to war is to re-deploy the bulk of our military force in the immemorial war against nature. On the other hand, we have Dave Foreman and others arguing that the war that humans are waging against the environment is unjust, and that the earth must be defended. However, both Foreman and James share a commitment to non-violence, although they may disagree about what counts as violence. In both cases, the pacifism is laudable, but the idea that humans are at war with nature is a bad one.

In this paper, I begin (in section 2) by distinguishing three different roles that analogies, such as the war analogy, might play in environmental philosophy and environmental policy making: a rhetorical function, a justificatory function, and a heuristic function. I also show how the war analogy might figure in an argument for ecosabotage. I then proceed to show (in section 3) that relevant differences between war and our relationship to the environment, not to mention counteranalogies, severely weaken the positive argument for ecosabotage. In section 4, I argue that pacifists and activists ostensibly committed to the use of non-violent tactics, such as James and Foreman, respectively, have very good reasons *not* to invoke the war analogy for purposes of justification or persuasion. In the concluding section 5, I show how the war analogy might nevertheless have an interesting and fruitful, if limited role to play in environmental philosophy, as a source of novel moral claims and insights. Although I recommend the war analogy for limited, heuristic purposes, I do not think that it should be used for purposes of justification, persuasion, or recruitment.

2. THE ROLE OF ANALOGY IN ENVIRONMENTAL PHILOSOPHY: SOME GENERAL CONSIDERATIONS

Let's begin with some further examples of the ways in which analogy and metaphor shape our thinking about the environment.

(1) Species resemble rivets in an airplane. If you remove just one rivet, nothing will happen. However, if you remove enough rivets, the plane will malfunction. If we want to avoid a plane crash—or ecosystemic collapse—we had better stop popping rivets (Ehrlich and Ehrlich 1981, pp. xi–xiv).

(2) Ecosystems, like organisms, can fall ill and even die. The science of ecology is a normative science, like medicine. Just as medicine has the goal of promoting the health of individual people, ecology has the goal of promoting the health of ecosystems. The ecologist's relation to the natural world resembles the doctor-patient relation (See, e.g. Norton 1988; as well as Leopold 1949).

(3) Rare plants and animals are artistic masterpieces. Our duty to protect endangered species derives from our more general duty to protect great works of art (Russow 1981; Slobodkin 2003, p. 139).

(4) Causing the extinction of a species is like killing a person. Both mean destroying something that is unique and irreplaceable. Extinction by natural causes resembles death by natural causes; anthropogenic extinction is analogous to murder (Rolston 1989).

(5) Causing the extinction of a species that scientists have not yet had a chance to study is like ripping out a page from the book of nature before anyone has had a chance to read it (Slobodkin 2003, pp. 139–40).

(6) Human beings are like parasites, and we are gradually killing our host—the earth.

(7) The goal of restoration ecology is to restore our relationship to the environment. That relationship has positive value of its own, much like a relationship between people—e.g. between siblings. Ecological restoration is like trying to re-establish contact with a sibling or an old friend with whom you have fallen out of touch (Light 2000).

(8) Our relationship to the environment is an oppressive one, rather like the relationship between men and women in a patriarchal society (Warren 1990).

(9) The human population on earth is like a bomb. Population growth during the last century has been explosive (Ehrlich 1968).

This list is by no means intended to be exhaustive. In some cases, the metaphors (e.g. population explosion) are so familiar that we may not even notice them. In other cases, such as (3) and (4), the analogies figure in important ethical arguments. In at least one case, (8), the analogy is central to an important philosophical movement, ecofeminism.

These examples of analogical thinking about the environment fall neatly into three groups. (1), (3), (5), and (9) all involve some sort of comparison between non-human nature and human artifacts—bombs, books, rivets, airplanes, and artworks. We can call these *artifactual analogies.* (2), (4), (7), and (8), on the other hand, involve comparisons of the human-environment relationship to various human-human relationships—between doctors and patients, killers and victims, friends, and between oppressors and oppressed. The proposition that our relationship to the environment is one of warfare, conquest, and armed occupation also falls in this group of *social analogies.* Finally, both (2) and (6) are *biological analogies.* In (2), ecosystems are compared to organisms, while in (6) the human-environment relationship is compared to the parasite-host relationship. It is ironic, to say the least, that we should find it so natural to use social and artefactual analogies to help us to think about our relationship to nature.

We could, in principle, call upon these analogies to serve any of three distinct functions: a rhetorical function, a justificatory function, and a heuristic function. I will briefly discuss each of these in turn, before proceeding to inquire which of these roles, if any, analogy *ought* to play in environmental philosophy. For purposes of this paper, I will not distinguish between analogy and metaphor; that distinction will not matter much, since both analogies and metaphors can serve these three functions.

First, we might want to use these analogies for purposes of persuasion. For example, if we wanted to persuade legislators to enact stricter laws protecting endangered species, or to provide funding for ecological restoration projects, we might do well to compare species to artistic treasures, or to compare attempts to restore our relationship to nature to attempts to re-establish ties with long lost friends. The analogies have persuasive power in part because they are so vivid. Of course, we could avoid the colorful language and just say that species are valuable, and that ecological restoration is important, but environmentalists are less likely to persuade people if they use such drab language. When Derrick Jensen and George Draffan decided to give their book about deforestation the title, *Strangely Like War*, their intent can only have been rhetorical.

Second, we might want to form analogical arguments in the strict philosophical sense, rather like the traditional argument from design. Here is an example of an analogical argument for strategic monkeywrenching:

P1. Human beings' relationship to the environment is like an armed occupation. And in the few places where untainted wilderness remains, that relationship is like war.

P2. In this conflict, humans are the aggressors, while the environment is the innocent victim (i.e., humans are to the environment as Iraq was to Kuwait in 1990).

P3. In war, it is morally permissible for a third party to offer military assistance to an innocent victim of aggression.

C. Therefore, it is morally permissible for monkeywrenchers and eco-saboteurs to take measures to protect the environment, including the destruction of property.

Young (2001) suggests that the best argument for monkeywrenching is a utilitarian one. However, proponents of monkeywrenching themselves seem to prefer the appeal to the analogy with war.

Edward Abbey gives a slightly different argument in his "Foreward!" to Dave Foreman's manual for monkeywrenchers:

> If a stranger batters your door down with an axe, threatens your family and yourself with deadly weapons, and proceeds to loot your home of whatever he wants, he is committing what is universally recognized—by law and morality—as a crime. In such a situation the householder has both the right and the obligation to defend himself, his family, and his property by what ever means are necessary. This right and this obligation is universally recognized, justified, and even praised by all civilized human communities. Self-defense against attack is one of the basic laws not only of human society but of life itself, not only of human life but of all life.
>
> The American wilderness, what little remains, is now undergoing exactly such an assault. (1987, p. 7)

Here Abbey draws an analogy not with war but with home invasion. Rather than saying that humans are at war with nature, he suggests that humans are breaking, entering, threatening, and looting. The crucial premise of Abbey's argument is very similar to P3 above: Not only do people have a right to defend their homes, loved ones, and property against attack, but third parties may also intervene to protect the innocent victims. Is Abbey's argument any more or less promising than the argument based on the analogy with war? In section 4, I discuss some specific problems with the war analogy that do not apply to Abbey's analogy with home invasion. For that reason, Abbey's argument seems more promising. On the other hand, in section 3 I raise some general problems with the evaluation of analogical arguments that apply with equal force to the war analogy and the home invasion analogy. The trouble is that there are some relevant differences between our relationship to the natural environment and home invasion. For example, one cannot, strictly speaking, invade a home that one has lived in all along. Moreover, one important reason why it is wrong for someone to enter my home and loot it is that doing so violates my property rights. It is not clear, however, that humans are violating anyone else's property rights when they "loot" the non-human environment. Anyone wishing to defend Abbey's argument would need to explain why these disanalogies are less important than the similarities between home invasion and our relationship to the environment. In the remainder of the paper, I will focus mainly on the war analogy, with the understanding that the home invasion analogy, though perhaps a bit more promising, raises many of the same difficulties.

If the above argument by analogy with war were any good, it might enable monkeywrenchers to answer some of the standard objections to what they do. "It is simply wrong to destroy other people's property." But there is a war going on, and everyone concedes that it is permissible to attack and destroy the aggressor's weapons and munitions, so as to make it more difficult for the other side to prosecute the war. "It is wrong to break the law." But the laws themselves are implements of war that facilitate destruction and conquest of the environment. I will discuss some of the problems with this kind of argument in a moment; for now, the point is just that in addition to serving a rhetorical function, analogies can also be used with the aim of justifying moral claims and policy recommendations.

Third and finally, philosophers of science have traditionally distinguished between the context of discovery and the context of justification. Historians and some philosophers of science have questioned the usefulness of this distinction. In practice—that is, when we look at a particular bit of reasoning, or at a particular episode in the history of science—it can be extraordinarily difficult, if not impossible, to discern whether that reasoning belongs to the context of discovery or justification. Yet in principle, the distinction makes sense: when we draw a conclusion from a set of premises, we may be trying to justify that conclusion, or we may be discovering that conclusion—i.e. formulating it for the first time, without worrying yet whether there are any good reasons for believing it. Now there might well be certain moral judgments and policy recommendations that *would never have occurred to anyone* but for their consideration of one of the analogies listed above. For example, the monkeywrenchers' proposition that we are entitled to take certain measures (including destroying property) in order to defend the environment against humans might never have occurred to anyone but for their consideration of the analogy with war, or perhaps the analogy with home invasion. In other words, in one context P1, P2, and P3 might be offered as reasons for believing C; in another context, we might discover C by inferring it from P1, P2, and P3. Insofar as we use an analogy as a source of novel insights, hypotheses, and recommendations, the analogy serves a heuristic function.

To sum up: Any of the analogies considered thus far may serve one or more of three distinct functions: (1) persuasion; (2) justification; and (3) discovery. Which, if any, of these functions *should* analogies play in environmental philosophy? Here we may be able to make some progress by looking at the history of science.

One of the most striking things about the history of science is the prevalence of analogical thinking. One of the earliest theories of vision, for example (one that we may owe to Empedocles) depended upon a comparison of sight to touch. According to this theory, the eye sends out visual rays which "touch" the objects in the visual field and communicate information about them back to the eye (Park 1997, p. 35). Scientists frequently use analogies in either of two situations: (a) something poorly understood, such as eyesight, is compared to something that is better understood, such as touch; and (b) something unobservable is compared to something observable. As an example of (b), consider the analogy between the origin of the universe and an explosion. Perhaps nothing like the big bang theory could have occurred to anyone if gunpowder had never been invented. Or to take an example from the life sciences, the cranial crests of some dinosaurs puzzled scientists for many years, but today the consensus view is that the hollow crests were adaptations for making hooting and honking noises. This hypothesis was first arrived at by a German scientist who noticed similarities between the structure of one specimen's skull and a medieval musical instrument called a krumhorne (Turner 2000). It would be easy to multiply the examples.

In science, there may well be no consensus view about the proper role of analogy. Indeed, scientists may well use different analogies for different purposes. A good example of this is Darwin: There is good reason to think that Darwin discovered the principle of natural selection by drawing an analogy between the economic realm (as described by Malthus) and the biological realm. Most commentators agree that Darwin uses the analogy between artificial and natural selection as part of the justification for his conclusion that existing species have descended, with modification, from common ancestors. No doubt Darwin also uses this analogy for rhetorical purposes. The problem of figuring out just when a particular scientist is using a particular analogy for a particular purpose poses difficult challenges to historians.

Nevertheless, there are good methodological reasons why scientists should adhere to the following maxims:

Ml. Use analogies liberally in order to help formulate novel hypotheses about either (a) especially complex and challenging phenomena, or (b) things that we cannot observe.

M2. Rather than trying to justify one's theory or hypothesis by making an argument by analogy, it is always better to try to derive predictions from the hypothesis or theory and test those predictions against the observational and/or experimental evidence.

M3. Do not use analogies to persuade others in the scientific community to accept a theory or hypothesis. Ideally, we should want others to accept the hypothesis in question on the basis of the experimental and/or observational evidence.

I take it that these maxims are fairly uncontroversial. One way to justify Ml would be to point to all of the historical successes—all the theories that were inspired by analogies and later survived rigorous observational and experimental testing. M2 just reflects scientists' basic commitment to empiricism, and to the idea that observation and experiment are (or should be) the final arbiter of theory choice. Finally, M3 reflects scientists' commitment to rationality, and to the idea that the scientific community ought to be swayed by evidence rather than by rhetoric. More could be said in defense of all three of these maxims, but since I assume that most readers will find them to be plausible, I will not take time to defend them here. What I want to do, rather, is to use this analogy with science to help generate a novel claim about environmental philosophy: Perhaps environmental philosophers, and environmentalists more broadly, should treat analogies in roughly the same way that scientists do: We might think that analogies do have an appropriate role to play in environmental philosophy, but that this role has to do with the context of discovery rather than the context of justification. The suggestion, in other words, is that in environmental philosophy, analogies should perform the function of discovery, but not the functions of persuasion and justification. This proposal applies to the analogy with war there may be nothing wrong with using the analogy to generate novel moral claims and/or policy recommendations that we then submit for further consideration. But as in science, it may be a bad idea to use that analogy either to justify or to persuade.

3. WHY THE WAR ANALOGY FAILS TO JUSTIFY ECOSABOTAGE

Consider once again the earlier argument in favor of strategic monkeywrenching. Whether that argument is any good depends upon the first premise:

P1. Human beings' relationship to the environment is like an armed occupation. And in the few places where untainted wilderness remains, that relationship is like war.

In order to evaluate this or any other argument by analogy, we need to ask: How similar are the primary subject (in this case, our relationship to the environment) and the analog (war, conquest, and/or armed occupation)? Assuming that we can arrive at a substantive agreement about what the similarities and differences between the primary subject and the analog *are,* then we must pose the further question whether the agreed-upon similarities are the ones that matter. With this in mind, let's look more carefully at the warfare analogy, beginning with the similarities and then moving on to consider the differences.

Our relationship to the environment does resemble a state of war in a number of interesting ways: (1) Our relationship to the environment is a destructive one, and the level of destructiveness has risen with the development of new technologies—including everything from chemical pesticides to automobiles. Indeed, some of the technologies that have had a destructive impact on the environment, from rifles to nuclear weapons, are themselves military technologies. (2) Just as we distinguish between combatants and noncombatants in war, so we also distinguish wilderness areas that may be developed and certain protected areas that we deem "off limits." In the United States, for example, National Parks, National Wildlife Refuges, and other protected areas enjoy a status resembling non-combatant immunity. (3) Human civilization began in a number of geographically restricted areas, such as the Nile and Indus valleys, China, the Eastern Mediterranean, and Mesoamerica, and it has been gradually expanding ever since. This encroachment of human civilization against wilderness areas resembles a war of imperial conquest and territorial expansion. (4) Finally, some of the features of human agriculture—especially the large-scale industrial agriculture that emerged in the second half of the twentieth century—resemble warfare. Humans make it a policy to exterminate their competitors, including everything from wolves and other animals that prey upon our livestock, to weeds that compete with our crops for sunlight and nutrients. Similarly, many wars are conducted with the aim of eliminating competition for land, natural resources, and so forth (for a more detailed discussion of the similarities and differences between agriculture and war, see Surgey 1989). In short, just about anyone can agree that our relationship to non-human nature bears *some* degree of resemblance to war. This is hardly surprising, since any two items taken at random will be found to resemble each other in some respects. The question is whether these similarities are important enough to support any interesting conclusions.

Our relationship to the environment also differs from war in some important and obvious ways: (1) The relation of being at war is, fundamentally, a relation that holds between human social groups. Thus, we might say that two countries are at war, or that a government is embroiled in a war with a guerilla organization. The point is that war is a social phenomenon. But our relationship to the environment is not a social relationship, for the obvious reason that the environment is not a person or a social group. (2) War is symmetrical. If X is at war with Y, then Y is at war with X (i.e. it is impossible for the United States to be at war with Iraq, unless Iraq is also at war with the US). Yet many of those who seem to think that humans are at war with the environment do *not* think that the environment is at war with us. Environments, planets, species, and ecosystems are not the sorts of things that can wage war. (3) In war, people typically do not cross over into enemy territory for recreational purposes. Yet large numbers of people visit wilderness areas in order to relax, enjoy themselves, and engage in recreational activities. (4) Typically, we would want to say that if X is at war with Y, then X could exist even if Y did not. With reference to the recent US invasion of Iraq, we would want to say that the United States could exist even if Iraq did not. But no such claim is true of our relationship to the environment. Humans could not exist at all unless the environment existed. Proponents of the war analogy would surely acknowledge the existence of these and other differences. The question, once again, is: How important are they? Are these differences important enough to undermine the argument for strategic monkeywrenching?

Suppose, for the sake of argument, that we arrive at a substantive agreement concerning the similarities and differences between our relationship to the environment and war. A disagreement of emphasis persists. Proponents of strategic monkeywrenching argue that the similarities outweigh the differences, while the critics insist that the differences are such as to render P1 implausible. It seems that we are at an impasse. However, in order to justify strategic monkeywrenching, the proponent of ecosabotage must break this impasse. If the earlier argument for monkeywrenching is to be any good at all, it must be possible to explain why the similarities are more important than the differences. The explanation must not be circular. It would not help to argue that the similarities between war and our relationship to the environment matter more than the differences because those similarities have interesting ethical consequences. To do so would be to beg the question against anyone who doubts that the similarities have any interesting ethical consequences because the differences matter more.

How, then, might one go about trying to show that the similarities between the primary subject (our relationship to the environment) and the analog (war between human groups) matter more than the differences? One strategy might be to argue that war has an essence. Suppose we could discover the essential features of war, so that we could say, "Having features *F*l, *F*2,....*F*n is both necessary and sufficient for something's counting as war." For the moment, we need not worry about what those features might be, or about the metaphysical problem of what accounts for the distinction between essential and merely accidental features. The point is that if we knew the essence of war, then we would be in a position to say whether the similarities or differences between the primary subject and the analog matter more, because the essential features of war would be the ones that matter. If our relationship to the environment had these essential features, then we could reasonably conclude that we *are* at war with nature, and that the differences—in this case, the differences between our relationship to the environment and wars, such as the US invasion of Iraq—are merely accidental.

This essentialist maneuver is one way to argue, in a non-circular fashion, that the similarities outweigh the differences. Yet even if we could agree that war has an essence at all, it is hard to see how we could ever resolve disagreements about which features are the essential ones. Some disputants will argue that we are *not* at war with nature, because *F* is essential to war, and our relationship to the environment lacks this feature. Others who think that we *are* at war with nature will grant that our relationship to the environment lacks feature *F* while denying that feature *F* is essential to war. But notice what has happened: The debate about whether the similarities or the differences are most important will show up all over again, though in slightly different terms, as a debate about whether certain features are essential to war. The essentialist move transmutes the initial disagreement without actually settling anything. While there may be other ways of explaining why the similarities between the primary subject and the analog are more important than the differences, this essentialist move is not too promising.

Of course, there might be other ways of justifying monkeywrenching that make no appeal to the war analogy. For instance, Young (2001) shows how a number of objections to monkeywrenching can be answered from a utilitarian perspective. All I want to claim here is that the war analogy does not add anything to the justification of monkeywrenching.

4. WHY THOSE WHO OPPOSE WAR AND/OR VIOLENCE SHOULD NOT USE THE WAR ANALOGY FOR RHETORICAL OR JUSTIFICATORY PURPOSES

It is awfully tempting to use the war analogy in order to recruit people to the cause. Groups such as *Earth First!* and the Earth Liberation Front rely heavily on this analogy. For example, in one Primer on *Earth First!* that can be found online, we learn that "over the last several hundred years, human civilization has declared war on large mammals . . ." and that humans are conducting a "blitzkrieg against the natural world. . . ." Although more mainstream environmental organizations tend to use less inflammatory rhetoric, they do frequently invite people to become "defenders" of wilderness, the environment, or endangered species. Yet the analogy with warfare poses some dangers that other analogies (e.g. that between ecology and medicine) do not. There are three dangers in particular that we should worry about. It is on account of these three dangers that we should avoid using the war analogy for rhetorical or justificatory purposes. A pacifist like James, and a person who, like Foreman, claims to be committed to nonviolence, ought to appreciate these dangers most of all.

First, each one of us is involved in a host of relationships—with other people, with institutions, with our local environments, and so on. If we wanted to, we could think of many of those relationships in terms of warfare. To give an example: If I wanted to, I could proceed on the assumption that the college where I teach is at war with its peer institutions. Or I could proceed on the assumption that my department is at war with other departments on campus. Or I could conceive of a philosophical dispute with colleagues as a kind of war. In many of these cases, it would probably be easy to find *some* small similarity between the relationship in question and warfare, just as one can do in the case of the relationship between humans and the environment. However, there are two very good reasons to resist the temptation to think of any of these relationships in terms of war. To begin with, we must try to avoid getting used to thinking of ourselves as being at war. Someone who is too ready to think of himself as being involved in a war (whether a war on drugs, a war with other departments on campus, or whatever) is someone who is *warlike.* One need not be a pacifist at all in order to recognize that being warlike is a vice. The reason why being warlike is properly thought of as a vice is that the warlike person is someone who will not be reluctant enough to support a war. Even if a war is rationally and legally justifiable, people's support for it should be reluctant, simply because it is a war. The warlike person, however, will have a more difficult time thinking of war as a last resort. Those who have gotten in the habit of using the analogy with warfare to think about the various relationships in which they are involved are liable to find it more difficult to think about real wars in a critical and detached way.

A second reason why we should avoid using the analogy with war to think about any of the relationships in which we are involved is that doing so will have the effect of making those relationships more like war. If the analytic philosophers within a department were to begin thinking of themselves as being at war with their colleagues who do continental philosophy, or vice versa, then relations between the two groups would become more and more acrimonious and less collegial over time. When an analogy plays the justificatory and rhetorical roles, it inevitably influences behavior; people who think of themselves as being at war will tend to behave at times as if they were at war. In other words, they will become more warlike, and their relationships will become more like war.

The third major problem with the war analogy, as it figures in some contemporary environmental thought, is that it promotes what Lisa Gerber calls the vice of misanthropy. Any environmentalist who supposes that humans are at war with nature will naturally be tempted to think of humans as the enemy. One's enemy is the proper object of hatred if anyone is, and Gerber defines misanthropy as a "mistrust, hatred, and disgust of humankind" (2002, p. 41). One minor shortcoming of Gerber's illuminating discussion of misanthropy in the environmental movement is that she does not notice the conceptual connection between misanthropy and the war analogy. She draws a useful distinction between self-righteous misanthropy and self-hating misanthropy. Foreman, surely, is a good example of the self-righteous misanthrope. Humans are involved in an unjust war against nature, but it is possible to become righteous by resisting on behalf of non-human nature. Gerber argues that misanthropy is a vice, first because it tends to produce hopelessness and despair, and second because the misanthrope sees humans "as a despicable mass" rather than as "individuals capable of moral and social change" (2002, p. 42). The goal toward which environmentalists should direct all their efforts is something like this: We want humans to flourish, and to live good lives, but not at the expense of future generations or other forms of life on earth. How can we make any progress toward achieving this goal if we proceed on the assumption that humans are the enemy?

5. CONCLUSION: MAKING PEACE WITH NON-HUMAN NATURE

I have argued that it is a bad idea to think of ourselves as being at war with nature, but I want to suggest in closing that the war analogy may not be all that bad, because it may have a useful heuristic role to play. It is fashionable and fruitful to compare ecology to medicine. Perhaps we could also learn something by comparing restoration ecology to peace and conflict resolution. We might think of restoration ecology as part of a broader attempt to establish peace between humans and the environment. Environmental legislation and policy decisions could also be thought of as part of this larger "peace process." The thought that we are presently at war with non-human nature could well lead us to think about what it might mean to make peace with the natural world.

Notice that making peace with nature would not mean leaving it alone to develop on its own, untouched by humans. When two countries make peace with one another in the international arena, this usually means that they continue to interact with one another, but in a different way. If anything, they seek to strengthen the cultural, political, and economic ties that bind them together, so that over time, each becomes more dependent on the other. Here, then, is just one example of an ethical claim that owes its inspiration to the war analogy: making peace with the environment does not mean leaving the environment alone, as preservationists recommend, but rather cultivating and strengthening all of the various ties between ourselves and the environment (analogous to the cultural, political, and economic ties between countries). I will leave it to others to pursue these leads. The point is simply that there is nothing wrong with using the war analogy as a tool for generating novel moral claims. If we think of ecological science and environmental policymaking as part of a larger project of bringing about peace between humans and nature, we avoid the pitfalls discussed in section 4: There is no danger here of promoting vices such as warlikeness and misanthropy. Furthermore, there is far less danger of arousing people's passions in ways that do not contribute to the goals of the environmental movement. Moreover, if we do not try to use the analogy for justificatory purposes, then the significant differences between war and our relationship to the environment will not be a problem. Those obvious differences should not prevent us from using the analogy as a source of novel ethical insights and policy recommendations.

NOTES

An early version of this paper was presented to colleagues and students at the Goodwin-Niering Center for Conservation Biology and Environmental Studies, Connecticut College, November 6, 2003. I am also grateful to Thomas Young for his help in improving this paper.

REFERENCES

Ehrlich, Paul. *The Population Bomb*. New York: Ballantine Books, 1968.

Ehrlich, Paul, and Anne Ehrlich. *Extinction: The Causes and Consequences of the Disappearance of Species*. New York: Random House, 1981.

Foreman, Dave. *Ecodefense: A Field Guide to Monkeywrenching*. Tucson, AZ: Ned Ludd, 1987.

Gerber, Lisa. "What Is So Bad About Misanthropy?" *Environmental Ethics* 24 (2002): 41–55.

James, William. "The Moral Equivalent of War." In *The Writings of William James: A Comprehensive Edition*, edited by John J. McDermott, 660–71. Chicago: University of Chicago Press, 1910/1977.

Jensen, Derrick, and George Draffan. *Strangely Like War: The Global Assault on Forests*. White River Junction, VT: Chelsea Green Publishing, 2003.

Leopold, Aldo. "The Land Ethic," in *A Sand County Almanac*. Oxford: Oxford University Press, 1949.

Light, Andrew. "Ecological Restoration and the Culture of Nature: A Pragmatic Perspective." In *Restoring Nature: Perspectives from the Social Sciences and Humanities*, edited by P. Gobster and P. Hall, 49–70. Washington: Island Press, 2000.

———. "Taking Environmental Ethics Public." In *Environmental Ethics: What Really Matters, What Really Works*, edited by David Schmidtz and Elizabeth Willott, 556–66. Oxford: Oxford University Press, 2002.

Norton, Bryan. "What Is a Conservation Biologist?" *Conservation Biology* 2 (1988): 237–38.

Park, D. *The Fire Within the Eye: A Historical Essay on the Nature and Meaning of Light*. Princeton, NJ: Princeton University Press, 1997.

Quinn, Daniel. *Ishmael*. New York: Bantam/Turner, 1992.

Rolston, Holmes III. "Duties to Endangered Species." In *Philosophy Gone Wild*, 206–20. Buffalo, NY: Prometheus Books, 1989.

Russow, Lily-Marlene. "Why Do Species Matter?" *Environmental Ethics* 3 (1981): 101–12.

Slobodkin, Lawrence B. *A Citizen's Guide to Ecology*. Oxford: Oxford University Press, 2003.

Surgey, John. "Agriculture: A War on Nature?" *Journal of Applied Philosophy* 6 (1989): 205–207.

Turner, Derek. "The Functions of Fossils: Inference and Explanation in Functional Morphology." *Studies in History and Philosophy of Biology and Biomedical Sciences* 31 (2000): 193–212.

Warren, Karen. "The Power and the Promise of Ecological Feminism." *Environmental Ethics* 12 (1990): 125–46.

Young, Thomas. "The Morality of Ecosabotage." *Environmental Values* 10 (2001): 385–93.

ENVIRONMENTAL ETHICS

*HOLMES ROLSTON III**

ON BEHALF OF BIOEXUBERANCE

The old ethic focused on the welfare of only one species; a new ethic must regard the welfare of the several million species that constitute evolving life on Earth.

The signs at a subalpine campground in the Rocky Mountain Rawah Range suggest a new kind of caring about plants. For years the trailside signs there read, "Please leave the flowers for others to enjoy." But recently, the wasted wooden signs were replaced by newly cut ones: "Let the flowers live!" Perhaps the intent was only to send a subtle psychological message, but I suspect a shifting ethic—respect for plants, replacing what was before only respect for persons.

There is similar evidence at the Indiana Dunes National Lakeshore, along Lake Michigan. The dunes were the site of the earliest studies in ecology, and their preservation required a long, bitter environmental fight. A major argument for saving the dunes was to preserve them as a playground for Chicagoans. But a recent Park Service poster depicts a clump of marram grass, sand and the lake and offers the injunction, "Let it be!" There seems now in the Park Service a caring for grass and dunes, something beyond mere maintenance of a lakeshore playground.

Such responses reflect a new ethic—one that adds a respect for plants to a respect for people. Not coincidentally, this comes at a time of enormous human-caused changes in the natural world. Previously, humans did not have much power to spoil ecosystems and cause extinctions, or much knowledge about what they were inadvertently doing. But today humans have more understanding than ever before of the biological processes, more predictive power to foresee the intended

*Holmes Rolston III, *Philosophy Gone Wild* (Prometheus Books, 1986) and *Environmental Ethics* (Temple University Press, to be released this fall).

and unintended results of their actions, and more power to reverse the undesirable consequences. We know many floristic natural histories; we find that willy-nilly we have a vital role in whether these stories continue.

We are appreciating vitality in the biological world, one that precedes and overleaps our personal or cultural presence. And with this new appreciation comes a deeper sense of responsibility.

Ethicists say that in *Homo sapiens* one species has appeared that not only exists but ought to exist. But why say this exclusively of a late-coming, highly developed form? Why not extend this duty more broadly to other species? Only the human species contains moral agents, but the paradox is that humankind, the single moral species, acts only in its collective self interest toward all the rest. Perhaps conscience ought not be used to exempt every other form of life from consideration.

In understanding why humans ought to let wildflowers and marram grass be, we need to see that to care about plant species is not to adopt some vague, subjective intuitions of romantic humans who fancy curious plants. To the contrary, it is to be quite nonanthropocentric and objective about botanical processes that take place independently of human preferences.

THE SHIFT AWAY FROM HUMANISM

In the past, the reasons given for preserving rare plants have routinely been humanistic: *Please leave the plants for others to use.* People have a strong obligation not to harm other people, and a weaker, though important, duty to benefit others. Given the many ways that humans use plants—agriculturally, industrially, medically, recreationally, aesthetically, scientifically, as cultural symbols, as environmental indicators, as part of their life-support system—humans are significantly affected by their flora. Even rare plants had value to the extent that they were part of this plant world that benefits humans.

In this human-oriented view, we would not say that the needless destruction of a plant species was cruel, but we might say that it was callous. We would not be concerned about what the plants felt, but about what the human destroyers did not feel. We would not be valuing sensitivity in plants, but censuring insensitivity in persons. And we might go on to ask what properties in plants a person should be sensitive to.

But when we look past a concern for people, when we try to articulate an ethic to explain the deeper, naturalistic, reasons to let rare plants be, we get lost in unfamiliar territory. We find that all the familiar moral landmarks are gone. We are not addressing humans, or culture, or moral agents; we are not considering animals that are close kin, or can suffer or experience anything, or that are sentient. Plants are not "valuers," with preferences that can be satisfied or frustrated.

In moving toward a new ethic, what we find ourselves caring about are "only" plants, and plants can't "care," so why should we? It seems odd to assert that rare flowers or species have rights or moral standing, or need our sympathy; odd to ask that we should consider their point of view.

Moreover, we are not caring about individual plants, but rather about species. To an even greater degree than individuals, species don't "care."

In addition, 98 percent of the species that have ever inhabited Earth are extinct. Evidently nature doesn't care about species, so why should we? Finally, why should we care about rare plants—what has their rarity to do with their value?

None of these elements—flora, species, ecosystems, wilderness, or rarity—has figured within the coordinates of prevailing ethical systems. In fact, ethics and biology have had uncertain relations. An often-heard argument forbids moving from what *is* the case (a description of botanical facts) to what *ought to be* (an ethical prescription of duty). Philosophers accuse anyone who argues in this way of committing what they call the naturalistic fallacy.

THE PLANT WAY OF CARING

A living plant, though lacking a brain or neural center, has a controlling program that enables it to maintain itself. The plant control program is coded in the DNA, the informational molecules. Through this program, the plant composes and recomposes itself, maintaining order against disordering tendencies and checking against its performance in the world via feedback loops. The genetic set distinguishes between what is and what ought to be—that is, it is a normative system.

Each plant develops and maintains a botanical identity, posting a boundary between itself and its environment. An acorn becomes an oak; the oak stands on its own.

Plants do not, of course, have ends in view, they do not have goals. And a plant is not a moral system—there are no moral agents in nonhuman nature. But a plant, unlike, say, a rock, is an evaluative system, selecting resources for itself.

From one perspective, a plant's activity is just biochemistry—the whir and buzz of proteins and other organic molecules. But from an equally valid (and equally objective) perspective, the activity is a valued state; the plant life is not merely *biological* but, given the way the plant defends itself, the life is *vital.*

Hence, to the assertion that plants don't care, the response is that plants do care—using botanical standards. They defend their lives, an intrinsic value, in the only form of caring available to them.

If we attach value to life defended (rather than to human preferences), then we must attach value to plants, because plants defend their lives as good-in-themselves. To say that there is no value involved because this activity is controlled by the genome and not by a conscious brain is something like saying that there is neither information nor life in the plant.

A plant is engaged in the biological conservation of its identity and kind. Conservation biologists and others ought therefore to respect plants for what they are: projects in conservation biology. This view aligns ethics with objective biology.

THE ARGUMENT FOR SPECIES

Although we can see why we might respect individual plants, what about species? In one view, a species is a useful fiction, like a center of gravity or a statistical average. Species might be only classes of convenience, or, like lines of latitude or contours, devices for mapping the world. Indeed, taxonomists insist on appending to the plant's Latin name the name of the "author" who, they say, "erected" the taxon.

Even Darwin wrote, "I look at the term species, as one arbitrarily given for the sake of convenience to a set of individuals closely resembling one another." Botanists are divided whether Illinois' Kankakee mallow, *Iliamna remota*, and Virginia's *Iliamna corei*, both rare, are distinct species. Perhaps all that exists objectively are the individual mallow plants; whether there are two species or one is a fuss over what labels to use.

Against this, though, is the claim that there are specific forms of life maintained in their ecosystems over time. Evolutionary lines develop into diverse kinds of life, each with a more or less distinct integrity, a breeding population, and a gene pool.

G. G. Simpson says, "An evolutionary species is a lineage (an ancestral-descendant sequence of populations) evolving separately from others and with its own unitary evolutionary role and tendencies." Niles Eldredge and Joel Cracraft insist that species are "discrete entities in time as well as space."

In this view, the idea of species does not seem arbitrary or fictitious at all, but rather, as certain as anything we believe about the empirical world, even though at times taxonomists revise the theories and taxa with which they map these forms. Species are like mountains and rivers, objectively there to be mapped. The edges of these natural groups will sometimes be fuzzy, to some extent discretionary; one species will slide into another over evolutionary time and in some cases actual speciation is now in progress. But the various criteria biologists use for defending species (descent, reproductive isolation, morphology, gene pool) provide evidence that species are really there.

What survives for a few months, years or decades is the individual plant; what survives for millennia is the kind, or species. Life is therefore something passing through the individual as much as something the individual possesses on its own. A species is a dynamic life form preserved in historical lines and has a vitality that persists genetically over millions of years, overleaping short-lived individuals.

Further, reproduction can be looked on as the means by which a species defends itself. This does not mean that a species has a controlling center, any more than a plant has a brain; but the species, like the individual, is a survival process. Both conserve botanical identity over time.

An ethic about plants sees that the species is a bigger event than the individual. In a sense the species level is more appropriate for moral concern since the species is a more comprehensive survival unit than the organism.

When an individual rhododendron dies, another one replaces it. The deaths of individual rhododendrons are even necessary if the species is to persist: Seeds are dispersed and replacement rhododendrons grow elsewhere in the forest. As landscapes change or succession shifts, later replacements, including mutants, provide a steady turnover. Thus the species improves in fitness or adapts to a changing climate or to competitive pressures. Tracking its environment over time, the species is conserved and modified. As this process is extended overtime, certain species are unable to adapt, there are natural extinctions, with re-speciation and a normal turnover—and the generative process continues unabated.

But with human-caused extinctions, this process stops. Such extinction shuts down the generative processes and is a kind of superkilling. It kills forms beyond individuals. It kills "essences" beyond "existences," the "soul" as well as the "body." It kills collectively, not just distributively. To kill a particular plant is to stop a life of a few years, while other lives of such kind continue unabated and the possibilities for the future are unaffected. To kill a particular species is to shut down a story of many millennia, and to leave no future possibilities.

A consideration of species strains any ethic focused on individuals, much less on sentience or persons. But, though it revises what was formerly thought logically permissible or ethically binding, the result can be a biologically sounder ethic. The species line is fundamental. It is more important to protect this integrity than to protect individuals. The appropriate survival unit is the appropriate level of moral concern.

"Ought species *x* to exist?" is a single element in the collective question, "Ought life on Earth to exist?" The answer to the question about one species is not always the same as the answer to the bigger question, but since life on Earth is an aggregate of many species, the two are sufficiently related that the burden of proof lies with those who wish deliberately to extinguish a species and simultaneously to care for life on Earth.

Humans ought not to play the role of murderers. The duty to species can be overridden, for example with pests or disease organisms. But a prima facie duty stands nevertheless.

THE ARGUMENT FOR EVOLVING ECOSYSTEMS

On evolutionary time scales, species, like individuals, are ephemeral. But the speciating process is not. Persisting through vicissitudes for two-and-a-half billion years, species evolution is about as long-continuing as anything on Earth can be.

Ecosystems are biotic communities, kept in dynamic evolution over time by selection pressures toward an optimally satisfactory fit for each species. Each species defends only its own kind, but the ecosystem coordinates kinds, through a spontaneously evolving order that arises when many such species interact. That order exceeds in richness, beauty and dynamic stability the order of any of the component parts. Species reproduce their own kind; evolutionary ecosystems produce new kinds. Bioexuberance, both diversity and complexity, is conserved while it is increased.

Ecosystems are the context of speciation. Neither individual nor species stands alone; both are embedded in an ecosystem, and in that sense it is even more important to save evolutionary ecosystems than to save species. Species are what they are where they are. The comprehensive ecosystem too is a vital survival unit.

It might seem that for humans to terminate plant species now and again is quite natural—after all, plants become extinct all the time. But when human culture supplants nature, extinction is radically different. Natural extinction is the key to the future because in nature, a species dies when it has become unfit in its habitats, and other species appear in its place. Artificial extinction closes off the future because it shuts down speciation.

One can say that nature doesn't care about species, and in a way that is true. But it does not follow that there is nothing in nature promoting, conserving, elaborating species, or that we should not care about this. "Caring" is perhaps not the appropriate language to describe the natural processes by which Earth conserves life, overleaping species, starting from zero to elaborate a biota of several million species. But nature does seem to generate species with remarkable fertility and extravagance in Earth's several billion years of creative struggle.

We hardly yet have a complete theoretical account of this richness of life, but bioscience gives us this certainty: The evolutionary odyssey is prolific, that is, pro-life. We ought to admire the process as much as the product.

VALUING THE RARE

Rarity per se is not a valuable property. Rarity simply means few individuals of this kind exist. We do not, or should not, value plants or plant encounters just because they are rare.

That a plant is naturally rare may seem to suggest its insignificance in an ecosystem. But naturally rare species, as much as common species, signify exuberance in nature: Each is a unique expression of the potential that drives evolution. Some rare plants may be en route to natural extinction, but it does not follow that most rare plants have less biological competence than common species. On the contrary, endemics or specialized species—like the grape fern *Botrychium pumicola*, which grows only on pumice at high elevations in the Cascade Mountains—may competently occupy restricted niches.

A rare flower is a botanical achievement, a bit of brilliance, a problem resolved, a threshold crossed. An endemic species, perhaps one specialized for an unusual habitat, represents a rare discovery in nature (in addition to the adventure that humans experience in finding it). Rare species ornament the display of life. Together, the myriad species make Earth a garden.

Some rare plants live on the cutting edge of adaptability; some are relics of the past. Either way they offer promise and memory of an inventive natural history. Even more poignantly than the common, they provide both a liberal and a conservative sign, evidence of life persisting in struggling beauty, flourishing, pushing on at the edge of perishing. The rare flowers—if one is open to a wider, more philosophical perspective—offer a moment of perennial truth.

Rare species have proved their right to life through being tested by natural selection. These examples of biological right-to-life, of adaptive fitness in an ecosystem, generate at least a presumption in the humans who encounter them that these are good kinds, good right where they are, and therefore that it is right for humans to let them be, to let them evolve. That leaves plants, species, and process all in place.

When humans make once-common plants artificially rare, biological vitality is lost. When humans extinguish species, they stop the story. That makes humans misfits in the system, because they bring death without survivors into Earth's prolific exuberance. Life is a many-splendored thing; extinction of the rare dims its luster.

Several billion years worth of creative toil, several million species of teeming life, have been handed over to the care of this late-coming species in which mind has flowered and morals have emerged. Ought not those of this sole moral species do something less self-interested than count all the produce of an evolutionary ecosystem as resources? Such an attitude hardly seems ethically adequate.

There is something overspecialized about an ethic that regards the welfare of only one of several million species as an object of duty. It is an ethic no longer functioning in, or suited to, the changing environment. There is something morally naive about living in a reference frame where one species takes itself as absolute and values everything else relative to its utility.

The old signs, "leave the flowers for others to enjoy," reflected a humanistic ethic. The new, naturalistic signs invite a change of reference frame. Love the flora too. Let it be. Let life flower!

EDWARD O. WILSON

BIOPHILIA AND THE CONSERVATION ETHIC

Biophilia, if it exists, and I believe it exists, is the innately emotional affiliation of human beings to other living organisms. Innate means hereditary and hence part of ultimate human nature. Biophilia, like other patterns of complex behavior, is likely to be mediated by rules of prepared and counterprepared learning—the tendency to learn or to resist learning certain responses as opposed to others. From the scant evidence concerning its nature, biophilia is not a single instinct but a complex of learning rules that can be teased apart and analyzed individually. The feelings molded by the learning rules fall along several emotional spectra: from attraction to aversion, from awe to indifference, from peacefulness to fear-driven anxiety.

The biophilia hypothesis goes on to hold that the multiple strands of emotional response are woven into symbols composing a large part of culture. It suggests that when human beings remove themselves from the natural environment, the biophilic learning rules are not replaced by modern versions equally well adapted to artifacts. Instead, they persist from generation to generation, atrophied and fitfully manifested in the artificial new environments into which technology has catapulted humanity. For the indefinite future more children and adults will continue, as they do now, to visit zoos than attend all major professional sports combined (at least this is so in the United States and Canada), the wealthy will continue to seek dwellings on prominences above water amid parkland, and urban dwellers will go on dreaming of snakes for reasons they cannot explain.

Were there no evidence of biophilia at all, the hypothesis of its existence would still be compelled by pure evolutionary logic. The reason is that human history did not begin eight or ten thousand years ago with the invention of agriculture and villages. It began hundreds of thousands or millions of years ago with the origin of the genus *Homo*. For more than 99 percent of human history people have lived in hunter-gatherer bands totally and intimately involved with other organisms. During this period of deep history, and still farther back, into paleo-hominid times, they depended on an exact learned knowledge of crucial aspects of natural history. That much is true even of chimpanzees today, who use primitive tools and have a practical knowledge of plants and animals. As language and culture expanded, humans also used living organisms of diverse kinds as a principal source of metaphor and myth. In short, the brain evolved in a biocentric world, not a machine-regulated world. It would be therefore quite extraordinary to find that all learning rules related to that world have been erased in a few thousand years, even in the tiny minority of peoples who have existed for more than one or two generations in wholly urban environments.

The significance of biophilia in human biology is potentially profound, even if it exists solely as weak learning rules. It is relevant to our thinking about nature, about the landscape, the arts, and mythopoeia, and it invites us to take a new look at environmental ethics.

How could biophilia have evolved? The likely answer is biocultural evolution, during which culture was elaborated under the influence of hereditary learning propensities while the genes prescribing the propensities were spread by natural selection in a cultural context. The learning rules can be inaugurated and fine-tuned variously by an adjustment of sensory thresholds, by a quickening or blockage of learning, and by modification of emotional responses. Charles Lumsden and I (1981; 1983; 1985) have envisioned biocultural evolution to be of a particular kind, gene-culture coevolution, which traces a spiral trajectory through time: a certain genotype makes a behavioral response more likely, the response enhances survival and reproductive fitness, the genotype consequently spreads through the population, and the behavioral response grows more frequent. Add to this the strong general tendency of human beings to translate emotional feelings into myriad dreams and narratives, and the necessary conditions are in place to cut the historical channels of art and religious belief.

Gene-culture coevolution is a plausible explanation for the origin of biophilia. The hypothesis can be made explicit by the human relation to snakes. The sequence I envision, drawn principally from elements established by the art historian and biologist Balaji Mundkur, is this:

1. Poisonous snakes cause sickness and death in primates and other mammals throughout the world.
2. Old World monkeys and apes generally combine a strong natural fear of snakes with fascination for these animals and the use of vocal communication, the latter including specialized sounds in a few species, all drawing attention of the group to the presence of snakes in the near vicinity. Thus alerted, the group follows the intruders until they leave.

3. Human beings are genetically averse to snakes. They are quick to develop fear and even full-blown phobias with very little negative reinforcement. (Other phobic elements in the natural environment include dogs, spiders, closed spaces, running water, and heights. Few modern artifacts are as effective—even those most dangerous, such as guns, knives, automobiles, and electric wires.)
4. In a manner true to their status as Old World primates, human beings too are fascinated by snakes. They pay admission to see captive specimens in zoos. They employ snakes profusely as metaphors and weave them into stories, myth, and religious symbolism. The serpent gods of cultures they have conceived all around the world are furthermore typically ambivalent. Often semihuman in form, they are poised to inflict vengeful death but also to bestow knowledge and power.
5. People in diverse cultures dream more about serpents than any other kind of animal, conjuring as they do so a rich medley of dread and magical power. When shamans and religious prophets report such images, they invest them with mystery and symbolic authority. In what seems to be a logical consequence, serpents are also prominent agents in mythology and religion in a majority of cultures.

Here then is the ophidian version of the biophilia hypothesis expressed in briefest form: constant exposure through evolutionary time to the malign influence of snakes, the repeated experience encoded by natural selection as a hereditary aversion and fascination, which in turn is manifested in the dreams and stories of evolving cultures. I would expect that other biophilic responses have originated more or less independently by the same means but under different selection pressures and with the involvement of different gene ensembles and brain circuitry.

This formulation is fair enough as a working hypothesis, of course, but we must also ask how such elements can be distinguished and how the general biophilia hypothesis might be tested. One mode of analysis, reported by Jared Diamond in this volume, is the correlative analysis of knowledge and attitude of peoples in diverse cultures, a research strategy designed to search for common denominators in the total human pattern of response. Another, advanced by Roger Ulrich and other psychologists, is also reported here: the precisely replicated measurement of human subjects to both attractive and aversive natural phenomena. This direct psychological approach can be made increasingly persuasive, whether for or against a biological bias, when two elements are added. The first is the measurement of heritability in the intensity of the responses to the psychological tests used. The second element is the tracing of cognitive development in children to identify key stimuli that evoke the responses, along with the ages of maximum sensitivity and learning propensity. The slithering motion of an elongate form appears to be the key stimulus producing snake aversion, for example, and preadolescence may be the most sensitive period for acquiring the aversion.

Given that humanity's relation to the natural environment is as much a part of deep history as social behavior itself, cognitive psychologists have been strangely slow to address its mental consequences. Our ignorance could be regarded as just one more blank space on the map of academic science, awaiting genius and initiative, except for one important circumstance: the natural environment is disappearing. Psychologists and other scholars are obligated to consider biophilia in more urgent terms. What, they should ask, will happen to the human psyche when such a defining part of the human evolutionary experience is diminished or erased?

There is no question in my mind that the most harmful part of ongoing environmental despoliation is the loss of biodiversity. The reason is that the variety of organisms, from alleles (differing gene forms) to species, once lost, cannot be regained. If diversity is sustained in wild ecosystems, the biosphere can be recovered and used by future generations to any degree desired and with benefits literally beyond measure. To the extent it is diminished, humanity will be poorer for all generations to come. How much poorer? The following estimates give a rough idea:

- Consider first the question of the *amount* of biodiversity. The number of species of organisms on earth is unknown to the nearest order of magnitude. About 1.4 million species have been given names to date, but the actual number is likely to lie somewhere between 10 and 100 million. Among the least-known groups are the fungi, with 69,000 known species but 6 million thought to exist. Also poorly explored are at least 8 million and possibly tens of millions of species of arthropods in the tropical rain forests, as well as millions of invertebrate species on the vast floor of the deep sea. The true black hole of systematics, however, may be bacteria. Although roughly 4,000 species have been formally recognized, recent studies in Norway indicate the presence of 4,000 to 5,000 species among the 10 billion individual organisms found on average in each gram of forest soil, almost all new to science, and another 4,000 to 5,000 species, different from the first set and also mostly new, in an average gram of nearby marine sediments.
- Fossil records of marine invertebrates, African ungulates, and flowering

plants indicate that on average each clade—a species and its descendants—lasts half a million to 10 million years under natural conditions. The longevity is measured from the time the ancestral form splits off from its sister species to the time of the extinction of the last descendant. It varies according to the group of organisms. Mammals, for example, are shorter-lived than invertebrates.

- Bacteria contain on the order of a million nucleotide pairs in their genetic code, and more complex (eukaryotic) organisms from algae to flowering plants and mammals contain 1 to 10 billion nucleotide pairs. None has yet been completely decoded.
- Because of their great age and genetic complexity, species are exquisitely adapted to the ecosystems in which they live.
- The number of species on earth is being reduced by a rate 1,000 to 10,000 times higher than existed in prehuman times. The current removal rate of tropical rain forest, about 1.8 percent of cover each year, translates to approximately 0.5 percent of the species extirpated immediately or at least doomed to much earlier extinction than would otherwise have been the case. Most systematists with global experience believe that more than half the species of organisms on earth live in the tropical rain forests. If there are 10 million species in these habitats, a conservative estimate, the rate of loss may exceed 50,000 a year, 137 a day.

Other species-rich habitats, including coral reefs, river systems, lakes, and Mediterranean-type heathland, are under similar assault. When the final remnants of such habitats are destroyed in a region—the last of the ridges on a mountainside cleared, for example, or the last riffles flooded by a downstream dam—species are wiped out en masse. The first 90 percent reduction in area of a habitat lowers the species number by one-half. The final 10 percent eliminates the second half.

It is a guess, subjective but very defensible, that if the current rate of habitat alteration continues unchecked, 20 percent or more of the earth's species will disappear or be consigned to early extinction during the next thirty years. From prehistory to the present time humanity has probably already eliminated 10 or even 20 percent of the species. The number of bird species, for example, is down by an estimated 25 percent, from 12,000 to 9,000, with a disproportionate share of the losses occurring on islands. Most of the megafaunas—the largest mammals and birds—appear to have been destroyed in more remote parts of the world by the first wave of hunter-gatherers and agriculture centuries ago. The diminuation of plants and invertebrates is likely to have been much less, but studies of archeological and other subfossil deposits are too few to make even a crude estimate. The human impact, from prehistory to the present time and projected into the next several decades threatens to be the greatest extinction spasm since the end of the Mesozoic era 65 million years ago.

Assume, for the sake of argument, that 10 percent of the world's species that existed just before the advent of humanity are already gone and that another 20 percent are destined to vanish quickly unless drastic action is taken. The fraction lost—and it will be a great deal no matter what action is taken—cannot be replaced by evolution in any period that has meaning for the human mind. The five previous major spasms of the past 550 million years, including the end-Mesozoic, each required about 10 million years of natural evolution to restore. What humanity is doing now in a single lifetime will impoverish our descendants for all time to come. Yet critics often respond, "So what? If only half the species survive, that is still a lot of biodiversity—is it not?"

The answer most frequently urged right now by conservationists, I among them, is that the vast material wealth offered by biodiversity is at risk. Wild species are an untapped source of new pharmaceuticals, crops, fibers, pulp, petroleum substitutes, and agents for the restoration of soil and water. This argument is demonstrably true—and it certainly tends to stop anticonservation libertarians in their tracks—but it contains a dangerous practical flaw when relied upon exclusively. If species are to be judged by their potential material value, they can be priced, traded off against other sources of wealth, and—when the price is right—discarded. Yet who can judge the *ultimate* value of any particular species to humanity? Whether the species offers immediate advantage or not, no means exist to measure what benefits it will offer during future centuries of study, what scientific knowledge, or what service to the human spirit.

At last I have come to the word so hard to express: spirit. With reference to the spirit we arrive at the connection between biophilia and the environmental ethic. The great philosophical divide in moral reasoning about the remainder of life is whether or not other species have an innate right to exist. That decision rests in turn on the most fundamental question of all: whether moral values exist apart from humanity, in the same manner as mathematical laws, or whether they are idiosyncratic constructs that evolved in the human mind through natural selection. Had a species other than humans attained high intelligence and culture, it would likely have fashioned different moral values. Civilized termites, for example, would support cannibalism of the sick and injured, eschew personal reproduction, and make a sacrament of the exchange and consumption of feces. The termite spirit, in short, would have been immensely different from the human spirit—horrifying to us in fact. The constructs of moral reasoning, in this evolutionary view, are the learning rules, the propensities to acquire or to resist certain emotions and kinds of knowledge. They have evolved genetically because they confer survival and reproduction on human beings.

The first of the two alternative propositions—that species have universal and independent rights regardless of how else human beings feel about the matter—may be true. To the extent the proposition is accepted, it will certainly steel the determination of environmentalists to preserve the remainder of life. But the species-right argument alone, like the materialistic argument alone, is a dangerous play of the cards on which to risk biodiversity. The independent-rights argument, for all its directness and power, remains intuitive, aprioristic, and lacking in objective evidence. Who but humanity, it can be immediately asked, gives such rights? Where is the enabling canon written? And such rights, even if granted, are always subject to rank-ordering and relaxation. A simplistic adjuration for the right of a species to live can be answered by a simplistic call for the right of people to live. If a last section of forest needs to be cut to continue the survival of a local economy, the rights of the myriad species in the forest may be cheerfully recognized but given a lower and fatal priority.

Without attempting to resolve the issue of the innate rights of species, I will argue the necessity of a robust and richly textured anthropocentric ethic apart from the issue of rights—one based on the hereditary needs of our own species. In addition to the well-documented utilitarian potential of wild species, the diversity of life has immense aesthetic and spiritual value. The terms now to be listed will be familiar, yet the evolutionary logic is still relatively new and poorly explored. And therein lies the challenge to scientists and other scholars.

Biodiversity is the Creation. Ten million or more species are still alive, defined totally by some 10^{17} nucleotide pairs and an even more astronomical number of possible genetic recombinants, which creates the field on which evolution continues to play. Despite the fact that living organisms compose a mere ten-billionth part of the mass of earth, biodiversity is the most information-rich part of the known universe. More organization and complexity exist in a handful of soil than on the surfaces of all the other planets combined. If humanity is to have a satisfying creation myth consistent with scientific knowledge—a myth that itself seems to be an essential part of the human spirit—the narrative will draw to its conclusion in the origin of the diversity of life.

Other species are our kin. This perception is literally true in evolutionary time. All higher eukaryotic organisms, from flowering plants to insects and humanity itself, are thought to have descended from a single ancestral population that lived about 1.8 billion years ago. Single-celled eukaryotes and bacteria are linked by still more remote ancestors. All this distant kinship is stamped by a common genetic code and elementary features of cell structure. Humanity did not soft-land into the teeming biosphere like an alien from another planet. We arose from other organisms already here, whose great diversity, conducting experiment upon experiment in the production of new life-forms, eventually hit upon the human species.

The biodiversity of a country is a part of its national heritage. Each country in turn possesses its own unique assemblages of plants and animals including, in almost all cases, species and races found nowhere else. These assemblages are the product of the deep history of the national territory, extending back long before the coming of man.

Biodiversity is the frontier of the future. Humanity needs a vision of an expanding and unending future. This spiritual craving cannot be satisfied by the colonization of space. The other planets are inhospitable and immensely expensive to reach. The nearest stars are so far away that voyagers would need thousands of years just to report back. The true frontier for humanity is life on earth—its exploration and the transport of knowledge about it into science, art, and practical affairs. Again, the qualities of life that validate the proposition are: 90 percent or more of the species of plants, animals, and microorganisms lack even so much as a scientific name; each of the species is immensely old by human standards and has been wonderfully molded to its environment; life around us exceeds in complexity and beauty anything else humanity is ever likely to encounter.

The manifold ways by which human beings are tied to the remainder of life are very poorly understood, crying for new scientific inquiry and a boldness of aesthetic interpretation. The portmanteau expressions "biophilia" and "biophilia hypothesis" will serve well if they do no more than call attention to psychological phenomena that rose from deep human history, that stemmed from interaction with the natural environment, and that are now quite likely resident in the genes themselves. The search is rendered more urgent by the rapid disappearance of the living part of that environment, creating a need not only for a better understanding of human nature but for a more powerful and intellectually convincing environmental ethic based upon it.

REFERENCES

I first used the expression "biophilia" in 1984 in a book entitled by the name (*Biophilia,* Harvard University Press). In that extended essay I attempted to apply ideas of sociobiology to the environmental ethic.

The mechanism of gene-culture coevolution was proposed by Charles J. Lumsden and myself in *Genes, Mind, and Culture* (Harvard University Press, 1981), *Promethean*

Fire (Harvard University Press, 1983), and "The Relation Between Biological and Cultural Evolution," *Journal of Social and Biological Structure* 8, no. 4 (October, 1985): 343–59. It represents an extension of theoretical population genetics in an effort to include the principles of cognition and social psychology.

Balaji Mundkur traced the role of snakes and mythic serpents in *The Cult of the Serpent: An Interdisciplinary Survey of Its Manifestations and Origins* (State University of New York Press, 1983).

I have reviewed the measures of global biodiversity and extinction rates in greater detail in *The Diversity of Life* (Harvard University Press, 1992).

In evaluating the environmental ethic I have been aided greatly by the writings of several philosophers, including most notably Bryan Norton (*Why Preserve Natural Diversity?* Princeton University Press, 1987), Max Oelschlaeger (*The Idea of Wilderness: From Prehistory to the Age of Ecology*, Yale University Press, 1991), Holmes Rolston III (*Environmental Ethics: Duties to and Values in the Natural World*, Temple University Press, 1988), and Peter Singer (*The Expanding Circle: Ethics and Sociobiology*, Farrar, Straus & Giroux, 1981).

ELLIOTT SOBER*

PHILOSOPHICAL PROBLEMS FOR ENVIRONMENTALISM

INTRODUCTION

A number of philosophers have recognized that the environmental movement, whatever its practical political effectiveness, faces considerable theoretical difficulties in justification.[1] It has been recognized that traditional moral theories do not provide natural underpinnings for policy objectives and this has led some to skepticism about the claims of environmentalists, and others to the view that a revolutionary reassessment of ethical norms is needed. In this chapter, I will try to summarize the difficulties that confront a philosophical defense of environmentalism. I also will suggest a way of making sense of some environmental concerns that does not require the wholesale jettisoning of certain familiar moral judgments.

Preserving an endangered species or ecosystem poses no special conceptual problem when the instrumental value of that species or ecosystem is known. When we have reason to think that some natural object represents a resource to us, we obviously ought to take that fact into account in deciding what to do. A variety of potential uses may be under discussion, including food supply, medical applications, recreational use, and so on. As with any complex decision, it may be difficult even to agree on how to compare the competing values that may be involved. Willingness to pay in dollars is a familiar least common denominator, although it poses a number of problems. But here we have nothing that is specifically a problem for environmentalism.

The problem for environmentalism stems from the idea that species and

---

ecosystems ought to be preserved for reasons additional to their known value as resources for human use. The feeling is that even when we cannot say what nutritional, medicinal, or recreational benefit the preservation provides, there still is a value in preservation. It is the search for a rationale for this feeling that constitutes the main conceptual problem for environmentalism.

The problem is especially difficult in view of the holistic (as opposed to individualistic) character of the things being assigned value. Put simply, what is special about environmentalism is that it values the preservation of species, communities, or ecosystems, rather than the individual organisms of which they are composed. "Animal liberationists" have urged that we should take the suffering of sentient animals into account in ethical deliberation.[2] Such beasts are not mere things to be used as cruelly as we like no matter how trivial the benefit we derive. But in "widening the ethical circle," we are simply including in the community more individual organisms whose costs and benefits we compare. Animal liberationists are extending an old and familiar ethical doctrine—namely, utilitarianism—to take account of the welfare of other individuals. Although the practical consequences of this point of view may be revolutionary, the theoretical perspective is not at all novel. If suffering is bad, then it is bad for any individual who suffers.[3] Animal liberationists merely remind us of the consequences of familiar principles.

But trees, mountains, and salt marshes do not suffer. They do not experience pleasure and pain, because, evidently, they do not have experiences at all. The same is true of species. Granted, individual organisms may have mental states; but the species—taken to be a population of organisms connected by certain sorts of interactions (preeminently, that of exchanging genetic material in reproduction)—does not. Or put more carefully, we might say that the only sense in which species have experiences is that their member organisms do: the attribution at the population level, if true, is true simply in virtue of its being true at the individual level. Here is a case where reductionism is correct.

So perhaps it is true in this reductive sense that some species experience pain. But the values that environmentalists attach to preserving species do not reduce to any value of preserving organisms. It is in this sense that environmentalists espouse a holistic value system. Environmentalists care about entities that by no stretch of the imagination have experiences (e.g., mountains). What is more, their position does not force them to care if individual organisms suffer pain, so long as the species is preserved. Steel traps may outrage an animal liberationist because of the suffering they inflict, but an environmentalist aiming just at the preservation of a balanced ecosystem might see here no cause for complaint. Similarly, environmentalists think that the distinction between wild and domesticated organisms is important, in that it is the preservation of "natural" (i.e., not created by the "artificial interference" of human beings) objects that matters, whereas animal liberationists see the main problem in terms of the suffering of any organism—domesticated or not. And finally, environmentalists and animal liberationists diverge on what might be called the *n + m question.* If two species—say blue and sperm whales—have roughly comparable capacities for experiencing pain, an animal liberationist might tend to think of the preservation of a sperm whale as wholly on an ethical par with the preservation of a blue whale. The fact that one organism is part of an endangered species while the other is not does not make the rare individual more intrinsically important. But for an environmentalist, this holistic property—membership in an endangered species—makes all the difference in the world: a world with n sperm and m blue whales is far better than a world with $n + m$ sperm and 0 blue whales. Here we have a stark contrast between an ethic in which it is the life situation of individuals that matters, and an ethic in which the stability and diversity of populations of individuals are what matter.[4]

Both animal liberationists and environmentalists wish to broaden our ethical horizons—to make us realize that it is not just human welfare that counts. But they do this in very different, often conflicting, ways. It is no accident that at the level of practical politics the two points of view increasingly find themselves at loggerheads. This practical conflict is the expression of a deep theoretical divide.

THE IGNORANCE ARGUMENT

"Although we might not now know what use a particular endangered species might be to us, allowing it to go extinct forever closes off the possibility of discovering and exploiting a future use." According to this point of view, our ignorance of value is turned into a reason for action. The scenario envisaged in this environmentalist argument is not without precedent; who could have guessed that penicillin would be good for something other than turning out cheese? But there is a fatal defect in such arguments, which we might summarize with the phrase *out of nothing, nothing comes*: rational decisions require assumptions about what is true and what is valuable (in decision-theoretic jargon, the inputs must be probabilities and utilities). If you are completely ignorant of values, then you are incapable of making a rational decision, either for or against preserving some species. The fact that you do not know the value of a species, by itself, cannot count as a reason for wanting one thing rather than another to happen to it.

And there are so many species. How many geese that lay golden eggs are there apt to be in that number? It is hard to assign probabilities and utilities precisely here, but an analogy will perhaps reveal the problem confronting this environmentalist argument. Most of us willingly fly on airplanes, when safer (but less convenient) alternative forms of transportation are available. Is this rational? Suppose it were argued that there is a small probability that the next flight you take will crash. This would be very bad for you. Is it not crazy for you to risk this, given that the only gain to you is that you can reduce your travel time by a few hours (by not going by train, say)? Those of us who not only fly, but congratulate ourselves for being rational in doing so, reject this argument. We are prepared to accept a small chance of a great disaster in return for the high probability of a rather modest benefit. If this is rational, no wonder that we might consistently be willing to allow a species to go extinct in order to build a hydroelectric plant.

That the argument from ignorance is no argument at all can be seen from another angle. If we literally do not know what consequences the extinction of this or that species may bring, then we should take seriously the possibility that the extinction may be beneficial as well as the possibility that it may be deleterious. It may sound deep to insist that we preserve endangered species precisely because we do not know why they are valuable. But ignorance on a scale like this cannot provide the basis for any rational action.

Rather than invoke some unspecified future benefit, an environmentalist may argue that the species in question plays a crucial role in stabilizing the ecosystem of which it is a part. This will undoubtedly be true for carefully chosen species and ecosystems, but one should not generalize this argument into a global claim to the effect that *every* species is crucial to a balanced ecosystem. Although ecologists used to agree that the complexity of an ecosystem stabilizes it, this hypothesis has been subject to a number of criticisms and qualifications, both from a theoretical and an empirical perspective.[5] And for certain kinds of species (those which occupy a rather small area and whose normal population is small) we can argue that extinction would probably not disrupt the community. However fragile the biosphere may be, the extreme view that everything is crucial is almost certainly not true.

But, of course, environmentalists are often concerned by the fact that extinctions are occurring now at a rate much higher than in earlier times. It is mass extinction that threatens the biosphere, they say, and this claim avoids the spurious assertion that communities are so fragile that even one extinction will cause a crash. However, if the point is to avoid a mass extinction of species, how does this provide a rationale for preserving a species of the kind just described, of which we rationally believe that its passing will not destabilize the ecosystem? And, more generally, if mass extinction is known to be a danger to us, how does this translate into a value for preserving any particular species? Notice that we have now passed beyond the confines of the argument from ignorance; we are taking as a premise the idea that mass extinction would be a catastrophe (since it would destroy the ecosystem on which we depend). But how should that premise affect our valuing the California condor, the blue whale, or the snail darter?

THE SLIPPERY SLOPE ARGUMENT

Environmentalists sometimes find themselves asked to explain why each species matters so much to them, when there are, after all, so many. We may know of special reasons for valuing particular species, but how can we justify thinking that each and every species is important? "Each extinction impoverishes the biosphere" is often the answer given, but it really fails to resolve the issue. Granted, each extinction impoverishes, but it only impoverishes a little bit. So if it is the *wholesale* impoverishment of the biosphere that matters, one would apparently have to concede that each extinction matters a little, but only a little. But environmentalists may be loathe to concede this, for if they concede that each species matters only a little, they seem to be inviting the wholesale impoverishment that would be an unambiguous disaster. So they dig in their heels and insist that each species matters a lot. But to take this line, one must find some other rationale than the idea that mass extinction would be a great harm. Some of these alternative rationales we will examine later. For now, let us take a closer look at the train of thought involved here.

Slippery slopes are curious things: if you take even one step onto them, you inevitably slide all the way to the bottom. So if you want to avoid finding yourself at the bottom, you must avoid stepping onto them at all. To mix metaphors, stepping onto a slippery slope is to invite being nickeled and dimed to death.

Slippery slope arguments have played a powerful role in a number of recent ethical debates. One often hears people defend the legitimacy of abortions by arguing that since it is permissible to abort a single-celled fertilized egg, it must be permissible to abort a fetus of any age, since there is no place to draw the line from 0 to 9 months. Antiabortionists, on the other hand, sometimes argue in the other direction: since infanticide of newborns is not permissible, abortion at any earlier time is also not allowed, since there is no place to draw the line. Although these two arguments reach opposite conclusions about the permissibility of abortions, they agree on the following idea: since there is no principled place to draw the line on the continuum from newly fertilized egg to fetus gone to term, one must treat all these cases in the same way. Either abortion is always permitted or it never is, since there is no place to draw the line. Both sides run their favorite slippery slope arguments, but try to precipitate slides in opposite directions.

Starting with 10 million extant species, and valuing overall diversity, the environmentalist does not want to grant that each species matters only a little. For having granted this, commercial expansion and other causes will reduce the tally to 9,999,999. And then the argument is repeated, with each species valued only a little, and diversity declines another notch. And so we are well on our way to a considerably impoverished biosphere, a little at a time. Better to reject the starting premise—namely, that each species matters only a little—so that the slippery slope can be avoided.

Slippery slopes should hold no terror for environmentalists, because it is often a mistake to demand that a line be drawn. Let me illustrate by an example. What is the difference between being bald and not? Presumably, the difference concerns the number of hairs you have on your head. But what is the precise number of hairs marking the boundary between baldness and not being bald? There is no such number. Yet, it would be a fallacy to conclude that there is no difference between baldness and hairiness. The fact that you cannot draw a line does not force you to say that the two alleged categories collapse into one. In the abortion case, this means that even if there is no precise point in fetal development that involves some discontinuous, qualitative change, one is still not obliged to think of newly fertilized eggs and fetuses gone to term as morally on a par. Since the biological differences are ones of degree, not kind, one may want to adopt the position that the moral differences are likewise matters of degree. This may lead to the view that a woman should have a better reason for having an abortion, the more developed her fetus is. Of course, this position does not logically follow from the idea that there is no place to draw the line; my point is just that differences in degree do not demolish the possibility of there being real moral differences.

In the environmental case, if one places a value on diversity, then each species becomes more valuable as the overall diversity declines. If we begin with 10 million species, each may matter little, but as extinctions continue, the remaining ones matter more and more. According to this outlook, a better and better reason would be demanded for allowing yet another species to go extinct. Perhaps certain sorts of economic development would justify the extinction of a species at one time. But granting this does not oblige one to conclude that the same sort of decision would have to be made further down the road. This means that one can value diversity without being obliged to take the somewhat exaggerated position that each species, no matter how many there are, is terribly precious in virtue of its contribution to that diversity.

Yet, one can understand that environmentalists might be reluctant to concede this point. They may fear that if one now allows that most species contribute only a little to overall diversity, one will set in motion a political process that cannot correct itself later. The worry is that even when the overall diversity has been drastically reduced, our ecological sensitivities will have been so coarsened that we will no longer be in a position to realize (or to implement policies fostering) the preciousness of what is left. This fear may be quite justified, but it is important to realize that it does not conflict with what was argued above. The political utility of making an argument should not be confused with the argument's soundness.

The fact that you are on a slippery slope, by itself, does not tell you whether you are near the beginning, in the middle, or at the end. If species diversity is a matter of degree, where do we currently find ourselves—on the verge of catastrophe, well on our way in that direction, or at some distance from a global crash? Environmentalists often urge that we are fast approaching a precipice; if we are, then the reduction in diversity that every succeeding extinction engenders should be all we need to justify species preservation.

Sometimes, however, environmentalists advance a kind of argument not predicated on the idea of fast approaching doom. The goal is to show that there is something wrong with allowing a species to go extinct (or with causing it to go extinct), even if overall diversity is not affected much. I now turn to one argument of this kind.

APPEALS TO WHAT IS NATURAL

I noted earlier that environmentalists and animal liberationists disagree over the significance of the distinction between wild and domesticated animals. Since both types of organisms can experience pain, animal liberationists will think of each as meriting ethical consideration. But environmentalists will typically not put wild and domesticated organisms on a par. Environmentalists typically are interested in preserving what is natural, be it a species living in the wild or a wilderness ecosystem. If a kind of domesticated chicken were threatened with extinction, I doubt that environmental groups would be up in arms. And if certain unique types of human environments—say urban slums in the United States— were "endangered," it is similarly unlikely that environmentalists would view this process as a deplorable impoverishment of the biosphere.

The environmentalist's lack of concern for humanly created organisms and environments may be practical rather than principled. It may be that at the level of values, no such bifurcation is legitimate, but that from the point of view of practical political action, it makes sense to put one's energies into saving items that exist in the wild. This subject has not been discussed much in the literature, so it is hard to tell. But I sense that the distinction between wild and domesticated has a certain theoretical importance to many environmentalists. They perhaps think that the difference is that we created domesticated organisms which would otherwise not exist, and so are entitled to use them solely for our own interests. But we did not create wild organisms and environments, so it is the height of presumption to expropriate them for our benefit. A more fitting posture would be one of "stewardship": we have come on the scene and found a treasure not of our making. Given this, we ought to preserve this treasure in its natural state.

I do not wish to contest the appropriateness of "stewardship." It is the dichotomy between artificial (domesticated) and natural (wild) that strikes me as wrong-headed. I want to suggest that to the degree that "natural" means anything biologically, it means very little ethically. And, conversely, to the degree that "natural" is understood as a normative concept, it has very little to do with biology.

Environmentalists often express regret that we human beings find it so hard to remember that we are part of nature—one species among many others—rather than something standing outside of nature. I will not consider here whether this attitude is cause for complaint; the important point is that seeing us as part of nature rules out the environmentalist's use of the distinction between artificial-domesticated and natural-wild described above. *If we are part of nature, then everything we do is part of nature, and is natural in that primary sense.* When we domesticate organisms and bring them into a state of dependence on us, this is simply an example of one species exerting a selection pressure on another. If one calls this "unnatural," one might just as well say the same of parasitism or symbiosis (compare human domestication of animals and plants and "slave-making" in the social insects).

The concept of naturalness is subject to the same abuses as the concept of normalcy. *Normal* can mean *usual* or it can mean *desirable*. Although only the total pessimist will think that the two concepts are mutually exclusive, it is generally recognized that the mere fact that something is common does not by itself count as a reason for thinking that it is desirable. This distinction is quite familiar now in popular discussions of mental health, for example. Yet, when it comes to environmental issues, the concept of naturalness continues to live a double life. The destruction of wilderness areas by increased industrialization is bad because it is unnatural. And it is unnatural because it involves transforming a natural into an artificial habitat. Or one might hear that although extinction is a natural process, the kind of mass extinction currently being precipitated by our species is unprecedented, and so is unnatural. Environmentalists should look elsewhere for a defense of their policies, lest conservation simply become a variant of uncritical conservatism in which the axiom "Whatever is, is right" is modified to read "Whatever is (before human beings come on the scene), is right."

This conflation of the biological with the normative sense of "natural" sometimes comes to the fore when environmentalists attack animal liberationists for naive do-goodism. Callicott writes:

> the value commitments of the humane movement seem at bottom to betray a world-denying or rather a life-loathing philosophy. The natural world as actually constituted is one in which one being lives at the expense of others. Each organism, in Darwin's metaphor, struggles to maintain its own organic integrity. . . . To live is to be anxious about life, to feel pain and pleasure in a fitting mixture, and sooner or later to die. That is the way the system works. If *nature as a whole is good, then pain and death are also good.* Environmental ethics in general require people to play fair in the natural system, The neo-Benthamites have in a sense taken the uncourageous approach. People have attempted to exempt themselves from the life death reciprocities of natural processes and from ecological limitations in the name of a prophylactic ethic of maximizing rewards (pleasure) and minimizing unwelcome information (pain). To be fair, the humane moralists seem to suggest that we should attempt to project the same values into the nonhuman animal world and to widen the charmed circle—no matter that it would be biologically unrealistic to do so or biologically ruinous if, per impossible, such an environmental ethic were implemented.
>
> There is another approach. Rather than imposing our alienation from nature and natural processes and cycles of life on other animals, we human beings could reaffirm our participation in nature by accepting life as it is given without a sugar coating . . .[6]

On the same page, Callicott quotes with approval Shepard's remark that "the humanitarian's projection onto nature of illegal murder and the rights of civilized people to safety not only misses the point but is exactly contrary to fundamental ecological reality: the structure of nature is a sequence of killings."[7]

Thinking that what is found in nature is beyond ethical defect has not always been popular. Darwin wrote:

> That there is much suffering in the world no one disputes.
>
> Some have attempted to explain this in reference to man by imagining that it serves for his moral improvement. But the number of men in the world is as nothing compared with that of all other sentient beings, and these often suffer greatly without any moral improvement. A being so powerful and so full of knowledge as a God who could create the universe, is to our finite minds omnipotent and omniscient, and it revolts our understanding to suppose that his benevolence is not unbounded, for what advantage can there be in the sufferings

> of millions of the lower animals throughout almost endless time? This very old argument from the existence of suffering against the existence of an intelligent first cause seems to me a strong one; whereas, as just remarked, the presence of much suffering agrees well with the view that all organic beings have been developed through variation and natural selection.[8]

Darwin apparently viewed the quantity of pain found in nature as a melancholy and sobering consequence of the struggle for existence. But once we adopt the Panglossian attitude that this is the best of all possible worlds ("there is just the right amount of pain," etc.), a failure to identify what is natural with what is good can only seem "world-denying," "lifeloathing," "in a sense uncourageous," and "contrary to fundamental ecological reality."

Earlier in his essay, Callicott expresses distress that animal liberationists fail to draw a sharp distinction "between the very different plights (and rights) of wild and domestic animals."[9] Domestic animals are creations of man, he says. "They are living artifacts, but artifacts nevertheless. . . . There is thus something profoundly incoherent (and insensitive as well) in the complaint of some animal liberationists that the 'natural behavior' of chickens and bobby calves is cruelly frustrated on factory farms. It would make almost as much sense to speak of the natural behavior of tables and chairs."[10] Here again we see teleology playing a decisive role: wild organisms do not have the natural function of serving human ends, but domesticated animals do. Cheetahs in zoos are crimes against what is natural; veal calves in boxes are not.

The idea of "natural tendency" played a decisive role in pre-Darwinian biological thinking. Aristotle's entire science—both his physics and his biology—is articulated in terms of specifying the natural tendencies of kinds of objects and the interfering forces that can prevent an object from achieving its intended state. Heavy objects in the sublunar sphere have location at the center of the earth as their natural state; each tends to go there, but is prevented from doing so. Organisms likewise are conceptualized in terms of this natural state model:

> . . . [for] any living thing that has reached its normal development and which is unmutilated, and whose mode of generation is not spontaneous, the most natural act is the production of another like itself, an animal producing an animal, a plant a plant.[11]

But many interfering forces are possible, and in fact the occurrence of "monsters" is anything but uncommon. According to Aristotle, mules (sterile hybrids) count as deviations from the natural state. In fact, females are monsters as well, since the natural tendency of sexual reproduction is for the offspring to perfectly resemble the father, who, according to Aristotle, provides the "genetic instructions" (to put the idea anachronistically) while the female provides only the matter.

What has happened to the natural state model in modern science? In physics, the idea of describing what a class of objects will do in the absence of "interference" lives on: Newton specified this "zero-force state" as rest or uniform motion, and in general relativity, this state is understood in terms of motion along geodesics. But one of the most profound achievements of Darwinian biology has been the jettisoning of this kind of model. It isn't just that Aristotle was wrong in his detailed claims about mules and women; the whole structure of the natural state model has been discarded. Population biology is not conceptualized in terms of positing some characteristic that all members of a species would have in common, were interfering forces absent. Variation is not thought of as a deflection from the natural state of uniformity. Rather, variation is taken to be a fundamental property in its own right. Nor, at the level of individual biology, does the natural state model find an application. Developmental theory is not articulated by specifying a natural tendency and a set of interfering forces. The main conceptual tool for describing the various developmental pathways open to a genotype is the norm of reaction. The norm of reaction of a genotype within a range of environments will describe what phenotype the genotype will produce in a given environment. Thus, the norm of reaction for a corn plant genotype might describe how its height is influenced by the amount of moisture in the soil. The norm of reaction is entirely silent on which phenotype is the "natural" one. The idea that a corn plant might have some "natural height," which can be augmented or diminished by "interfering forces," is entirely alien to post-Darwinian biology.

The fact that the concepts of natural state and interfering force have lapsed from biological thought does not prevent environmentalists from inventing them anew. Perhaps these concepts can be provided with some sort of normative content; after all, the normative idea of "human rights" may make sense even if it is not a theoretical underpinning of any empirical science. But environmentalists should not assume that they can rely on some previously articulated scientific conception of "natural."

APPEALS TO NEEDS AND INTERESTS

The version of utilitarianism considered earlier (according to which something merits ethical consideration if it can experience pleasure and/or pain) leaves the environmentalist in the lurch. But there is an alternative to Bentham's hedonistic utilitarianism that has been thought by some to be a foundation for environmentalism. Preference utilitarianism says that an object's having interests, needs, or preferences gives it ethical status. This doctrine is at the core of Stone's affirmative answer to the title question of his book *Should Trees Have Standing?*[12] "Natural objects can communicate their wants (needs) to us, and in ways that are not terribly ambiguous. . . . The lawn tells me that it wants water by a certain dryness of the blades and soil—immediately obvious to the touch—the appearance of bald spots, yellowing, and a lack of springiness after being walked on." And if plants can do this, presumably so can mountain ranges, and endangered species. Preference utilitarianism may thereby seem to grant intrinsic ethical importance to precisely the sorts of objects about which environmentalists have expressed concern.

The problems with this perspective have been detailed by Sagoff.[13] If one does not require of an object that it have a mind for it to have wants or needs, what is required for the possession of these ethically relevant properties? Suppose one says that an object needs something if it will cease to exist if it does not get it. Then species, plants, and mountain ranges have needs, but only in the sense that automobiles, garbage dumps, and buildings do, too. If everything has needs, the advice to take needs into account in ethical deliberation is empty, unless it is supplemented by some technique for weighting and comparing the needs of different objects. A corporation will go bankrupt unless a highway is built. But the swamp will cease to exist if the highway is built. Perhaps one should take into account all relevant needs, but the question is how to do this in the event that needs conflict.

Although the concept of need can be provided with a permissive, all-inclusive definition, it is less easy to see how to do this with the concept of want. Why think that a mountain range "wants" to retain its unspoiled appearance, rather than house a new amusement park?[14] Needs are not at issue here, since in either case, the mountain continues to exist. One might be tempted to think that natural objects like mountains and species have "natural tendencies," and that the concept of want should be liberalized so as to mean that natural objects "want" to persist in their natural states. This Aristotelian view, as I argued in the previous section, simply makes no sense. Granted, a commercially undeveloped mountain will persist in this state, unless it is commercially developed. But it is equally true that a commercially untouched hill will become commercially developed, unless something causes this not to happen. I see no hope for extending the concept of wants to the full range of objects valued by environmentalists.

The same problems emerge when we try to apply the concepts of needs and wants to species. A species may need various resources, in the sense that these are necessary for its continued existence. But what do species want? Do they want to remain stable in numbers, neither growing nor shrinking? Or since most species have gone extinct, perhaps what species really want is to go extinct, and it is human meddlesomeness that frustrates this natural tendency? Preference utilitarianism is no more likely than hedonistic utilitarianism to secure autonomous ethical status for endangered species.

Ehrenfeld describes a related distortion that has been inflicted on the diversity/stability hypothesis in theoretical ecology.[15] If it were true that increasing the diversity of an ecosystem causes it to be more stable, this might encourage the Aristotelian idea that ecosystems have a natural tendency to increase their diversity. The full realization of this tendency—the natural state that is the goal of ecosystems—is the "climax" or "mature" community. Extinction diminishes diversity, so it frustrates ecosystems from attaining their goal. Since the hypothesis that diversity causes stability is now considered controversial (to say the least), this line of thinking will not be very tempting. But even if the diversity/stability hypothesis were true, it would not permit the environmentalist to conclude that ecosystems have an interest in retaining their diversity.

Darwinism has not banished the idea that parts of the natural world are goal-directed systems, but has furnished this idea with a natural mechanism. We properly conceive of organisms (or genes, sometimes) as being in the business of maximizing their chances of survival and reproduction. We describe characteristics as adaptations—as devices that exist for the furtherance of these ends. Natural selection makes this perspective intelligible. But Darwinism is a profoundly individualistic doctrine. Darwinism rejects the idea that species, communities, and ecosystems have adaptations that exist for their own benefit. These higher-level entities are not conceptualized as goal-directed systems; what properties of organization they possess are viewed as artifacts of processes operating at lower levels of organization. An environmentalism based on the idea that the ecosystem is directed toward stability and diversity must find its foundation elsewhere.

GRANTING WHOLES AUTONOMOUS VALUE

A number of environmentalists have asserted that environmental values cannot be grounded in values based on regard for individual welfare. Aldo Leopold wrote in *A Sand County Almanac* that "a thing is right when it tends to preserve the integrity, stability, and beauty of the biotic community. It is wrong when it tends otherwise."[16] Cailicott develops this idea at some length, and ascribes to ethical environmentalism the view that "the preciousness of individual deer, *as of any other specimen*, is inversely proportional to the population of the species."[17] In his *Desert Solitaire*, Edward Abbey notes that he would sooner shoot a man than a snake.[18] And Garrett Hardin asserts that human beings injured in wilderness areas ought not to be rescued: making great and spectacular efforts to save the life of an individual "makes sense only when there is a shortage of people. I have not lately heard that there is a shortage of people."[19] The point of view suggested by these quotations is quite clear. It isn't that preserving the integrity of ecosystems has autonomous value, to be taken into account just as the quite distinct value of individual human welfare is. Rather, the idea is that the only value is the holistic one of maintaining ecological balance and diversity. Here we have a view that is just as monolithic as the most single-minded individualism; the difference is that the unit of value is thought to exist at a higher level of organization.

It is hard to know what to say to someone who would save a mosquito, just because it is rare, rather than a human being, if there were a choice. In ethics, as in any other subject, rationally persuading another person requires the existence of shared assumptions. If this monolithic environmentalist view is based on the notion that ecosystems have needs and interests, and that these take total precedence over the rights and interests of individual human beings, then the discussion of the previous sections is relevant. And even supposing that these higher-level entities have needs and wants, what reason is there to suppose that these matter and that the wants and needs of individuals matter not at all? But if this source of defense is jettisoned, and it is merely asserted that only ecosystems have value, with no substantive defense being offered, one must begin by requesting an argument: *why* is ecosystem stability and diversity the only value?

Some environmentalists have seen the individualist bias of utilitarianism as being harmful in ways additional to its impact on our perception of ecological values. Thus, Callicott writes:

> On the level of social organization, the interests of society may not always coincide with the sum of the interests of its parts. Discipline, sacrifice, and individual restraint are often necessary in the social sphere to maintain social integrity as within the bodily organism. A society, indeed, is particularly vulnerable to disintegration when its members become preoccupied totally with their own particular interest, and ignore those distinct and independent interests of the community as a whole. One example, unfortunately, our own society, is altogether too close at hand to be examined with strict academic detachment. The United States seems to pursue uncritically a social policy of reductive utilitarianism, aimed at promoting the happiness of all its members severally. Each special interest accordingly clamors more loudly to be satisfied while the community as a whole becomes noticeably more and more infirm economically, environmentally, and politically.[20]

Callicott apparently sees the emergence of individualism and alienation from nature as two aspects of the same process. He values "the symbiotic relationship of Stone Age man to the natural environment" and regrets that "civilization has insulated and alienated us from the rigors and challenges of the natural environment. The hidden agenda of the humane ethic," he says, "is the imposition of the anti-natural prophylactic ethos of comfort and soft pleasure on an even wider scale. The land ethic, on the other hand, requires a shrinkage, if at all possible, of the domestic sphere; it rejoices in a recrudescence of the wilderness and a renaissance of tribal cultural experience."[21]

Callicott is right that "strict academic detachment" is difficult here. The reader will have to decide whether the United States currently suffers from too much or too little regard "for the happiness of all its members severally" and whether we should feel nostalgia or pity in contemplating what the Stone Age experience of nature was like.

THE DEMARCATION PROBLEM

Perhaps the most fundamental theoretical problem confronting an environmentalist who wishes to claim that species and ecosystems have autonomous value is what I will call the *problem of demarcation.* Every ethical theory must provide principles that describe which objects matter for their own sakes and which do not. Besides marking the boundary between these two classes by enumerating a set of ethically relevant properties, an ethical theory must say why the properties named, rather than others, are the ones that count. Thus, for example, hedonistic utilitarianism cites the capacity to experience pleasure and/or pain as the decisive criterion; preference utilitarianism cites the having of preferences (or wants, or interests) as the decisive property. And a Kantian ethical theory will include an individual in the ethical community only if it is capable of rational reflection and autonomy. Not that justifying these various proposed solutions to the demarcation problem is easy; indeed, since this issue is so fundamental, it will be very difficult to justify one proposal as opposed to another. Still, a substantive ethical theory is obliged to try.

Environmentalists, wishing to avoid the allegedly distorting perspective of individualism, frequently want to claim autonomous value for wholes. This may take the form of a monolithic doctrine according to which the only thing that matters is the stability of the ecosystem. Or it may embody a pluralistic outlook according to which ecosystem stability and species preservation have an importance additional to the welfare of individual organisms. But an environmentalist theory shares with all ethical theories an interest in not saying that everything has autonomous value. The reason this position is proscribed is that it makes the adjudication of ethical conflict very difficult indeed. (In addition, it is radically implausible, but we can set that objection to one side.)

Environmentalists, as we have seen, may think of natural objects, like mountains, species, and ecosystems, as mattering for their own sake, but of artificial objects, like highway systems and domesticated animals, as having only instrumental value. If a mountain and a highway are both made of rock, it seems unlikely that the difference between them arises from the fact that mountains have wants, interests, and preferences, but highway systems do not. But perhaps the place to look for the relevant difference is not in their present physical composition, but in the historical fact of how each came into existence. Mountains were created by natural processes, whereas highways are humanly constructed. But once we realize that organisms construct their environments in nature, this contrast begins to cloud. Organisms do not passively reside in an environment whose properties are independently determined. Organisms transform their environments by physically interacting with them. An anthill is an artifact just as a highway is. Granted, a difference obtains at the level of whether conscious deliberation played a role, but can one take seriously the view that artifacts produced by conscious planning are thereby *less* valuable than ones that arise without the intervention of mentality.[22] As we have noted before, although environmentalists often accuse their critics of failing to think in a biologically realistic way, their use of the distinction between "natural" and "artificial" is just the sort of idea that stands in need of a more realistic biological perspective.

My suspicion is that the distinction between natural and artificial is not the crucial one. On the contrary, certain features of environmental concerns imply that natural objects are exactly on a par with certain artificial ones. Here the intended comparison is not between mountains and highways, but between mountains and works of art. My goal in what follows is not to sketch a substantive conception of what determines the value of objects in these two domains, but to motivate an analogy.

For both natural objects and works of art, our values extend beyond the concerns we have for experiencing pleasure. Most of us value seeing an original painting more than we value seeing a copy, even when we could not tell the difference. When we experience works of art, often what we value is not just the kinds of experiences we have, but, in addition, the connections we usually have with certain real objects. Routley and Routley have made an analogous point about valuing the wilderness experience: a "wilderness experience machine" that caused certain sorts of hallucinations would be no substitute for actually going into the wild.[23] Nor is this fact about our valuation limited to such aesthetic and environmentalist contexts. We love various people in our lives. If a molecule-for-molecule replica of a beloved person were created, you would not love that individual, but would continue to love the individual to whom you actually were historically related. Here again, our attachments are to objects and people as they really are, and not just to the experiences that they facilitate.

Another parallel between environmentalist concerns and aesthetic values concerns the issue of context. Although environmentalists often stress the importance of preserving endangered species, they would not be completely satisfied if an endangered species were preserved by putting a number of specimens in a zoo or in a humanly constructed preserve. What is taken to be important is preserving the species in its natural habitat. This leads to the more holistic position that preserving ecosystems, and not simply preserving certain member species, is of primary importance. Aesthetic concerns often lead in the same direction. It was not merely saving a fresco or an altar piece that motivated art historians after the most recent flood in Florence. Rather, they wanted to save these works of art in their original ("natural") settings. Not just the painting, but the church that housed it; not just the church, but the city itself. The idea of objects residing in a "fitting" environment plays a powerful role in both domains.

Environmentalism and aesthetics both see value in rarity. Of two whales, why should one be more worthy of aid than another, just because one belongs to an endangered species? Here we have the $n + m$ question mentioned in the introduction to this selection. As an ethical concern, rarity is difficult to understand. Perhaps this is because our ethical ideas concerning justice and equity (note the word) are saturated with individualism. But in the context of aesthetics, the concept of rarity is far from alien. A work of art may have enhanced value simply because there are very few other works by the same artist, or from the same historical period, or in the same style. It isn't that the price of the item may go up with rarity; I am talking about aesthetic value, not monetary worth. Viewed as valuable aesthetic objects, rare organisms may be valuable because they are rare.

A disanalogy may suggest itself. It may be objected that works of art are of instrumental value only, but that species and ecosystems have intrinsic value. Perhaps it is true, as claimed before, that our attachment to works of art, to nature, and to our loved ones extends beyond the experiences they allow us to have. But it may be argued that what is valuable in the aesthetic case is always the relation of a valuer to a valued object.[24]

When we experience a work of art, the value is not simply in the experience, but in the composite fact that we and the work of art are related in certain ways. This immediately suggests that if there were no valuers in the world, nothing would have value, since such relational facts could no longer obtain. So, to adapt Routley and

Routley's "last man argument," it would seem that if an ecological crisis precipitated a collapse of the world system, the last human being (whom we may assume for the purposes of this example to be the last valuer) could set about destroying all works of art, and there would be nothing wrong in this.[25] That is, if aesthetic objects are valuable only in so far as valuers can stand in certain relations to them, then when valuers disappear, so does the possibility of aesthetic value. This would deny, in one sense, that aesthetic objects are intrinsically valuable: it isn't they, in themselves, but rather the relational facts that they are part of, that are valuable.

In contrast, it has been claimed that the "last man" would be wrong to destroy natural objects such as mountains, salt marshes, and species. (So as to avoid confusing the issue by bringing in the welfare of individual organisms, Routley and Routley imagine that destruction and mass extinctions can be caused painlessly, so that there would be nothing wrong about this undertaking from the point of view of the nonhuman organisms involved.) If the last man ought to preserve these natural objects, then these objects appear to have a kind of autonomous value; their value would extend beyond their possible relations to valuers. If all this were true, we would have here a contrast between aesthetic and natural objects, one that implies that natural objects are more valuable than works of art.

Routley and Routley advance the last man argument as if it were decisive in showing that environmental objects such as mountains and salt marshes have autonomous value. I find the example more puzzling than decisive. But, in the present context, we do not have to decide whether Routley and Routley are right. We only have to decide whether this imagined situation brings out any relevant difference between aesthetic and environmental values. Were the last man to look up on a certain hillside, he would see a striking rock formation next to the ruins of a Greek temple. Long ago the temple was built from some of the very rocks that still stud the slope. Both promontory and temple have a history, and both have been transformed by the biotic and the abiotic environments. I myself find it impossible to advise the last man that the peak matters more than the temple. I do not see a relevant difference. Environmentalists, if they hold that the solution to the problem of demarcation is to be found in the distinction between natural and artificial, will have to find such a distinction. But if environmental values are aesthetic, no difference need be discovered.

Environmentalists may be reluctant to classify their concern as aesthetic. Perhaps they will feel that aesthetic concerns are frivolous. Perhaps they will feel that the aesthetic regard for artifacts that has been made possible by culture is antithetical to a proper regard for wilderness. But such contrasts are illusory. Concern for environmental values does not require a stripping away of the perspective afforded by civilization; to value the wild, one does not have to "become wild" oneself (whatever that may mean). Rather, it is the material comforts of civilization that make possible a serious concern for both aesthetic and environmental values. These are concerns that can become pressing in developed nations in part because the populations of those countries now enjoy a certain substantial level of prosperity. It would be the height of condescension to expect a nation experiencing hunger and chronic disease to be inordinately concerned with the autonomous value of ecosystems or with creating and preserving works of art. Such values are not frivolous, but they can become important to us only after certain fundamental human needs are satisfied. Instead of radically jettisoning individualist ethics, environmentalists may find a more hospitable home for their values in a category of value that has existed all along.

NOTES

1. Mark Sagoff, "On Preserving the Natural Environment," *Yale Law Review* 84 (1974): 205–38; J. Baird Callicott, "Animal Liberation: A Triangular Affair," *Environmental Ethics* 2 (1980): 311–38; and Bryan Norton, "Environmental Ethics and Nonhuman Rights," *Environmental Ethics* 4 (1982): 17–36.

2. Peter Singer, *Animal Liberation* (New York: Random House, 1975) has elaborated a position of this sort.

3. Occasionally, it has been argued that utilitarianism is not just *insufficient* to justify the principles of environmentalism, but is actually mistaken in holding that pain is intrinsically bad. Callicott writes: "I herewith declare in all soberness that I see nothing wrong with pain. It is a marvelous method, honed by the evolutionary process, of conveying important organic information. I think it was the late Alan Watts who somewhere remarks that upon being asked if he did not think there was too much pain in the world replied, 'No, I think there's just enough'" ("A Triangular Affair," p. 333). Setting to one side the remark attributed to Watts, I should point out that pain can be intrinsically bad and still have some good consequences. The point of calling pain intrinsically bad is to say that one essential aspect of experiencing it is negative.

4. A parallel with a quite different moral problem will perhaps make it clearer how the environmentalist's holism conflicts with some fundamental ethical ideas. When we consider the rights of individuals to receive compensation for harm, we generally expect that the individuals compensated must be one and the same as the individuals harmed. This expectation runs counter to the way an affirmative action program might be set up, if individuals were to receive compensation simply for being members of groups that have suffered certain kinds of discrimination, whether or not they themselves were victims of discrimination. I do not raise this example to suggest that a holistic conception according to which groups have entitlements is beyond consideration. Rather, my point is to exhibit a case in which a rather common ethical idea is individualistic rather than holistic.

5. David Ehrenfeld, "The Conservation of Non-Resources," *American Scientist* 64

(1976): 648–56. For a theoretical discussion see Robert M. May, *Stability and Complexity in Model Ecosystems* (Princeton, NJ: Princeton University Press, 1973).

6. Callicott, "A Triangular Affair," pp. 333–34 (my emphasis).

7. Paul Shepard, "Animal Rights and Human Rites," *North American Review* (Winter 1974): 35–41.

8. Charles Darwin, *The Autobiography of Charles Darwin* (London: Collins, 1876 and 1958), p. 90.

9. Callicott, "A Triangular Affair," p. 330.

10. Ibid.

11. Aristotle, *DeAnima*, p. 415a26.

12. Christopher Stone, *Should* Trees *Have Standing?* (Los Altos, CA: William Kaufmann, 1972), p. 24.

13. Sagoff, "Natural Environment," pp. 220–24.

14. Ibid.

15. Ehrenfeld, "The Conservation of Non-Resources," pp. 651–52.

16. Aldo Leopold, *A Sand County Almanac* (New York: Oxford University Press, 1949), pp. 224–25.

17. Callicott, "A Triangular Affair," p. 326 (emphasis mine).

18. Edward Abbey, *Desert Solitaire* (New York: Ballantine Books, 1968), p. 20.

19. Garrett Hardin, "The Economics of Wilderness," *Natural History* 78 (1969): 176.

20. Callicott, "A Triangular Affair," p. 323.

21. Ibid., p. 335.

22. Here we would have an inversion, not just a rejection, of a familiar Marxian doctrine—the labor theory of value.

23. Richard Routley and Val Routley, "Human Chauvinism and Environmental Ethics," *Environmental Philosophy, Monograph Series* 2, edited by D. S. Mannison, M. A. McRobbie, and R. Rowley (Philosophy Department, Australian National University, 1980), p. 154.

24. Donald H. Regan, "Duties of Preservation," *The Preservation of Species,* edited by B. Norton (Princeton , NJ: Princeton University Press, 1986), pp. 195–220.

25. Routley and Routley, "Human Chauvinism," pp. 121–22.

GREGORY M. MIKKELSON

TOWARD A GENERAL THEORY OF DIVERSITY AND EQUALITY

INTRODUCTION

Adam Smith founded the science of economics with his sprawling treatise *Wealth of Nations* (1776).[1] Not quite a century later, Charles Darwin begat the disciplines of evolutionary biology and ecology, with his more concise *Origin of Species* (1859).[2] Economics and ecology share the first three letters of their names, based on the Greek root "oikos," meaning "household." And there has been quite a bit of sharing over the years, of concepts like "producer" and "consumer," exponential increase, and optimal resource use. However, the two "eco-" fields also grew isolated from each other in certain key respects, to the detriment of both. While economists increasingly abstracted away from the biophysical processes that sustain economies, ecologists under-emphasized the role that *Homo sapiens* often plays in ecosystems (O'Neill and Kahn 2000).

Ecological economics is a new "trans-discipline" aimed to remedy some of the shortcomings in its parent sciences. This new field poses several fundamental challenges to mainstream economics, and also challenges ecologists to make their science more useful for solving the environmental problems created by a world economy run amok. In this chapter, I will introduce the basic conceptual framework of ecological economics, and suggest some lines of empirical inquiry to further articulate this framework. I will highlight a major philosophical aspect of the ecological-economic project, namely to change certain widespread ideas about the nature of science. Except where otherwise noted, my description of ecological economics is drawn from a recent textbook on the subject (Daly and Farley 2004).

THE CONCEPTUAL FRAMEWORK

Ecological economists ("ecologicals," for short) disagree with mainstream views about the basic relationship of the economy to the environment, about the proper questions to ask in economics, and about the degree to which economics should be connected to other spheres of inquiry. As we shall see in the next section, these disagreements about economics in particular reflect more general differences over the nature of science in general.

Ecologicals like to introduce their subject by pointing out that the human economy is only part of the more-than-human biosphere. In other words, the economy is a "wholly owned subsidiary" of the environment. Since the biosphere itself is not expanding, the economy cannot expand forever, at least in biophysical terms. The ecologicals accuse mainstream economists of glossing over this fact, and of treating the environment as just another part of the economy, rather than as the whole that contains and sustains the economy.

The world economy is already huge relative to this larger whole. By the 1980s, humans were appropriating 40 percent of all living material produced through photosynthesis on the land surface of the Earth (Vitousek, et al. 1986). And yet both the human population and per capita consumption continue to increase. The costs of this increase, e.g., in terms of resource depletion and pollution, may have already begun to outweigh the benefits. As a net result, world life expectancy at age one has now begun to decline (data from UNICEF 2005 and the UN Population Division 2005). And our fellow species are going extinct faster than at any time since a meteor wiped out the dinosaurs 65 million years ago (Leakey and Lewin 1996). For reasons like this, ecologicals insist that questions of *scale* must come first in economics: how much is enough, and how much is too much?

Questions of *distribution* come second for ecological economists. Since the 1930s, mainstream economists have downplayed the gap between the rich and the poor, to focus instead on overall economic growth as the way to alleviate poverty. Once the limits to such growth are recognized, however, this becomes a less viable option. And besides, one's economic status relative to others has much more of an effect on health and happiness than does one's absolute level of wealth or income (Sen 1993). Growth by itself does not alleviate relative poverty at all; improving economic equality does. Unfortunately, though, the world has become steadily more unequal over the past several decades, both within and between countries (Pitt Inequality Project 2006; World Bank 2006).

Ecologicals' third priority is *allocation*. While distribution has to do with how resources are divided up among different people, allocation has to do with how they are divided up among different processes or products. How much money and effort should go into producing corn vs. wheat, for example? Ecologicals join mainstream economists in celebrating the power of markets to solve many allocation problems. But they strongly dissent from the mainstream assumption that markets can adequately address scale and distribution as well. According to ecologicals, the latter require additional, democratic institutions and/or policies that come into play before the invisible hand of the market works its magic.

Finally, what are the ultimate purposes of markets, trade agreements, property rights, transportation policies, and other aspects of modern economic life? This question connects economics to ethics, which is about goals or ends such as excellence, happiness, fairness, freedom, etc. And what are the physical, chemical, and biological means available for achieving such ends? This question connects economics to the natural sciences. Ecologicals accuse mainstream economists of largely forgetting, or even disdaining, these connections. For example, rather than assessing the degree to which economic growth achieves important goals, economists seem to have turned it into an end in itself. Nor have they paid enough attention to its effects on natural resources.

THE LEGACY OF LOGICAL POSITIVISM IN MAINSTREAM ECONOMICS

The ecologicals' call to re-connect economics with both ethics and natural science harks back to some arguments by John Dewey in the early twentieth century. At that time, a school of thought called logical positivism was gaining preeminence within academic philosophy. Although the positivists valued unity among different sciences, they treated the unification process as a technical, even linguistic, matter of "reduction" (see below). In contrast, Dewey (1938) viewed the unity of science as a non-reductive "social problem," and his arguments called for the unity of not just science but of all knowledge, including ethics. The conceptual foundations of mainstream economics owe more to the positivists than to Dewey. In this section I will argue that these foundational premises exacerbate the "disconnection problems" diagnosed by the ecologicals.

The logical positivists subscribed to a strong form of ethical *subjectivism*, according to which moral claims are actually meaningless (Ayer 1936). They also espoused a type of *reductionism* entailing that "higher level" sciences like sociology and economics are reducible to—i.e., completely explicable in terms of—lower level sciences like biology, and ultimately to the lowest-level science of all, namely physics (Carnap 1938). Finally, the positivists endorsed a version of *instrumentalism*. Instrumentalism is the view that scientific theories are mere predictive tools or instruments that do not reflect, and are not accountable to, the "deep" nature of reality. The positivists' specific version of this doctrine was that claims about "unobservable" entities like atoms are, like ethical statements, literally meaningless (Neurath, et al. 1929).

Most philosophers have since rejected the versions of subjectivism, reductionism, and instrumentalism put forth by the logical positivists. But other versions still persist within both philosophy and science, and especially within mainstream economics (Mikkelson in press). Subjectivism—the belief that while science may be objective, ethics is by definition purely subjective—drove economists to divorce their subject from ethics beginning in the 1930s (Proctor 1991; Putnam 2002). Economics was to be a "positive" science, untainted by "normative" considerations. As many authors have pointed out, though, economists failed in the impossible task of purging value judgments. Instead, they covertly elevated certain values, such as the "efficiency" of economic allocation, above others, like the fairness of economic distribution, or the sustainability of economic scale.

While subjectivism has thus exacerbated the disconnection of economics from ethics, I would argue that a certain kind of reductionism has aggravated a disconnection problem within economics. The most popular version of reductionism among economists is so-called "methodological individualism" (Kincaid 1996; Udehn 2002). According to this doctrine, social patterns are to be explained in terms of individual behavior, rather than vice versa. While individualistic explanations thus highlight one kind of link between the "micro" and "macro" levels, they ignore the opposite kind, whereby social structures influence or constrain individual behavior. By thus foregoing an entire class of unifying links, I would argue that economists hinder their ability to understand the hitherto "mysterious" relationship between micro- and macro-economics (Daly and Farley 2004). In contrast, ecologicals try to take into account the reciprocal effects that individuals and their societies have on each other.

Finally, instrumentalism has arguably worsened the disconnect between economics and other sciences. Milton Friedman famously argued that economists can safely ignore questions about whether the assumptions built into their theories are "realistic," as long as they yield accurate predictions about certain variables, such as prices. For example, rather than worry about the combination of selfish and altruistic motives that actually underlies human behavior, economists "'assume' single-minded pursuit of pecuniary self-interest" (Friedman 1953). Mainstream economists dismiss ecology as casually as they do psychology, when they assume, for example, that "manmade" capital can adequately substitute for "natural capital." Ecologicals easily demonstrate the absurdity of this assumption by pointing out, e.g., that having more and better fishing boats does no good if there are no fish left in the ocean.

RESEARCH FRONTIERS

The conceptual framework of ecological economics has suggested many new lines of empirical research. Given the short history of this field, most of these research programs have only just gotten started.[3] In this section I will discuss three areas that deserve more consideration.

The first of these is the nature of the corporation. *Corporations* have amassed so many legal powers over the past couple of centuries that our current economic system has become fundamentally different from the free market advocated by Adam Smith. These powers have enabled corporations to become "externalizing machines" that channel the benefits of their endeavors to managers and shareholders, while shifting the costs and harms onto workers, consumers, the public at large, and the environment (Bakan 2004). Corporations thus re-distribute wealth to those who are already well off, and worsen pollution and other environmental problems.

Convincing governments to subsidize them is one way that corporations secure benefits for their managers and shareholders. Myers and Kent (2001) identify industries related to *automobiles* as by far the biggest recipients of what they call "perverse subsidies"—government giveaways that on balance do more harm than good. Besides causing far more "accidental" deaths than any of its alternatives, automobile transportation has greatly increased the scale of habitat destruction, urban sprawl, noise pollution, and many other health and environmental problems (Whitelegg and Haq 2003).

Corporate capitalism and automobile transportation thus have many effects on scale and distribution that, I submit, it would behoove ecological economists to study more closely. *Biodiversity* is a third topic that merits further attention. Above, I mentioned mass extinction as one result of the world economy's continually expanding scale. The biodiversity crisis can also be seen as a matter of distribution. Just as we can ask whether it is wise to let a relatively small number of people control most of the world's economic resources, we can also question whether it is right for one species (*Homo sapiens*) to use up so great a fraction of the world's ecological resources, thereby depriving others of access to what they need.

In addition to this conceptual link between economic equality and biological diversity, my colleagues and I have discovered an empirical relationship. Inequitable societies, besides having more health and social problems than equitable ones, also tend to have a greater proportion of their species headed toward extinction. This relationship between economic inequality and biodiversity loss holds up even after factoring out the influence of biophysical variables, human population size, and per capita income. And it characterizes both countries across the world, and states across the US (Mikkelson, et al. unpublished).

CONCLUSION

In this chapter, I have introduced the new field of ecological economics as an attempt to remedy some shortcomings of both ecology and economics. I focused on the ecologicals' critique of mainstream economics. Some of the same criticisms also apply to mainstream ecology, though to a lesser degree (see Mikkelson in press).

Ecological economists have called for greater attention to the overall scale of, and distribution of resources within, the economy. This, in turn, requires greater incorporation of both natural science and ethics into economic analysis. And it implies that economists will have to give up some deeply-held convictions about the nature of science. These include the ideas that "positive" and "normative" inquiry can and should be pursued separately, that wholes are reducible to their parts (e.g., societies to individuals), and that theories need not make realistic assumptions.

The goals of ecological economics are thus quite ambitious. In the last section above, I suggested some specific research projects that might help it to reach these goals. If the ecologicals succeed, they will achieve a revolution in both science and public policy.

REFERENCES

Ayer, A. J. *Language, Truth, and Logic*. London: Victor Gollancz, 1936.

Bakan, J. *The Corporation: The Pathological Pursuit of Profit and Power*. New York: Free Press, 2004.

Carnap, R. "Logical Foundations of the Unity of Science." *International Encyclopedia of Unified Science* 1 (1938): 42–62.

Daly, H. E., and J. Farley. *Ecological Economics: Principles and Applications*. Washington, DC: Island, 2004.

Darwin, C. *On the Origin of Species by Means of Natural Selection*. 1859.

Dewey, J. "Unity of Science as a Social Problem." *Encyclopedia of Unified Science* 1 (1938): 29–38.

Friedman, M. *Essays in Positive Economics*. Chicago: University of Chicago Press, 1953.

Kincaid, H. *Philosophical Foundations of the Social Sciences: Analyzing Controversies in Social Research.* New York: Cambridge University, 1996.

Leakey, R., and R. Lewin. *The Sixth Extinction: Biodiversity and Its Survival*. London: Weidenfeld and Nicolson, 1996.

Mikkelson, G. M. "Ecology." In *Companion to the Philosophy of Biology*, edited by D. L. Hull and M. Ruse. New York: Cambridge University Press, in press.

Mikkelson, G. M., A. Gonzalez, and G. D. Peterson. *Economic Inequality and Biodiversity Loss*. Unpublished.

Myers, N., and J. Kent. *Perverse Subsidies: How Misused Tax Dollars Harm the Environment and the Economy*. Washington, DC: Island Press, 2001.

Neurath, O., R. Carnap, and H. Hahn. *Wissenschaftliche Weltauffassung: Der Wiener Kreis*. Vienna, Austria: 1929.

O'Neill, R. V., and J. R. Kahn. "Homo economus as a keystone species." *BioScience* 50 (2000): 333–37.

Pitt Inequality Project. *Standardized Income Distribution Database*. Pittsburgh: University of Pittsburgh, 2006.

Proctor, R. N. *Value-Free Science? Purity and Power in Modern Knowledge*. Cambridge, MA: Harvard University Press, 1991.

Putnam, H. *The Collapse of the Fact/Value Dichotomy, and Other Essays*. Cambridge, MA: Harvard University Press, 2002.

Sen, A. "The Economics of Life and Death." *Scientific American* (May 1993): 40–47.

Smith, A. *An Inquiry into the Nature and Causes of the Wealth of Nations*. 1776.

Udehn, L. "The Changing Face of Methodological Individualism." *Annual Review of Sociology* 28 (2002): 479–507.

United Nations Children's Fund (UNICEF). *Monitoring the Situation of Children and Women*. New York: UNICEF, 2005.

United Nations Secretariat, Population Division of the Department of Economic and Social Affairs. *World Population Prospects: The 2004 Revision*. New York: United Nations, 2005.

Vitousek, P. M., P. R. Ehrlich, A. H. Ehrlich, and P. A. Matson. "Human Appropriation of the Products of Photosynthesis." *BioScience* 36 (1986): 368–73.

Whitelegg, J., and G. Haq, eds. *The Earthscan Reader in World Transport Policy and Practice*. London: Earthscan, 2003.

World Bank. *World Development Indicators 2006 Online*. Washington, DC: World Bank, 2006.

NOTES

1. At the time the science was called “political economy.”
2. Haeckel (1866), a fan of Darwin, coined the term “oekologie.”
3. The journal *Ecological Economics* got going fewer than 20 years ago.

WHEN LIFE BEGINS

REPORT OF THE SOUTH DAKOTA TASK FORCE TO STUDY ABORTION

SUBMITTED TO THE GOVERNOR AND LEGISLATURE OF SOUTH DAKOTA

DECEMBER 2005

2. WHAT HAS BEEN LEARNED FROM THE PRACTICE OF ABORTION SINCE THE *ROE V. WADE* DECISION

It can no longer be doubted that the unborn child from the moment of conception is a whole separate human being. During the 2005 legislative session, the South Dakota Legislature passed HB 1166 that expressly found that "all abortions, whether surgically or chemically induced, terminate the life of a whole, separate, unique, living human being." The Act amends SDCL 34-23A-10.1 to require a physician to disclose in writing to a pregnant mother "that the abortion will terminate the life of a whole, separate, unique, living human being." "Human being" is used in the biological sense as a whole member of the species *Homo sapiens* (SDCL 34-23A-1[4]).[1]

The Task Force received testimony from numerous experts who reiterated this fact. Of significance were detailed affidavits received from nationally and internationally renowned experts from a number of scientific and medical disciplines, who explained the scientific facts and information that establish the fact that abortion terminates the life of a human being. The Task Force concludes that the scientific evidence not only supports the Legislature's finding on this matter, but that it is indisputable.

Dr. David Fu-Chi Mark, a distinguished molecular biologist who has patented certain polymerase chain reaction technologies, provided a declaration (or affidavit) and explained that the new recombinant DNA technologies that have developed over the past twenty years provide scientific evidence about the unborn child's existence and early development and its ability to react to the environment and feel pain prior to birth. Dr. Mark stated that:

> [U]ntil the development of molecular biology and modern molecular biological techniques first began in the 1970's and exploding throughout the 1980's and 1990's, most scientific knowledge concerning human identity and human development prior to birth was based solely upon gross morphological observations and biochemical studies. . . . The new techniques developed through the exploding revolution over the past ten to eighteen years permits scientists to observe human existence and development at a molecular level, which is applicable in determining genetic uniqueness, genetic diseases and related information through the analysis of human genes well in advance of the old gross, anatomical observation. (Mark Declaration, P. 5, Par. 6.)

Dr. Mark explained nine different DNA technologies that, in essence, have "turned on the lights" for scientists over the past twenty years. (See Section II-B of this Report.)

A number of other nationally and internationally recognized experts also corroborated the findings of the Legislature found in HB 1166. Dr. Bruce Carlson, a renowned human embryologist and professor at University of Michigan Medical School, has published a text on Human Embryology used in medical schools in many nations. Dr. Carison, Dr. Mark, and the human geneticist, Dr. Marie Peeters-Ney, all emphasized the significance of the way that genetic information is expressed and the manner in which it is "pre-programmed" for life. Dr. Carlson stated, "The wholeness (or completeness) of the human being during the embryonic ages cannot be fully appreciated without an understanding of how the genetic information is packaged, and how the information becomes unfolded and cascades into visible structures" (Carlson Declaration, P.4, Par.12.).

The Task Force also reviewed a declaration from Dr. Bernard Nathanson, a board certified obstetrician and gynecologist, a Diplomat of the American Board of Obstetrics and Gynecology, and a Fellow of the American College of Obstetrics and Gynecology. Dr. Nathanson practiced medicine in New York for many decades and was personally responsible for approximately 75,000 abortions. Dr. Nathanson was also one of the founders of the National Association for the Repeal of the Abortion Laws (NARAL) in the United States in 1969. He stated that: "I was active among the pro-abortion community for a number of years, and I was actively involved in attempting, along with other abortion providers, to win public support for all forms of abortion" (Nathanson Declaration, Par. 1 to Par. 5).

Dr. Nathanson testified that the fact an abortion terminates the life of a living human being is generally known among obstetricians and scientists. However, Dr. Nathanson stated that abortion doctors and operators of abortion clinics often deny this fact for strategic reasons. He testified that he and other strategists for NARAL, for instance, adopted certain tactics to win the public perception that all forms of abortion should be and remain legal. Dr. Nathanson stated that one tactic was to suppress and denigrate all scientific evidence that supported the conclusions that a human embryo or fetus was a separate human being. He stated that he and others denied what they knew was true: "The abortion industry would routinely deny the undeniable, that is, that the human embryo and fetus is, as a matter of biological fact, a human being" (Nathanson Declaration, Par. 14). Dr. Byron Calhoun, a specialist in internal fetal medicine, also testified that it cannot be denied that the unborn child is a separate human being. Specifically, Dr. Nathanson stated that:

> The abortion procedure is an extraordinary one because in it the physician is proposing to terminate the life of one of his patients to whom the physician owes a legal and professional duty. The doctor has no legal authority to do so. Under normal circumstances, if he terminated the life of the unborn child, he would be guilty of a battery upon the mother, and, in fetal homicide states, such as South Dakota, he would be guilty of a homicide. The physician is given his authority to terminate the life of one of his patients only if he receives authority in the form of consent from the pregnant mother.
>
> In order for such a consent to be informed, at a minimum, the physician must be satisfied that the patient understood that the second patient was in existence, and that the procedure would terminate the life of her unborn child. These facts go directly to the risks, and effect the procedure would have on the second patient, but they also explain the nature of the procedure. The nature of the procedure is to terminate the life of the unborn child. Withholding these facts from the pregnant mother deprives her of the ability to make an informed decision for herself. Such informed written consent fails to meet the reasonable patient standard of disclosure and deprives the mother of her rights of self determination. (Nathanson Declaration, Par.10 and 11)

No credible evidence was presented that challenged these scientific facts. In fact, when witnesses supporting abortion were asked when life begins, not one would answer the question, stating that it would only be their personal opinion.

A number of physicians also testified that it has been recognized for some time that the doctor who has a pregnant mother as a patient has two separate patients—the mother and the child to whom he owes a professional and legal duty. Dr. Glenn Ridder, a physician who practices in Sioux Falls, South Dakota, testified that the physician has a duty to both patients, that disclosures about the risks or harms to the child must be made to the pregnant mother and that only she can make the decision for the child (Ridder Declaration, Par. 9).

Dr. Yvonne B. Seger, and Dr. Cynthia Davis, both of whom practice obstetrics and gynecology in Sioux Falls, South Dakota, also submitted statements that explained that the physician who has a pregnant woman as a patient has two separate patients—the mother and the child—and that the physician must make disclosures about the risks and effect any procedure would have on each (Davis Declaration, Par. 5; Seger Declaration, Par. 5).

Dr. Byron Calhoun, Maternal-Fetal Medical Specialist from Rockford, Illinois, explained that the unborn child is considered a separate patient in his or her own right. Dr. Calhoun also stated that there is no outcome data to support the idea that the mother's life is put at risk by allowing her to carry to term a critically ill baby. See also Harrison, M. R., M. S. Golbus, R. A. Filly, eds., *The Unborn Patient,* 2nd ed. (W. B. Saunders, Phil. 1991); American College of Obstetricians & Gynecologists, *Ethics in Obstetrics and Gynecology* 34, 2nd ed. (2004). (The maternal-fetal relationship is unique in medicine . . . because both the fetus and the woman are regarded as patients of the obstetrician.)

Dr. Mark Rosen, a Fetal Anesthesiologist practicing in San Francisco, California, explained that with the advances of modern medical techniques, fetal surgery is now performed on the unborn child in utero. The procedures include, among others, inserting fetal shunts, blood transfusions, muscle biopsies, and procedures to repair congenital diaphragmatic hernias.

The Task Force concludes the following:

1. That abortion terminates the life of a unique, whole, living human being;
2. That the physician performing an abortion terminates the life of one of the physician's patients to whom the physician owes a professional and legal duty;
3. That the authority for the physician to terminate the life of his or her patient rests exclusively upon the written consent of the pregnant mother, which, at the time it is signed, terminates the doctor's duty to the child; and
4. That the mother has an existing and important and beneficial relationship with her child that is irrevocably terminated by the abortion procedure.

B. The Body of Knowledge Concerning the Development and Behavior of the Unborn Child Which Has Developed Because of Technological Advances and Medical Experience Since the Legalization of Abortion

The Task Force received scientific information from highly credentialed scientists and medical practitioners. Information about the nature and development of the unborn child is now available that was not in existence at the time of *Roe v. Wade.*

1. The Development of the Field of Molecular Biology and Other Sciences

It has been known for the past five decades that human beings are biologically made up of molecular building blocks. The development of these building blocks is controlled by genetic material known as deoxyribonucleic acids (DNA) and ribonucleic acids (RNA). DNA contains genetic information, and RNA contains instructions for the synthesis of proteins.

The Task Force received a declaration prepared by Dr. David Fu-Chi Mark, who explained the modern developments in molecular biology, the information it has recently revealed, and the significance of that information. Dr. Mark is a nationally celebrated molecular biologist who has patented various polymerase chain reaction (PCR) techniques. In 1986, Dr. Mark was given the award of *Inventor of the Year*, which is a single award given across all disciplines of science and technology. That particular award was given to Dr. Mark for his work in obtaining a patent for Human Recombinant Interleukin-2 Muteins, which is used to treat cancer of the kidney and skin, and is still marketed internationally. He also obtained a patent for Human Recombinant Cysteine Depleted Interferon-B Muteins, which is a drug that is used to treat Multiple Sclerosis, and is also marketed internationally. These two drugs were developed by employment of new molecular biological techniques: DNA cloning, in vitro modification of DNA, and DNA sequencing.

Dr. Mark observed that until the development of molecular biology and modern molecular biological techniques,

> most scientific knowledge concerning human identity and human development prior to birth was based solely upon gross morphological observations and biochemical studies. Over the past [twenty] years there have been extraordinary scientific, medical and technological advances and discoveries which expose the

> rather rudimentary level of knowledge and ignorance of science, errors of fact and judgment concerning past scientific understanding of the child's existence as a human being, the child's early development and ability to react to the child's environment and feel pain prior to birth. The new techniques developed through the exploding revolution over the past [twenty years] permits scientists to observe human existence and development at a molecular level, which is applicable in determining genetic uniqueness, genetic diseases and related information through the analysis of human genes well in advance of the old gross, anatomical observation. (Mark Declaration, P. 5, Par. 6)

Dr. Mark described and explained in technical detail, with full citations to the relevant literature, nine of the many new major molecular biological technologies, and how they have been used to discover information about the unborn child:

1. Use of Restriction Endonuclease Enzymes: a technique discovered early in molecular biology that allowed scientists to use enzymes to cut pieces of DNA so that DNA can be manipulated in a test tube. This technique has bad great practical application (Mark P. 5–7, Par. 7A).
2. DNA Cloning: a technique first achieved in 1974 which allows a scientist to take a portion of DNA from a single cell, reproduce it, and make copies of it, allowing for the modern study of DNA and its reproduction. It was with the advent of development of DNA cloning in 1974 that molecular biology began in 1974 (Mark, P. 7–8, Par. 7B).
3. DNA Probe: a technology, first developed in 1979, that allows scientists to determine whether information contained in a certain gene is being expressed; study genome structure and identify sites of cytosine methylation (discussed below); and facilitate the development of DNA fingerprinting technology (Mark, P. 8–9, Par. 7C).
4. Southern Blot: a technique that permits the study of a single gene fragment. The importance of Southern Blot is the new ability to visualize the DNA of specific interest to the scientist, and it has led to the discovery and use of DNA fragmentation patterns visualized by Southern Blot as DNA fingerprints. DNA fingerprints, as discussed below, allows for the identification of DNA fragments both specific to the species *Homo sapiens*, and the specific individual member of the species (Mark, P. 10, Par. 7D).
5. Northern Blot: a technique that permits detection of messenger RNA (mRNA) in extremely small quantities of material. The importance of Northern Blot is the new ability of science to determine whether a spe-

cific gene is expressed in a particular tissue, which led to an understanding of the role of DNA methylation in regulating gene expression (Mark, P. 11, Par. 7E).

6. DNA Mapping: an important technique that allows scientists to determine if there are differences in DNA sequences, which provide science with the ability to detect abnormalities due to mutations in DNA, and to identify sites of DNA methylation (Mark, P. 11–12, Par. 7F).
7. DNA Fingerprinting: a technique first discovered in the mid-1980s by Alec Jeffries in Great Britain which gained wide application in the early to mid-1990s being introduced as evidence in American courts. It was learned by DNA mapping and Southern Blot analysis that the human genome contains many repetitive DNA sequences. Jeffries and his colleagues discussed in 1985 that, with the combined use of DNA mapping and Southern Blot, that a highly polymorphic DNA fragmentation pattern can be visualized. It was discovered that the highly variable DNA fragmentation patterns are characteristic of each individual human being, and the same pattern is found in all the cells of an individual. The significance of DNA fingerprinting is that it demonstrates the uniqueness of each human being, even at the first cell stage (Mark, P. 12, Par. 7G).
8. DNA Sequencing: the currently used rapid sequencing techniques were first developed in 1997. The importance of DNA sequencing is that from the gene code, science can better understand the functioning and development of the human being, including the ability to identify potential sites for DNA methylation. It also helps science determine the difference in genes in order to identify the nature of mutations (Mark, P. 12–14, Par. 7H).
9. Polymerase Chain Reaction (PCR): without PCR, DNA could not be analyzed from a single cell. The PCR technique was first invented in 1985 to rapidly amplify a segment of DNA up to a million fold from a very small amount of material. PCR greatly enhanced the ability of science to understand the uniqueness of each human being (Mark, P. 14, Par. 71).

2. Science Now Explains the Wholeness and Uniqueness of Every Human Being from Conception

DNA fingerprinting and the refinement of it by polymerase chain reaction (PCR) techniques developed in the mid-1990s have proven that each human being is totally unique immediately at fertilization. Dr. Mark explained that the invention and widespread use of the DNA techniques such as restriction enzymes, DNA cloning, DNA sequencing, and Southern Blotting provided scientists with the ability to clone human DNA and study the organization of the genes encoded by DNA. This resulted in many discoveries in the mid-1980s leading to the finding that individual species DNA bands can be observed as a fingerprint of an individual human being. A child's DNA fingerprint is completely unique.

The invention of the PCR techniques has led to further refinements of the DNA fingerprinting techniques, which has given science the ability to obtain a human being's DNA fingerprinting—and therefore his or her identity—from a single cell.

There can no longer be any doubt that each human being is totally unique from the very beginning of his or her life at fertilization (Mark, P. 19–21).

The significance of methylation of cytosine was unknown until 1985. It has a profound significance in understanding the wholeness or completeness of a human being immediately following conception. Cytosine is one of the four base components of DNA. Methylation of cytosine, just as other methods of gene regulation, is a natural method by which genetic information is periodically silenced or activated for purposes of human development. Understanding how the genetic information contained in each human being's DNA is activated and how that information is programmed for life is essential to understanding that the human being is whole and complete at fertilization.

A human being at an embryonic age and that human being at an adult age are naturally the same, the biological differences are due only to the differences in maturity. Changes in methylation of cytosine demonstrate that the human being is fully programmed for human growth and development for his or her entire life at the one cell age (Mark, P. 21–25).

Although the material messenger RNA initially present in the fertilized egg can provide the basic functions necessary to transcribe the child's DNA in the initial one or two cell divisions immediately following fertilization, these messenger RNAs are quickly degraded and lost after the first two rounds of cell division, and the housekeeping genes in the child's own DNA are transcribed into messenger RNA at that point. This newly synthesized RNA directs the program of global demethylation of genes so that they can be activated to replenish the functions lost after the degradation of the maternal RNA. Modern molecular biology has discovered that by the third cell division (long before implantation) all control of growth and development are established by the child's DNA. This means that immediately after conception, all programming for growth of the human being is self-contained (Mark, P. 26).

At the pre-implantation age, the child synthesizes a platelet activating factor (PAF) (discovered by O'Neil in 1991), beginning at the one cell age, that enhances the child's ability to implant into his or her mother's uterine wall; and at 7.5 days old, before implantation into the uterus, the child begins to produce an enzyme (IDO) that inhibits the mother's immune system from attacking and rejecting the child (discovered by Mann, et al in 1998) (Mark, P. 25–26).

Molecular biology has also revealed information about chemical reactions within the unborn child that assist the child to adjust to its environment and defend itself from painful stimuli. The role of substance P in pain transmission through activation of a sub-population of the primary afferent c nerve fibers has been recently understood and documented in the 1990s. The presence of substance P, known to be a pain transmitter, was not observed in the human being during gestation until the late 1980s. The discoveries concerning neuropeptides, enkephalin and beta-endorphin, pain modulators, natural pain inhibitors, as found in the unborn child, is discussed in Section II-H of this Report, along with studies done in the 1990s that measured fetal plasma cortical and beta endorphin responses to painful stimuli given to the fetus in utero (Mark, P. 27, 28, Par. 9 to 11).

NOTE

1. South Dakota Deputy Attorney General John Guhin addressed the Task Force on August 1, 2005, and reported that a preliminary injunction was entered enjoining the enforcement of HB 1166 in order to preserve the status quo while the case filed by Planned Parenthood challenging the Act is pending in the Federal District Court. Mr. Githmn reported that the state filed an appeal from the entry of the preliminary injunction. Members of the Task Force have subsequently learned that the trial in that case is not yet scheduled and it is anticipated that it will likely be held in the mid to later half of 2006.

It was of some interest to members of the Task Force to learn that while the Federal District Court entered the Order imposing the preliminary injunction, the Court did so on the basis that the Court sought to protect Planned Parenthood's First Amendment right of free speech which they asserted in that case. The Court recognized there was a conflict between these asserted rights of the abortion providers and the interests of the pregnant mothers which were sought to be protected by the Act. We note that the Court attempted to weigh the harm to the personal rights of the abortion providers against the harms to the interests of the pregnant mothers, and the Court chose to protect the interests of the abortion providers as a way to preserve the status quo while the case is litigated in the District Court.

SOUTH DAKOTA WOMEN'S HEALTH AND HUMAN LIFE PROTECTION ACT (HB 1215)*

SIGNED INTO LAW BY
SOUTH DAKOTA GOV. MIKE ROUNDS

MARCH 6, 2006

BE IT ENACTED BY THE LEGISLATURE OF THE STATE OF SOUTH DAKOTA:

Section 1. The Legislature accepts and concurs with the conclusion of the South Dakota Task Force to Study Abortion, based upon written materials, scientific studies, and testimony of witnesses presented to the task force, that life begins at the time of conception, a conclusion confirmed by scientific advances since the 1973 decision of *Roe v. Wade*, including the fact that each human being is totally unique immediately at fertilization. Moreover, the Legislature finds, based upon the conclusions of the South Dakota Task Force to Study Abortion, and in recognition of the technological advances and medical experience and body of knowledge about abortions produced and made available since the 1973 decision of *Roe v. Wade*, that to fully protect the rights, interests, and health of the pregnant mother, the rights, interest, and life of her unborn child, and the mother's fundamental natural intrinsic right to a relationship with her child, abortions in South Dakota should be prohibited. Moreover, the Legislature finds that the guarantee of due process of law under the Constitution of South Dakota applies equally to born and unborn human beings, and that under the Constitution of South Dakota, a pregnant mother and her unborn child, each possess a natural and inalienable right to life.

South Dakota Legislature. http://news.findlaw.com/nytimes/docs/abortion/sdabortionlaw06.html

Section 2. That chapter 22–17 be amended by adding thereto a NEW SECTION to read as follows:

> No person may knowingly administer to, prescribe for, or procure for, or sell to any pregnant woman any medicine, drug, or other substance with the specific intent of causing or abetting the termination of the life of an unborn human being. No person may knowingly use or employ any instrument or procedure upon a pregnant woman with the specific intent of causing or abetting the termination of the life of an unborn human being.
>
> Any violation of this section is a Class 5 felony.

Section 3. That chapter 22–17 be amended by adding thereto a NEW SECTION to read as follows:

> Nothing in section 2 of this Act may be construed to prohibit the sale, use, prescription, or administration of a contraceptive measure, drug or chemical, if it is administered prior to the time when a pregnancy could be determined through conventional medical testing and if the contraceptive measure is sold, used, prescribed, or administered in accordance with manufacturer instructions.

Section 4. That chapter 22–17 be amended by adding thereto a NEW SECTION to read as follows:

> No licensed physician who performs a medical procedure designed or intended to prevent the death of a pregnant mother is guilty of violating section 2 of this Act. However, the physician shall make reasonable medical efforts under the circumstances to preserve both the life of the mother and the life of her unborn child in a manner consistent with conventional medical practice.
>
> Medical treatment provided to the mother by a licensed physician which results in the accidental or unintentional injury or death to the unborn child is not a violation of this statute.
>
> Nothing in this Act may be construed to subject the pregnant mother upon whom any abortion is performed or attempted to any criminal conviction and penalty.

Section 5. That chapter 22–17 be amended by adding thereto a NEW SECTION to read as follows:

> Terms used in this Act mean:

1. "Pregnant," the human female reproductive condition, of having a living unborn human being within her body throughout the entire embryonic and fetal ages of the unborn child from fertilization to full gestation and child birth;
2. "Unborn human being," an individual living member of the species, homo sapiens, throughout the entire embryonic and fetal ages of the unborn child from fertilization to full gestation and childbirth;
3. "Fertilization," that point in time when a male human sperm penetrates the zona pellucida of a female human ovum.

Section 6. That § 34-23A-2 be repealed.

Section 7. That § 34-23A-3 be repealed.

Section 8. That § 34-23A-4 be repealed.

Section 9. That § 34-23A-5 be repealed.

Section 10. If any court of law enjoins, suspends, or delays the implementation of a provision of this Act, the provisions of sections 6 to 9, inclusive, of this Act are similarly enjoined, suspended, or delayed during such injunction, suspension, or delayed implementation.

Section 11. If any court of law finds any provision of this Act to be unconstitutional, the other provisions of this Act are severable. If any court of law finds the provisions of this Act to be entirely or substantially unconstitutional, the provisions of § 34-23A-2, 34-23A-3, 34-23A-4, and 34-23A-5, as of June 30, 2006, are immediately reeffective.

Section 12. This Act shall be known, and may be cited, as the Women's Health and Human Life Protection Act. An Act to establish certain legislative findings, to reinstate the prohibition against certain acts causing the termination of an unborn human life, to prescribe a penalty therefor, and to provide for the implementation of such provisions under certain circumstances.

JANE MAIENSCHEIN

HOW AND WHEN DOES A LIFE BEGIN?

For decades, bioethicists and medical practitioners have focused on determining what to count as the end of a life. The traditional answer for centuries was that when a person stops breathing, that signals the end of that life and of that living person. For a variety of reasons, including legal concerns about mistakes, that definition gave way to heart death as defining the end of life: when the heart stops, the person ends. But, then, when Christian Barnard showed in 1967 that he could successfully transplant a human heart from one person to another and society embraced that exciting new procedure, we needed another new definition. If the heart was defined a being alive, we needed a new way to define a person as dead. Brain-death served that purpose, and current legal definitions of death focus on the end of brain function.[1]

Understanding what counts as the end of life depends on the social and medical context. The definition is socially determined, drawing on underlying metaphysical assumptions about what it is that exists, epistemological assumptions about how we can know, and also ethical and social assumptions and values. In the case of end of life discussions, biological considerations must also play a central role. What we know about the functioning of the human heart or brain is equally important information in shaping the social decisions.

Precisely the same is true of efforts to define the beginning of a life. Metaphysical assumptions about what a person is, epistemological assumptions about how we can best know, and ethical and social assumptions about what we should value mix together. Unlike the end of life case, however, beginning of life discussions have been dominated and shaped by ethical, social, and especially religious assumptions, and have paid far less attention to biological understandings of life. Furthermore, they have typically failed to acknowledge the many different senses of life and even of "a life." Our answer will necessarily not be a simple definition but rather will embrace a plurality of answers depending on the details of the question.

This essay looks at beginning of life discussions in order to illuminate the underlying assumptions involved, with the goal of bringing biological understanding more centrally into the picture. Current views of life as beginning at conception are part of a centuries-long preformationist tradition that emphasizes predeterminism and materialism. This is paradoxical, since the leading advocates of the view today are driven by religious convictions that life involves soul as well as material and that typically defy a materialistic sort of determinism. We need to unpack the mix of assumptions in order to reflect wisely and to make sound social policy.

TRADITION

Tradition shapes our assumptions and constrains our thinking. By the seventeenth century, two traditions already co-existed with competing definitions of beginning of life and the form and function that constitute life. Aristotle had laid out a clear, logical, and empirically reasonable interpretation of life as a process.[2] The final, formal, material, and efficient causes work together to cause a person. There is no instant of creation, but rather a process, guided by the causes. It takes time and interaction of the causes for a human to come into being, that is for the potential to be actualized. This is a gradual process, with form and function emerging over time, and is called epigenesis. Early epigenesists assumed the existence of something beyond the material to guide the emergence of form and function, and that something was typically assumed to be some sort of non-material vital force or cause. The causes work together to bring about the epigenetic emergence.

The Roman Catholic Church readily absorbed the Aristotelian ideas and added a Creator to put the epigenetic developmental process into action and provide a soul in addition to the material body. Though the Christian concept of soul was quite different from Aristotle's philosophical soul-in-terms-of-causes, it proved easy to interpret Aristotle in Christian and epigenetic terms. For early Catholics, an individual life emerged gradually and became a person at the moment of metaphysical ensoulment when the spiritual soul became one with the emerging form and function of the body. Jewish and Muslim traditions held similar epigenetic views, with each life emerging gradually and reaching presumed personhood only as the biological organism reached the point of ensoulment, typically taken to occur at forty days.[3]

In contrast, seventeenth century materialists rejected the metaphysical components such as the soul and vital causes. These participants in what was later called the Scientific Revolution demanded explanations in terms of material and motion. They rejected what they saw as unwarranted metaphysical assumptions that anything non-material like a soul exists. We can only know what we can experience empirically through our senses, they insisted, and we must explain such phenomena in material terms. This, they concluded, meant that the life of an individual organism begins at the material beginning. Instead of a gradual coming into being, the adult form and function must be there from the start. It must be preformed and predetermined. Once biological knowledge made it clear by the mid-nineteenth century that the epigenetic biological process of developing an organism begins with the egg and gradually keeps changing every stage after that, the preformationist tradition held that life begins at fertilization.[4] Remember that they were motivated by the assumption that the form and function must be in the fertilized egg from this beginning, and thus the life of the individual organism is both preformed and predetermined.

In 1869, Pope Pius IX declared that life begins at "conception," thereby changing centuries of Catholic tradition. Conception was taken as meaning fertilization or the coming together of sperm and egg cells, which was taken as being a moment even though it is actually a rather lengthy process that takes time and involves many changes. The Pope's reasons are not publicly documented, and it is not clear whether the Pope was influenced by accumulating biological understanding of the fertilization process or whether he was motivated by metaphysical or other political considerations. Most likely, his intention was to stop attempts to prevent pregnancies and to stop abortions. What is clear is that the Church changed its mind, and that this has had significant social impact.[5]

Given the knowledge of the day and a desire to exert greater control over social matters, Pope Pius IX's decision was not unreasonable. Yet it is ironic. In changing the Church position, he embraced the preformationist view that life and the person begins all at once in the fertilized egg, a view held by materialists for reasons that rejected the presence of a soul or any such non-material entities. Thus Pope Pius IX was allying himself with the materialists, preformationists, and those who traditionally viewed life and the person as defined by material alone.

That seems odd. But the situation becomes even more odd when we realize that developmental biologists today reject the concept that an individual organism begins at fertilization in any but a very rudimentary material sense. Form, and its function, emerges only gradually, over time, that is epigenetically. In their role as developmental biologists, today's biological epigenesists are materialists or epigenesists. Insofar as any developmental biologists accept the existence of something beyond the material, that is a metaphysical or religious assumption and lies outside the scientific understanding of life.

Note that I refer to developmental biologists here. By contrast, geneticists have often failed to understand fully the epigenetic emergence of form and function and, especially in recent decades, have often sounded as if they did believe in a form of predeterminism. In their enthusiasm for the power of genetics, some have sounded as though they are embracing a form of preformation (and even predestination), as if genes alone transmit and define destiny. This view is misguided: understanding the gradual processes of the emergence of form and function during development shows why.

In summary, since 1869 the Catholic Church has adopted a view of life that is, ironically, traditionally preformationist and predeterminist and that has historically denied the existence of anything other than the material. Developmental biologists, in contrast, hold a view that is traditionally epigenetic and emergentist and that until 1900 required the existence of some vital force to turn material into a life and a person. What are the assumptions, evidence, and arguments underlying each view, and what are the implications?

ASSUMPTIONS, EVIDENCE, AND ARGUMENT

Of course the Roman Catholic Church and other religions are not driven by scientific understanding alone, or even primarily. While some churches may hold a preference for being consistent with scientific knowledge as far as is possible, most religions embrace something beyond the material world that makes up the subject of science. Religious evidence and argument are thought to come from such sources as revelation and belief; epistemic warrant comes from values within religions and need not be logical nor even rational. Therefore, any complaints by scientists that something "makes no sense" scientifically carry no force for religious interpreters.

This means that claims about what defines a person are bound to compete when some groups appeal to the material, empirical, logical, and rational values of science, others appeal to spiritual, revelatory, and religious values, and the two lead to different conclusions. When this happens, there are competing claims for authority. Even if we reject the authority of specific religious values for social decisions, those values often intersect with secular social and moral values and therefore play a role in social decision-making.

I focus here on understanding the scientific study of development rather than the religious interpretations. By the middle of the nineteenth century, researchers had shown in animals (not yet humans, since human development occurs inside the mother and remains invisible—at least until the recent advent of in vitro fertilization made it observable in some cases) that an individual organism begins as an egg cell. Fertilization of the egg cell by a sperm cell initiates a series of events that lead, gradually, to differentiation and to complex form and function.[6]

Experimental embryological research in the last quarter of the nineteenth century and into the twentieth century showed that the early stages of development across many species occur gradually and with considerable response to changing environmental conditions. Cell lineage studies tracked the changes in each cell and showed that, while the cell divisions are fairly regular and predictable in most species, they are also more or less able to respond to changes. Individuals of species such as sea urchins are quite "regulatory" and able to compensate for disruptions, accidents, and other changes in the normal pattern of cell division and still develop normally. In contrast, individuals in other species are more "mosaic" or "determinate" and respond to disruptions either by dying or developing abnormally. The assumption in the late nineteenth century was that the "higher" the organism, the more determinate it was, though this assumption soon proved to be wrong. At the time little was known about heredity, and the focus remained on development and differentiation.[7]

The regular steps observed during development in most species under normal conditions led researchers to describe the progress in developmental stages. At first, much of that research was oriented toward discovering and demonstrating presumed evolutionary relationships, or denying those relationships and attempting to show natural relationships based on underlying types.

By the beginning of the twentieth century, however, embryologists had already turned to study of embryos in their own right, asking: what are the stages of development for each species? They assumed that each member of a species would follow a common pattern, with minor variations based primarily on environmental conditions (though by the 1920s and 1930s the possibility of a hereditary or genetic contribution had to be considered as one of the many contributing factors in the developmental process). Embryologists, nonetheless, continued to focus on embryos and left it largely to others to look at the unobserved and still largely hypothetical genes.

Ross Granville Harrison was a typical embryologist, still insisting in the 1950s that Yale did not need a geneticist but rather should stick with empirical verifiable biological research such as embryology.[8] Harrison laid out the standard stages of frog development in a clear way that he passed on to his graduate students. Each had a photographic copy of the standard stages, which they could then use for comparison with experimental examples to see what difference experimental manipulation made in the results. If they surgically removed some cells, for example, or transplanted them from one part of an embryo to another, what was the result and how did it compare with the normal control embryo?

Frogs (through research carried out especially by Harrison and Hans Spemann and their students), a number of marine invertebrates (through research carried out especially at the Marine Biological Laboratory in Woods Hole, Massachusetts, and at the Naples Zoological Station), and even human developmental stages received close attention.

What Harrison called the embryological "gold rush" of new information about development showed the power of scientific investigation. Careful observation, combined with experimental control of variables to produce additional information, provided a strong biological basis for interpreting organismal development. It was clear by 1900 that each individual animal undergoes a developmental process that starts with the egg and fertilization, when it already contains hereditary material concentrated in the nucleus and chromosomes. Eggs then gradually undergo cell division after cell division eventually to yield differentiated parts and eventually the form and function characteristic of the species. The entire process shows just how little the initial egg and fertilization stages contribute.

HUMAN EMBRYOLOGY

Aside from embryologists interested in comparative development of different types of animals, humans are understandably most interested in human development. Yet unlike with frogs or sea urchins, we cannot see the early stages before birth because the mother protects the embryo (defined as the early developmental stages up to the eight week stage) and fetus (defined as the eight week stage to birth) from view. Thus, interpreting human embryonic development has not been easy. The only source of specimens for study came from abortions or the pregnant woman's death and, beyond the blastocyst stage, this remains the case. Aborted fetuses especially had long been the subject of fascination and curiosity. Many museums held jars of preserved specimens and, indeed, still hold them though many old specimens have been placed into storage out of the public view. The selections remained largely accidental and largely pathological until the end of the nineteenth century, however, so it was difficult to establish the course of normal development.

In the 1870s, Wilhelm His began systematically reconstructing the stages of human development, an enterprise joined by Franz Keibel in Germany. They sought to collect and model as many human developmental stages as possible, as Nick Hopwood has discussed in his excellent and beautifully illustrated *Embryos in Wax*. As Hopwood explains, establishing each normal stage and then accurately identifying the sequence held many challenges.[9] In fact, His and his contemporaries had no source of information or specimens from the earliest developmental stages. Such stages are too small, not yet recognizable as embryos, and the mothers in most cases did not even know they were pregnant. In addition, as Hopwood points out, there is a good deal of interpretation involved in understanding individual specimens, many of which were available only because they were abnormal from the beginning.

Nonetheless, the collection that His began was the largest and most scientifically useful by far, and Keibel's and Curt Elze's 1908 *Normentafel* set the standard. American anatomist Franklin Paine Mall visited Leipzig to study with His and began his own collection of human embryos once he returned to the United States and took up his position at the Johns Hopkins University Medical School. In 1913, Mall received his first grant, for $15,000, from the Carnegie Institution of Washington to catalog his existing collections to combine the specimens from different sources, and to preserve the collection in fireproof facilities.

Mall began what he called "a vigorous campaign" to collect new specimens, reportedly reaching half the US physicians and many internationally. The collection grew rapidly in both number and quality of specimens. The next year, Mall hired Keibel from Germany to help prepare and model the embryos, and continued to receive increasing Carnegie funding. At Mall's death in 1917, the Carnegie Institution of Washington continued support, and in 1991 the collection moved to the National Museum of Health and Medicine of the Armed Forces Institute of Pathology in Washington, DC. There, the now vast collection of specimens, records, models, and files were made available for researchers and public exhibits. The collection also forms the central core of material for the "Human Embryology Digital Library and Collaboratory Support Tools," funded by the National Institutes of Health.[10] The current collection director Adrianne Noe reports that the public finds the collections fascinating and educational, and that she receives thanks rather than complaints about both the virtual exhibit and the physical exhibits of embryos and fetuses.[11]

Yet even this marvelous collection cannot reveal any details of the very earliest stages of development, when all that exists is a few cells. That had to wait until the 1970s when physician Patrick Steptoe and embryologist Robert Edwards succeeded with what came to be called in vitro fertilization.

IVF AND EARLY HUMAN DEVELOPMENT

What in vitro fertilization did, quite literally, was to take the egg, the fertilization process, and the fertilized egg and early pre-implantation embryo out of the mother and into the glass dish. This made it visible. Researchers could actually observe an egg cell and a sperm cell uniting during fertilization, as had been predicted based on knowledge from other species. Then the cell divides: one into two, two into four, and four into eight, though in humans not eight into sixteen in the same regular way. Up to the eight-cell stage, the cells are biologically identical for all practical purposes. Each is totipotent, or in other words has the capacity to become a whole organism.

This situation begins to change, gradually, after the eight-cell stage, and this fact about the biological cellular material is important for our interpretation of when an individual organism begins. Together, the eight cells make up one whole. They will, under normal circumstances, become one individual. In the laboratory, however, technicians can—and do under some circumstances in fertility clinics—remove one or two of the cells, leaving the other seven or six to make up one whole. The one or two cells can be tested genetically, which is the main reason they are sometimes removed following voluntary protocols established by leading fertility clinics.

Each cell at this stage, or apparently any combination of up to eight cells together, can become one whole. In principle, it would be possible to separate any combination of the eight cells and produce twins or triplets or up to octuplets in the lab. Since it has been done often, we know this is possible with mice, and we have every reason to believe it is possible in humans though it is considered unethical to experiment with humans just for the sake of satisfying our curiosity about what is possible. However, we have reason to think that this is precisely what happens in perfectly normal cases of identical twins, triplets, quadruplets, and so on.

Each of the eight cells has the same set of genetic material in the chromosomes, of course, but there seems to be no gene expression up to this point. The cells are just mechanically dividing into more and more small cells. The total cell mass does not increase appreciably, and the only significant biological action comes from cells signaling to the others in some mechanical and/or chemical way that they need to work together as a whole.

Up to this point, one can become two or more, and two or more can become one. In other words, one cell can divide into two, four, and eight that can be separated to yield twins or up to octuplets. It is also possible for two cells to be combined into one individual. Even two fertilized eggs cells can come together and make up a hybrid (or chimera) which then contains the genetic material from two different sets of cells. In humans this happens naturally, though very rarely. In mice, however, researchers have combined two, four, and even eight cells from different eggs and each of these combinations has proven capable of working together to produce a whole and apparently normal mouse.[12]

This shows that in these early cell stages what counts as an individual living organism is still very flexible. Changing environmental conditions produce changing developmental responses, and we see that the early developing embryo carries tremendous capacity to regulate itself as a whole, though it is not yet formed and the cells are not yet differentiated into different cell types. What counts as an individual organism is highly accidental and malleable at this point. Therefore, from the biological point of view, it is very odd to consider these stages of early cell division as a person or indeed even as a defined organism.

In fact, there is very good evidence that most fertilized eggs and even most early stages of pre-implantation embryos never go on to become implanted. Even of those implanted, many do not continue to develop. Some estimate that at least ⅔ of naturally fertilized eggs never become babies, and perhaps an even higher percentage fail to pass through all the stages of development to birth. If the biological life of an individual human begins at fertilization, even more do not make it beyond more than a few cell divisions before dying. This seems a tremendous waste and a very odd interpretation of meaningful life. Thus it is far more reasonable and scientifically accurate to focus on later developmental stages when the fetus is actually viable as what we should consider a human individual organism.

Stanford developmental biologist Irving Weissman lobbies for wider understanding of what others have argued for at least 25 years. He wants to call the stages before implantation a "pre-embryo," as President George W. Bush did during his August 9, 2001, televised speech concerning stem cell research. Weissman wants to make clear for public discussion, as Bush accepted implicitly by his choice of words, that developmentally this early bunch of cells has divided a number of times but has not yet been differentiated nor even implanted in the mother. Weissman wants to call the pre-embryo what others would call the pre-implantation embryo and has urged the National Academy of Sciences to adopt his definition. The important point is that there is a widely recognized difference between this early stage and what appears even a few days later.[13]

The only reasons anyone would insist that these earliest stages define the individual organism that may come later are metaphysical: (a) they assert that since there is material continuity of the later with these earlier stages it must be the same ontological entity; or (b) they assume a sort of preformationism that holds that the chromosomal complement defines the later adult; or (c) for other religious or non-material reasons that invoke the existence of some sort of soul, vital force, or other religiously postulated defining factor.

SO, HOW AND WHEN DOES A HUMAN LIFE BEGIN?

Early stages are, biologically, just a bunch of material cells. There is as yet no gene action, no differentiation, and absolutely nothing that could be called an individual organized organism. Within the blastula, for the first time, cells begin the simplest differentiation. A single layer of cells gathers around the surface of the blastula, and this layer will become the placenta. All the other cells gather inside, where they multiply and compact tightly together into an inner cell mass. These cells are the embryonic stem cells and each seems to be pluripotent, meaning that none can become a whole organism by itself but that each can become any kind of cell. Which kind of cell it actually becomes depends on the environmental conditions in which it grows—how many cells, what culture medium, and other factors that we do not fully understand.

These pluripotent human embryonic stem cells are the subject of public fuss since 1998, when they were first isolated and when the promises for possible stem cell therapies first emerged—against the background of ethical concerns that harvesting the cells requires destroying a human embryo. So here we are at that question: when does the individual organism begin? Is the blastocyst stage of the pre-implantation embryo an individual human organism already?

Biologically, no. The genes are not yet being expressed, the cells are not differentiated, and the whole has no organization and no form even remotely resembling the organism that it will become in either structure or function. Only for those preformationists who insist on other than empirical biological grounds that fertilization marks the beginning of the individual can the blastocyst be considered more than a biological collection of cells at this point. To emphasize "potential" for what might come later is to read in more than is biologically there.

Up through the blastula stage, embryonic cells can grow in the lab, in the glass dish, and do not require any exchange of nutriment or waste with the mother. This happens just at the end of the blastula stage as the stage called gastrulation begins. This is the point at which the cells begin for the first time to divide into the three germ layers, each of which will give rise to different kinds of cells. The embryo is, for the first time, beginning to become organized in a very early way, one still open to considerable malleability in response to changing environmental conditions. It is at the blastula stage and before gastrulation begins that the pre-implantation embryo must be implanted (or frozen for later implantation) or it will die.

For our purposes, what is important is that there is a long-recognized series of documented developmental stages for humans just as for frogs or sea urchins or mice or whatever. These stages correspond to gradually increasing organization and differentiation. Just as we can clearly and precisely identify an embryo as an undefined and undifferentiated blastula, an early stage gastrula, an eight week fetus with rudimentary form, or a later stage fetus with considerable organic function, so perhaps we can identify stages of emerging organization and individuality. The biological progress of stages makes clear that we have nothing like a defined human organism in the early stages. It makes little sense to call the blastula a human being in any robust sense. Only a preformationist who believes that the early stages are predestined to become a human and therefore already deserve that name could suggest otherwise. I note again how very odd it is that traditionally those holding this view in the past were the materialists seeking to avoid vitalism or metaphysics, while those who hold this view now often do so for religious or metaphysical reasons.

WHAT FOLLOWS?

Consider a block of marble. There it sits. With the right sculptor, it may become a beautiful sculpture, perhaps a human figure. Yet do we want to say that the marble material is already the sculpture? Even if the sculptor has the plan in mind and has begun the very first scratches into the surface, we surely do not want to pretend that we have a sculpture already. In saying that the sculpture already exists within the marble and that it is a matter of removing the excess, even Michelangelo does not persuade us that a completed form is already literally and materially there except in his mind. True, we might say that we have a sculpture even at some stage before it is finally and fully formed, but surely not in the earliest developmental stages. We need a much more concrete and actualized form that has resulted from the formative sculpting process.

Science alone cannot answer the question when a person begins, since that is necessarily at least in part a social definition. But biology can give us guidance about what biological capacity each developmental stage has, which provides some guidance to what meaning we might want to assign.

As with the marble, the egg is not a human individual organism, not even after it is fertilized and becomes the zygote. Nor is the two cell stage. With twinning or other splitting up to the eight-cell stage, the egg may even become more than one human organism. Even the blastocyst is not an organism yet, but rather a bunch of raw material cells that have not yet experienced gene expression and are still unformed. The implanted blastocyst becomes one with the mother, and takes on a different process of gastrulation and other stages that begin gene expression and finally lead to the expression of form. Only after eight weeks of human development do we have all the organ and major parts differentiated sufficiently that the form is there, at least rudimentarily. At this stage, the embryo becomes a fetus. This is the first point at which we might reasonably declare that we have a relatively defined human organism from a biological perspective.

Yet form alone is not enough for a human. The form must digest and respire in order to live. There are good reasons, therefore, to consider the viable fetus as a proper beginning of an individual human life, though there may be room for social negotiation between the beginning of the fetal stage at eight weeks and viability some time later. And there is considerable room for social negotiation about what it means to have such a fetal organism, whether we want to call it a person, and what meaning and status it might have.

That is not our charge here, which is to emphasize that there are no good biological reasons to define early developmental stages such as the blastocyst stage, with its much-discussed embryonic stem cells, as a differentiated human organism. Only the preformationists, those who already are looking ahead at the potential for something that they believe is coming later, insist that the earliest stages are already individual humans—or already sculptures. Biologically, an individual human life begins only gradually, over time, and only well into the fetal stages does it acquire form, function, and human-ness. Biologically, development of an individual organism takes place only gradually through predictable and meaningful developmental stages, as undifferentiated cells become organized into an organism of a certain form and function. Any other answer requires us to invoke metaphysics or religion, not science.

NOTES

1. See, for example, Eelco F. M. Wijdicks, ed., *Brain Death* (Baltimore: Lippincott, Williams, and Wilkins, 2001); or Stuart J. Youngner, Robert M. Arnold, and Renie Schapiro, eds., *The Definition of Death: Contemporary Controversies* (Baltimore: John Hopkins University Press, 2002).

2. Aristotle, *Generation of Animals*, trans. A. L. Peck (Cambridge, MA: Harvard University Press, 1979).

3. Jane Maienschein, *Whose View of Life: Embryos, Cloning, and Stem Cells* (Cambridge, MA: Harvard University Press, 2003) provides much greater detail and background on some of the discussions here. See especially chapter 1 on early church interpretations. There is much discussion about St. Thomas Acquinas's interpretations of ensoulment, for example, including the suggestion that males are ensouled at 40 days and females at 80 days.

4. Shirley A. Roe, *Matter, Life, and Generation* (Cambridge: Cambridge University

Press, 1981); Peter Bowler, "Preformation and Pre-existence in the Seventeenth Century: A Brief Analysis," *Journal of the History of Biology* (1971) 4: 221–44; Jane Maienschein, "Competing Epistemologies and Developmental Biology," in *Biology and Epistemology*, edited by Richard Creath and Jane Maienschein (Cambridge: Cambridge University Press, 2000), pp. 122–37.

5. It is important to note that Pope Pius IX presided over what has been called Vatican I, a period of intense activity designed to restore what was regarded as the conservative basis to the Church. This is undoubtedly relevant to the particular decision about conception.

6. E. S. Russell, *Interpretation of Development and Heredity: A Study in Biological Method* (Clarendon Press, 1930) provides a nice, even if dated, introduction to the issues and discoveries of the 19th century.

7. See, for example, Maienschein, *Whose View*, chapters 2 and 3 for a summary of this research.

8. In the Ross Granville Harrison Archives, Yale University, he comments to his friend Edwin Grant Conklin that he saw no point in hiring a geneticist. J. P. Trinkaus confirmed, personal conversation, that when he first went to Yale as a young professor, this attitude continued. Embryology was seen as the proper study of development. Genetics was about inheritance and a different subject.

9. Nick Hopwood, *Embryos in Wax: Models from the Ziegler Studio* (Cambridge: Whipple Museum of the History of Science, 2002); Hopwood, "Producing Development: The Anatomy of Human Embryos and the Norms of Wilhelm His," *Bulletin of the History of Medicine* 74 (2000): 29–79.

10. See, for example, as of August 19, 2006, http://netlab.gmu.edu/visembryo.htm.

11. Adrianne Noe, "The Human Embryo Collection," in Jane Maienschein, Marie Glitz, and Garland Allen, eds., *The Centennial History of the Carnegie Institution of Washington Department of Embryology* (Cambridge: Cambridge University Press, 2005), pp. 21–61; and see other essays in that volume. Franz Keibel and Curt Elze, with contributions from Ivar Broman, I. August Hammar, and Julius Tandler, *Normentafel zur Entwicklungsgeschichte des Menschen* 8, ed. Franz Keibel (Jena: Gustav Fischer, 1908); Franz Keibel and Franklin P. Mall, eds., *Manual of Human Embryology* (Philadelphia: Lippincott, 1910).

12. Beatrice Mintz did much of the early research on hybrid mice, as discussed in Ann B. Parson, *The Proteus Effect: Stem Cells and their Promise for Medicine* (Washington, DC: Joseph Henry Press, 2004).

13. Irving Weissman has worked to persuade lawmakers and the National Academies of Science in numerous presentations on the subject of early embryos. The term "pre-embryo" has been used for at least 25 years; President Bush, on stem cell research, August 9, 2001, http://www.whitehouse.gov/news/releases/2001/08/20010809-2.html.

H.R. 810: STEM CELL RESEARCH ENHANCEMENT ACT OF 2005

To amend the Public Health Service Act to provide for human embryonic stem cell research.

Representative Michael Castle (R-DE) introduced the bill February 15, 2005. It passed the House and the Senate, with 97 percent of Democrats supporting the bill and 66 percent of Republicans opposing it.

Be it enacted by the Senate and House of Representatives of the United States of America in Congress assembled.

SECTION 1. SHORT TITLE.

This Act may be cited as the "Stem Cell Research Enhancement Act of 2005."

SECTION 2. HUMAN EMBRYONIC STEM CELL RESEARCH

Part H of title IV of the Public Health Service Act (42 U.S.C. 289 et seq.) is amended by inserting after section 498C the following:

(a) IN GENERAL—Notwithstanding any other provision of law (including regulation or guidance), the Secretary shall conduct and support research that utilizes human embryonic stem cells in accordance with this section (regardless of the date on which the stem cells were derived from a human embryo).

(b) ETHICAL REQUIREMENTS—Human embryonic stem cells shall be eligible for use in any research conducted or supported by the Secretary if the cells meet each of the following:

(1) The stem cells were derived from human embryos that have been donated from in vitro fertilization clinics, were created for the purposes of fertility treatment, and were in excess of the clinical need of the individuals seeking such treatment.

(2) Prior to the consideration of embryo donation and through consultation with the individuals seeking fertility treatment, it was determined that the embryos would never be implanted in a woman and would otherwise be discarded.

(3) The individuals seeking fertility treatment donated the embryos with written informed consent and without receiving any financial or other inducements to make the donation.

(c) GUIDELINES—Not later than 60 days after the date of the enactment of this section, the Secretary, in consultation with the Director of NIH, shall issue final guidelines to carry out this section.

(d) REPORTING REQUIREMENTS—The Secretary shall annually prepare and submit to the appropriate committees of the Congress a report describing the activities carried out under this section during the preceding fiscal year, and including a description of whether and to what extent research under subsection (a) has been conducted in accordance with this section.

STATEMENT OF ADMINISTRATION POLICY H.R. 810 STEM CELL RESEARCH ENHANCEMENT ACT (REP. CASTLE [R] DE AND 200 COSPONSORS)

The Administration strongly opposes Senate passage of H.R. 810, which would use Federal taxpayer dollars to support and encourage the destruction of human life for research. The bill would compel all American taxpayers to pay for research that relies on the intentional destruction of human embryos for the derivation of stem cells, overturning the President's policy that funds research without promoting such ongoing destruction. *If H.R. 810 were presented to the President, he would veto the bill.*

The President strongly supports medical research and worked with Congress to dramatically increase resources for the National Institutes of Health (NIH). This Administration is the first to provide Federal funds for human embryonic stem cell research and has done so without encouraging the destruction of human embryos. The President's policy permits the funding of research using embryonic cell lines created prior to August 9, 2001, the date his policy was announced, along with stem cell research using other kinds of cell lines. Scientists can therefore explore the potential applications of such cells, but the Federal government does not offer incentives or encouragement for the destruction of human life.

Over the past five years, more than $90 million has been devoted to embryonic stem cell research through the NIH. However, this bill would provide Federal funding for the first time for a line of research that involves the intentional destruction of living human embryos for the derivation of their cells. Destroying nascent human life for research raises serious ethical problems, and many millions of Americans consider the practice immoral.

The Administration believes that government has a duty to use the people's money responsibly, both supporting important public purposes and respecting moral boundaries. Every year since 1995, Congress has upheld this balance on a bipartisan basis by prohibiting Federal funds for research in which an embryo is destroyed. The Administration's policy upholds this same principle.

H.R. 810 seeks to replace the Administration's policy with one that uses Federal dollars to offer a prospective incentive for the destruction of human embryos. Embryonic stem cell research is at an early stage of basic science and has never yielded a therapeutic application in humans. While no treatments or cures have been developed from embryonic stem cell research, there are therapies and promising treatments from adult stem cells and other forms of non-embryonic stem cells.

Alternative types of human stem cells—drawn from adults, children, and umbilical-cord blood without doing harm to the donors—have already achieved therapeutic results in thousands of patients with many different diseases. Researchers are now also developing promising new techniques to produce stem cells just as versatile as those derived from human embryos, but not requiring the use of embryos. The Administration believes that the availability of alternative sources of stem cells further counters the case for compelling the American taxpayer to encourage the ongoing destruction of human embryos for research.

Moreover, private sector support and public funding by several States for this line of research, which will add up to several billion dollars in the coming few years, argues against any urgent need for an additional infusion of Federal funds, which would not approach such figures even if completely unrestricted. Whatever one's view of the ethical issues or the state of the research, the future of this field does not require a policy of Federal subsidies offensive to the moral principles of millions of Americans.

H.R. 810 advances the proposition that the Nation must choose between science and ethics. The Administration, however, believes it is possible to advance scientific research without violating ethical principles—both by enacting the appropriate policy safeguards and by pursuing the appropriate scientific techniques. H.R. 810 is seriously flawed legislation that would undo current safeguards, and provide a disincentive to pursuing new techniques that do not raise ethical concerns.

CHRISTOPHER A. PYNES*

WHY IS THERE A STEM CELL DEBATE? AND HOW TO DEPOLITICIZE IT

I. INTRODUCTION

There is a constant mix of confusion and certainty in the stem cell debate. Opponents think the reasons why stem cells should not be used in scientific research are clear and simple; while proponents think it is clear and simple that one should use stem cells in scientific research. What is going on in this debate? Why are both sides so certain of their views? In what follows I shall aim at three basic tasks: (i) clear up confusion about the science of stem cells; (ii) explain the sources of some of the more problematic stem cells issues that foster misunderstanding and unnecessary debate; and (iii) try to move the stem cell debate forward from its current state of deadlock between the two groups.

II. TERMS, DEFINITIONS, AND STEM CELL BIOLOGY

In an effort to make the moral debate about stem cells more focused, it will be useful to explain the biological facts. A human ovum/egg has to be intercepted in some way by a human sperm for those gametes to progress through the stages of human development, and give rise to the properties of different kinds of human cells. There are several stages of human development, but the following four are particularly important for this discussion:

*Department of Philosophy and Religious Studies, Western Illinois University, Macomb, IL 61455, CA-Pynes@wiu.edu

Stages of Human Development:

Zygote—a fertilized human ovum before cellular division
Blastocyst—the result of cellular division of a zygote from day 2 to week 2
Embryo—week three to week eight of human development
Fetus—after week eight of human development

Now, the intermediate stages between blastocyst and embryo are the domain of stem cell biology, and the area that most of the debate centers around as these are the cells most useful for research. Blastocysts, the result of cellular division, are generally not allowed to grow for longer than two weeks, when used for research, for this is when typical implantation into a uterus occurs in a natural setting. Since it is the early stages of human development that are in question, and these balls of cells are neither embryos nor fetuses yet, it has lead some people, like President Bush, into referring to them as "pre-embryos."[1] But why is it that these "pre-embryo" cells or blastocyst cells are so important to scientists and opponents of stem cell research?

As cells develop they lose some of their early abilities. Cells can have one of three cellular abilities or properties depending on what stage of development they have reached.

Types of Cell Properties

Multipotent—can become a defined set of cells of the human organism
Pluripotent—can become most but not all cells of the human organism
Totipotent—can become any of the cells of the human organism

Obviously totipotent, or pre-embryonic, cells are most useful for research into the cures of degenerative diseases like Alzheimer and Parkinson's because they can become any of the cells of the human body. The stem cell debate, however, has focused on the relative advantages to using adult stem cells over pre-embryonic stem cells for scientific research. This would be acceptable if each type of stem cell didn't have different biological properties. Exactly how does the cellular property of each kind of stem cell affect the perception of the moral permissibility of using it in scientific research? That is, why are opponents of embryonic stem cell research so keen on adult cells and down on totipotent, pre-embryonic stem cells? Potentiality!

III. THREE KINDS OF STEM CELLS & POTENTIALITY

The term "stem cell" is used in the debate to refer to everything from embryonic stem cells (typically cells from things that are not actually embryos, but "pre-embryos") to stem cells found in umbilical chord blood to adult stem cells. Many opponents of embryonic stem cell use in scientific research claim that what ought to be used are stem cells from umbilical chord blood or adult stem cells, instead of embryonic or pre-embryonic stem cells. If scientists can do all the same things with all three types of cells, then there is no need to use stem cells from embryos or pre-embryos, which typically requires the destruction of the organism. One might wonder what kind of in principle, moral distinction might be made between embryonic stem cells, chord blood stem cells, and adult stem cells. One might also wonder why it would matter which version of the cells are used in scientific research for if all three types of cells can do the same thing for scientific research, then the impermissibility of one should make the use of them all impermissible. So, it would seem that this kind of unprincipled distinction between stem cell types and moral permissibility for scientific research cannot hold up to close inspection. So what is really going on here?

There are differences between the kind of stem cells found in umbilical chord blood, embryonic stem cells, and adult stem cells.[2] Stem cells gained from chord blood are pluripotent, and adult stem cells are multipotent, but only embryonic stem cells are totipotent. So, there does seem to be good reason to think that there is an in principle distinction between the three kinds of stem cells. Only embryonic stem cells (given the right environment and conditions) have the potential to give rise to a new, complete human life. It is because undifferentiated stem cells have the ability to become any type of cell in the body that they are such a useful scientific tool. Can the issue of the cell's potential, however, be overcome in a morally satisfying way? I think it can, and there are at least two ways to argue that it can be overcome.

First, one could argue that there really isn't any potential in these cells, at least the kind that matters morally as differentiation[3] has not occurred in the embryonic stem cell type—the cells are still unspecialized. The objection to the use of the totipotent (embryonic) stem cells is that they have the potential to give rise to a complete full-fledged human being and that kind of potential should not be destroyed. Do the kinds of cells that can be generated by fertilizing an ovum in the lab have the potential to become full-fledged humans? In one sense, absolutely they do. But if they are not implanted into a uterus, then they in fact do not have the potential to become a full-fledged, adult human. So in a second sense, absolutely they do not have the potential. It is here that there seems to be confusion between potentiality and possibility. Clearly potentiality involves possibility, but it has to be more than just mere possibility. For example, we can all agree that I was once a possible first round NFL draft pick, but I no longer am a potential first round NFL draft pick even though it is still in some sense possible.[4] So just because something is possible doesn't mean that that something is potential or in other words: it is false that possibility entails potentiality. But I don't think that this will comfort anyone that isn't already sympathetic to the use of totipotent (embryonic) stem cells for scientific research.

Second, one could even grant that potentiality of the cells to become adults is there but it does not warrant the kind of moral assessment that many people think it should. The issue seems to hinge on the notion of potentiality we are using. It is one thing to state that something has potential, and in a sense everything has potential, but it is quite another to say that potential matters morally. In the case of stem cells' potentiality it does not seem to matter morally. It appears as if confusion between two types of questions about potentiality and stem cells has crept into the debate:

(A) Are stem cells the kind of thing that is worthy of moral status? and

(B) Are stem cells (think blastocyst) that will become a person worthy of moral status?[5]

If (B) is answered in the affirmative, that does not entail the truth of (A). And if (A) is answered in the negative, that does not mean that (B) has to be. And a denial of (B) entails a denial of (A), and affirming (A) entails an affirmation of (B). These kinds of potential arguments are easy to explain in a more natural example. We won't let 17-year-old people vote in presidential elections. If the person replies that they are a potential 18-year-old and thus deserve the right to vote, we still don't allow them to vote. This seems a natural and appropriate manner in which to handle potentiality claims, and many kinds of examples are useful in this regard: driving, drinking, and marriage all seem to work. Potentiality isn't relevant to the assessment of moral status in these cases, and thus can be eliminated from the moral debate when it comes to stem cells.[6] This being said, it seems more likely that the real source of the stem cell debate positions comes analogously from the abortion debate. We shall now turn our attention to that very issue.

IV. STEM CELLS, ABORTION, AND FERTILITY TREATMENTS

Most of the debate about stem cell use focuses on embryonic (totipotent) stem cells; and thus is linked, I shall argue, inappropriately to the abortion debate, when it should, more appropriately, be linked to the issue of infertility treatments. To begin, the vitriol between the pro-life and pro-choice groups feeds into the stem cell debate and the misunderstanding about the difference between embryonic stem cells and aborted fetuses. Using embryonic stem cells is not abortion. In fact, since most of the fetal tissue from aborted fetuses is already differentiated, and thus is only multipotent or pluripotent, their usefulness is limited, unlike the embryonic totipotent cells. This is precisely why embryonic (totipotent) stem cells are created in the lab using human gametes, or in other words, not all embryonic stem cells come from embryos. As such, the stem cell debate should be more closely linked to the fertility treatment issue, rather than the abortion debate since the cells of pre-implantation, pre-embryos are most useful to stem cell science, and they are found in infertility treatments.[7]

If we find fertility treatments to be morally acceptable and as a part of a culture of life, then we have to accept that there are going to be fertilized eggs, even pre-embryos about 14 days old, left over from fertility treatments. These pre-embryos can be adopted, as President Bush likes to point out, but in many cases they are considered bio-waste after the couple decides that they don't want additional children, treatment or after a certain amount of storage time has elapsed. If the cells of these pre-embryos can provide scientific information that can save the lives of actual human beings, then it is part of the culture of life that these cells be used in just such a way. This is the view of the embryonic stem cell research proponent. This isn't an issue about abortion, but an issue about how to create and save lives in the best possible way with the resources we have.

The question comes to this: what kind of obligation do we have to human cells that have the potential to become adult humans given the right set of circumstances, but will not become adult humans because those circumstances will never occur? Now clearly there are those that don't want a factory of human embryos created and then destroyed for scientific research. This goes against a sense of human dignity and sensibility that everyone shares. And this is why scientists only allow fertilized ovum to be grown for fourteen days. Ultimately we need to realize that the abortion issue and the stem cell debate are not the same thing; and even if one has a moral objection to abortion, it does not mean that one ought to have an objection to embryonic (totipotent) stem cell research out of some moral consistency claim. There isn't a moral consistency relation between the two.

To make it clear, the use of aged fertilized eggs (blastocysts) for their stem cells is not the same thing as aborting a fetuses and using their parts for scientific research. The former is less likely to be a future human than the latter. And just because, as President Bush points out, they may be adopted, used, and given birth to, doesn't mean that they have moral status when they won't be adopted, used, and given birth to. This would be to confuse questions (A) and (B) from section III above. Finally a culture of life that supports the use of fertility treatment for people that want to have children should be willing to support that same culture of life for those individuals that are ill and in need of scientific cures that could result from stem cell research.

In the last 30 years there have been over 3 million IVF babies born. Let's imagine a case of infertility where the loss of fertilized eggs was just one for every ten children born. If this loss were the norm, then there would have been 300,000 fertilized eggs that went unclaimed or used over the last 30 years, which is 10,000 a year, more than enough to take care of stem cell lines for research. What is to be done with these fertilized eggs especially given the fact that the number for left over pre-embryos is closer to two orders of magnitude greater? Ultimately, if we think that infertility treatments are morally permissible, then we have to accept the consequences of the treatment, namely, left over fertilized eggs. If we accept these consequences, then it is unreasonable to not use these pre-embryo cells for scientific research instead of just incinerating them or letting them sit in liquid nitrogen until they are unusable for any purpose. It is for this reason that advocates of stem cell research are, at their core, part of a culture of life. They want to make those that are living with painful, debilitating diseases live lives that are worth living; not at the expense of others that might live, but through the process of science.

V. PRESIDENT BUSH AND THE STEM CELL DEBATE

If one looks closely at the information President Bush has put out regarding stem cells, one can see why the debate has raged on in such a divisive way.[8] Three things stand out as particularly troubling when it comes to the President's portrayal of the stem cell debate. President Bush states:

(i) that he is the first President to fund stem cell research (with a balanced approach);
(ii) that stem cell research destroys embryos; and
(iii) that stem cell research isn't illegal and that there are enough stem cell lines to do the work that scientists need to do.

As for (i) it is important to remember our history. While President Bush was the first president to fund certain kinds of stem cell research he shouldn't be lauded so loudly for this accidental fact of history, for stem cells weren't discovered until November 1998. At that point the country was in the grip of the Monica Lewinsky scandal and the impeachment trial of President Clinton. This lasted from early 1998 until the final Senate vote in February 1999. Then in April of 1999 the country was transfixed by the Columbine, Colorado, killing of twelve students and one teacher by fellow students. Then the country was in the midst of presidential election politics where little legislation gets done. The fact is that this president is the first to have a real opportunity to fund stem cell research, and when given the opportunity to do so, did it in a way that was not conducive to good science.

When it comes to claim (ii) the President continues to confuse people by changing the vocabulary of the stem cell discussion from one of "pre-embryos" to "embryos" when talking about stem cell research. He muddles the waters even more when he introduces the notion of adoption and shows up to veto a bill with adopted IVF children in tow claiming a culture of life requires the protection of the embryos (he changed his terms again) that will never become adult individuals.[9]

Finally, we will see how the claims found in (iii) generate problems. Yes, it is true that stem cell research isn't illegal. Since so much scientific research is funded by the federal government, however, if the funding isn't provided, then the much needed scientific research on stem cells will very likely not be done, and if it is done at all will be delayed so that those that need the cures will be gone and have suffered unnecessarily. Moreover, the fact that there are some stem cell lines that are used for research doesn't justify the President's claim that his funding is balanced. Many of the stem cell lines the President funded are contaminated with mice DNA and other foreign materials that make them useless for research; or if they can be used at all, it is prohibitively expensive to clean up the cell lines, extend the time to research cures, and provide instances where the research could be compromised by the extra steps to clean the cell lines. New stem cell lines need to be created, preserved, and used to move forward in so many areas of research. The President is either wrong or misrepresenting the issue when he makes claims (i)–(iii) above, which fosters an unnecessary debate and politicizes scientific research.

VI. CONCLUSION

The confusion and politicized nature of the stem cell debate derives from a number of issues: confusion about biology and the terms used, a link with the abortion debate, and a political agenda by our current President. If we understand that pre-embryos and the totipotent stem cells they have are not embryos and that they will never be used to give rise to adult humans, then we realize that we don't have the same obligations to them that we have to actual individuals. If we realize that scientific research is a part of the culture of life, then we can see that using the remaining fertilized eggs from IVF and other ART (assisted reproductive treatments) can give us knowledge to help actual people with real diseases. This is why the proponent of embryonic stem cell research is struck by the fact that there is a debate at all. The way to resolve the debate is to resolve the misunderstandings and depoliticize the funding process. I hope that I have done just that in support of embryonic stem cell research.

NOTES

1. See the White House Web site link for one such example from August 9, 2001, in a speech on stem cell research, http://www.whitehouse.gov/news/releases/2001/08/20010809-2.html. The President has since stopped using this term because it undercuts his desire to limit stem cell research. More on this below.

2. From here forward, I will be referring to all early stem cells that are totipotent as "embryonic stem cells" even though most of these cells that are to be used come from pre-embryos that are created for infertility treatments or assisted reproductive treatments (ART). It would be nice if we had a neutral term to refer to them instead of "embryonic" which connotes baby or person to most people. What is important about these cells is that they are totipotent, not that they are embryonic for they are not from embryos, but from blastocysts and morulas. The multiple senses of "embryonic" as "from an embryo" and as "undeveloped" is what gives rise to this confusion and is a major source of the deadlock in the debate.

3. Differentiation occurs when the cell changes and begins to perform a particular function. This process occurs as human development does, and the totipotent, multipotent, pluripotent changes are referred to as differentiation.

4. The same could have been said of Maurice Clarett before the NFL draft in 2005. Or if you like, one could make the same point about being a military recruit or any number of things. After a certain age, one is no longer a potential military recruit, but it is still possible to be a recruit depending on the state of the world and a given military's readiness.

5. By "moral status" all I mean is that property that makes something worthy of moral respect or to be treated like a moral patient even though in all senses, these cells are not moral agents at the time in question. Being a moral agent does, however, make one immediately a moral patient.

6. I should point out that voting and drinking ages can be arbitrary and seemingly unfair, but just because they are in a sense arbitrary and unfair doesn't mean that there isn't

a need to draw a line somewhere. This is also something that I think affects the kind of thinking for opponents of stem cell use. Just about any line could be considered arbitrary and somewhat unfair, but a line has to be drawn, and it can reasonably be drawn at some places rather than others.

7. President Bush has done just this when he linked stem cells use with fertility treatment by promoting the adoption of pre-embryos left over from fertility treatments. In his speech on July 19, 2006, the President equated adoption of pre-embryos with the culture of life and how these potential children are worthy of respect. You can find the news release here: http://www.whitehouse.gov/news/releases/2006/07/20060719-3.html.

8. For example see the White House press release from July 17, 2006: "Setting the Record Straight: President Bush's Stem Cell Policy Is Working," which can be found online here: http://www.whitehouse.gov/news/releases/2006/07/20060717-5.html. Or the President's stem cell "fact sheet" from July 19, 2006, which can be found online here: http://www.whitehouse.gov/news/releases/2006/07/20060719-6.html.

9. See President Bush's veto speech of July 19, 2006, for more details. It can be found online here: http://www.whitehouse.gov/news/releases/2006/07/20060719-3.html.

GOD AND BIOLOGY

THE FIRST BOOK OF MOSES, CALLED GENESIS

CHAPTER 1

In the beginning God created the heaven and the earth.

2. And the earth was without form, and void; and darkness was upon the face of the deep. And the Spirit of God moved upon the face of the waters.

3. And God said, Let there be light: and there was light.

4. And God saw the light, that it was good: and God divided the light from the darkness.

5. And God called the light Day, and the darkness he called Night. And the evening and the morning were the first day.

6. And God said, Let there be a firmament in the midst of the waters, and let it divide the waters from the waters.

7. And God made the firmament, and divided the waters which were under the firmament from the waters which were above the firmament: and it was so.

8. And God called the firmament Heaven. And the evening and the morning were the second day.

9. And God said, Let the waters under the heaven be gathered together unto one place, and let the dry land appear: and it was so.

10. And God called the dry land Earth; and the gathering together of the waters called he Seas: and God saw that it was good.

11. And God said, Let the earth bring forth grass, the herb yielding seed, and the fruit tree yielding fruit after his kind, whose seed is in itself, upon the earth: and it was so.

12. And the earth brought forth grass, and herb yielding seed after his kind, and the tree yielding fruit, whose seed was in itself, after his kind: and God saw that it was good.

13. And the evening and the morning were the third day.

14. And God said, Let there be lights in the firmament of the heaven to divide the day from the night; and let them be for signs, and for seasons, and for days, and years,

15. And let them be for lights in the firmament of the heaven to give light upon the earth: and it was so.

16. And God made two great lights; the greater light to rule the day, and the lesser light to rule the night: he made the stars also.

17. And God set them in the firmament of the heaven to give light upon the earth,

18. And to rule over the day and over the night, and to divide the light from the darkness: and God saw that it was good.

19. And the evening and the morning were the fourth day.

20. And God said, Let the waters bring forth abundantly the moving creature that hath life, and fowl that may fly above the earth in the open firmament of heaven.

21. And God created great whales, and every living creature that moveth, which the waters brought forth abundantly, after their kind, and every winged fowl after his kind: and God saw that it was good.

22. And God blessed them, saying, Be fruitful, and multiply, and fill the waters in the seas, and let fowl multiply in the earth.

23. And the evening and the morning were the fifth day.

24. And God said, Let the earth bring forth the living creature after his kind, cattle, and creeping thing, and beast of the earth after his kind: and it was so.

25. And God made the beast of the earth after his kind, and cattle after their kind, and every thing that creepeth upon the earth after his kind: and God saw that it was good.

26. And God said, Let us make man in our image, after our likeness: and let them have dominion over the fish of the sea, and over the fowl of the air, and over the cattle, and over all the earth, and over every creeping thing that creepeth upon the earth.

27. So God created man in his own image, in the image of God created he him; male and female he created he them.

28. And God blessed them, and God said unto them, Be fruitful, and multiply, and replenish the earth, and subdue it: and have dominion over the fish of the sea, and over the fowl of the air, and over every living thing that moveth upon the earth.

29. And God said, Behold, I have given you every herb bearing seed, which is upon the face of all the earth and every tree in the which is the fruit of a tree yielding seed; to you it shall be for meat.

30. And to every beast of the earth, and to every fowl of the air, and to every thing that creepeth upon the earth, wherein there is life, I have given every green herb for meat: and it was so.

31. And God saw every thing that he had made, and, behold, it was very good. And the evening and the morning were the sixth day.

CHAPTER 2

Thus the heavens and the earth were finished, and all the host of them.

2. And on the seventh day God ended his work which he had made; and he rested on the seventh day from all his work which he had made.

3. And God blessed the seventh day, and sanctified it: because that in it he had rested from all his work which God created and made.

4. These are the generations of the heavens and of the earth when they were created, in the day that the LORD God made the earth and the heavens,

5. And every plant of the field before it was in the earth, and every herb of the field before it grew: for the LORD God had not caused it to rain upon the earth, and there was not a man to till the ground,

6. But there went up a mist from the earth, and watered the whole face of the ground.

7. And the LORD God formed man of the dust of the ground, and breathed into his nostrils the breath of life; and man became a living soul.

8. And the LORD God planted a garden eastward in Eden, and there he put the man whom he had formed.

9. And out of the ground made the LORD God to grow every tree that is pleasant to the sight, and good for food the tree of life also in the midst of the garden, and the tree of knowledge of good and evil.

10. And a river went out of Eden to water the garden; and from thence it was parted, and became into four heads.

11. The name of the first is Pishon: that is it which compasseth the whole land of Hāv´ĭ-läh, where there is gold;

12. And the gold of that land is good: there is bdellium and the onyx stone.

13. And the name of the second river is Gī´hŏn: the same is it that compasseth the whole land of Ethiopia.

14. And the name of the third river is Hĭ´dĕ-kĕl: that is it which goeth toward the east of Assyria. And the fourth river is Eû-phrā´tēs

15. And the LORD God took the man, and put him into the garden of Eden to dress it and to keep it.

16. And the LORD God commanded the man, saying, Of every tree of the garden thou mayest freely eat:

17. But the tree of the knowledge of good and evil, thou shalt not eat of it: for in the day that thou eatest thereof thou shalt surely die.

18. And the LORD God said, It is not good that the man should be alone; I will make a helper suitable for him.

19. And out of the ground the LORD God formed every beast of the field, and every fowl of the air; and brought them unto Adam to see what he would call them: and whatsoever Adam called every living creature, that was the name thereof.

20. And Adam gave names to all cattle, and to the fowl of the air, and to every beast of the field; but for Adam there was not found a suitable helper for him.

21. And the LORD God caused a deep sleep to fall upon Adam, and he slept: and he took one of his ribs, and closed up the flesh instead thereof;

22. And the rib, which the LORD God had taken from man, made he a woman, and brought her unto the man.

23. And Adam said, This is now bone of my bones, and flesh of my flesh: she shall be called Woman, because she was taken out of Man.

24. Therefore shall a man leave his father and his mother, and shall cleave unto his wife: and they shall be one flesh.

25. And they were both naked, the man and his wife, and were not ashamed.

JOE CAIN

RETHINKING ATTACKS ON EVOLUTION: LESSONS FROM THE 1925 SCOPES TRIAL

Fundamentalist Christians deny evolution and attack its teaching in schools. Why? This has nothing to do with the facts of nature or the arguments of scientists. No fossil or experiment will change their minds. At the same time, their denial involves neither ignorance nor prejudice. It relates to much deeper cultural values and a century-long dispute between traditional and modernist world views. In the 1925 Scopes Trial, evolution was a proxy for this much wider dispute in America. That dispute began long before and continued long after the Scopes Trial. Since the 1920s, evolution has remained a focal point for these deep cultural differences. If the people involved in this dispute are serious about resolving their differences, they would be wise to shift to the real source of their disagreements.

BACKGROUND

A famous courtroom drama took place in Dayton, Tennessee, during the summer of 1925. This was the trial of John Scopes for teaching evolution.

In March 1925, Tennessee created a law "prohibiting the teaching of the Evolution Theory." This made illegal the teaching of "any theory that denies the story of the Divine Creation of man as taught in the *Bible*, and to teach instead that man has descended from a lower order of animals." In a plan to test the constitutionality of this law, Scopes volunteered to be arrested and charged with its violation.

Strong legal teams came together for both prosecution and defense. Both teams also received volunteer assistance from celebrities. To Scopes' defense came Clarence Darrow, one of America's best defense attorneys. He was famous for defending alleged murderers and subversives. For the prosecution came William Jennings Bryan, three-time candidate for US President, populist hero, and widely-loved defender of traditional Christian faith. Bryan had campaigned against evolution for years. Media outlets used the presence of these celebrities to create a frenzy, promising the trial would be far more than a legal test. Dayton become the site for a clash of titanic proportions: Bryan versus Darrow, good versus evil, religion versus reason, God versus Darwin.

By presenting the Scopes Trial in terms of polar-opposite, the sensationalist media did a disservice to both sides and to their readers. Most scientists comfortably reconciled science and religion. Most religious leaders had no difficulty with modern scientific ideas. The genuine differences existing between science and religion were left undiscussed.

JOHN SCOPES

With so many celebrities and legal scholars circulating around this case, Scopes was the least important person at his own trial. Born in Paducah, Kentucky, he developed a strong interest in the physical sciences before graduating in law at the University of Kentucky in 1924. He took work as a teacher to earn money for more legal training. Dayton's high school hired Scopes to coach their football team and to provide teaching in chemistry, physics, and algebra. Scopes and Dayton were linked only late in the summer 1924. Dayton's previous coach had left on short notice. The school was desperate. Young and inexperienced, Scopes seemed to have more in common with his students than with their parents.

Scopes was well-liked around town. He had a casual demeanor and made friends easily. In his autobiography, Scopes spoke warmly of Dayton before the trial. He took pains to represent fellow townspeople as welcoming and hospitable (Scopes, 1967). Scopes also complained about the Trial in his autobiography, stressing its "carnival" atmosphere, his admiration for the defense team, and his "utter contempt" for Bryan.

After the trial, Scopes was offered his teaching job in Dayton again. However, he chose to leave Dayton—no doubt to the relief of all concerned. Scopes declined lucrative offers to exploit his popularity through speaking tours, advertising sponsorships, and even appearances in Hollywood films. Instead, Scopes accepted a scholarship for graduate studies in geology at the University of Chicago. He never completed his doctoral studies. Financial problems forced Scopes back to work.

In 1927, Scopes began a 3-year contract as a petroleum geologist for Gulf Oil in Venezuela. He enjoyed the chance to escape his new-found notoriety. In 1929, Scopes was forced home by a severe case of blood poisoning. After recovering, Scopes was soon back in Venezuela, when he married Mildred Walker, daughter of an American businessman. For his new wife, Scopes was baptized Roman Catholic. In 1930, Scopes took two prospecting trips on the Orinoco towards Colombia. On the second, he said, Gulf Oil wanted him to illegally prospect over the border. Scopes refused and was fired. The married couple returned to the United States.

In the early 1930s, Scopes attempted graduate studies again but did not see them to completion. In 1940, he returned to economic geology, working for United Gas Corporation. He served mainly in Louisiana until his retirement in 1963.

Throughout his life, Scopes regularly received proposals to profit from the 1925 trial. He consistently refused, shunning publicity and interviews. He ended his silence in 1960, with the release of the film, *Inherit the Wind* (discussed below).

THE LEGAL SIDE OF THE SCOPES TRIAL

Tennessee was one of several states restricting the teaching of evolution during the 1920s. The connections drawn in such legislation is revealing. In 1922 Kentucky created a law forbidding the teaching of agnosticism, Darwinism, atheism, and evolution as it pertained to humans. Florida made similar restrictions in 1925. Civil rights experts followed these enactments closely. The American Civil Liberties Union (ACLU) wanted to test their constitutionality in court. When Tennessee put its anti-evolution law into effect in March 1925, they canvassed Tennessee for a volunteer to challenge the new law in a test case. Scopes stepped forward.

The legal case, *State of Tennessee v. John Thomas Scopes*, is easily lost in the media circus that engulfed Dayton in 1925. In legal terms, Scopes was charged with committing a minor offense. In the courtroom, however, both prosecution and defense used their arguments to raise issues of much larger importance.

PROSECUTION

Defending the law and prosecuting Scopes and a law-breaker seemed straightforward. States had the power to prescribe curriculum, the prosecution argued, as well as to decide what should and should not be taught in the classroom. This right came from the State's support of education. As chief financier, it could direct how public moneys were spent. In this case the legislature could prohibit schools from spending their money in ways it thought inappropriate. In this case, the prohibition was against curriculum which seemed to them to conflict with the Bible. From the prosecution's point-of-view, Scopes was an employee of the State. When working in that capacity, he was required to uphold the policies of his employer. This particular policy did not interfere with anyone's freedom to worship. Outside the school, Scopes was free to defend whatever views he chose, and the law did not force anyone to accept certain beliefs. The law being challenged simply prevented teachers from using state funds for curriculum that criticized or undermined Christian faith.

DEFENSE

The defense challenged Tennessee's anti-evolution law on many levels. Part of this challenge focused on technical points of the law itself, its precise language, and the process of its creation. Their constitutional challenge had three parts.

First, they emphasized rights and liberties. One fundamental liberty, they argued, was a right to worship according to one's own conscience. Forcing Scopes to teach in ways consistent only with the King James Bible was the same as compelling him to defend it (worse, to defend one particular interpretation of that text). This unjustly restricted his liberties on belief and speech. It made Tennessee an advocate of some religious views rather than others. Reciprocally, the law made a crime out of certain kinds of learning, especially science. It restricted Scopes' academic freedom and right to free speech when it prohibited him from telling his students about the views of modern science.

As a second strategy, the defense focused on legal notions of jurisdiction and competency: who was eligible to decide on a matter of science or to engage in the discussion? In passing the anti-evolution law, they argued, Tennessee's legislature used the power of the State to pronounce on matters of scientific theory and fact. This was outside the jurisdiction of government—just as no legislature could dictate the value of pi or define the law of gravity. Neither were lawmakers experts on the merits of scientific facts and theories. In this case, scientists and science teachers were experts, not the State.

Third, in setting evolution and the Bible into contradiction, the anti-evolution law created a false dichotomy. Many scientists believed evolutionary processes were, in fact, the mechanisms God, as lawgiver, used to carry out Creation. In this view, the history of life either was unfolding a Divine plan or following divine law within the context of a clockwork universe. Thus, the defense argued, the Tennessee law was nonsense: evolution did not occur *instead of* Creation (as the law represented them); rather, evolution *was* the Creation. The defense was ready to have half a dozen prestigious scientists testify to this view during the trial.

DAY-BY-DAY IN DAYTON

The Scopes Trial lasted eight days between 10 and 21 July 1925. Most trials are dull proceedings interspersed with dramatic moments, and the Scopes Trial was no exception. At regular intervals insults and insinuations flew across the room. Jokes were told to the crowd. Presentations were made with the flourish of actors on stage, drawing applause and hisses from the audience. Some speeches were eloquent and brought the crowd to its feet. Other moments provided a feast for tabloid writers. For instance, at one point Clarence Darrow was cited for contempt while complaining the judge had shown prejudice against the defense. Everyone in town scrambled for a seat in the jury box, but the jury saw little of the courtroom fireworks. Most of the real action took place during procedural motions and in debates about admissibility of evidence. In these discussions, the jury was excluded for fear of prejudicing their conclusions. Transcripts of the trial are available (Rhea County Historical Society, 1978); many newspapers printed daily accounts of the proceedings.

Day One was devoted to technical matters about Scopes' original indictment, the State's outline of its case, and jury selection. All lawyers were trying to be careful about procedure because they wanted the law tested on it merits alone. In Day Two, the defense moved to have the law summarily ruled unconstitutional. Arguments on this motion, which encapsulated each side's basic approach to the case, lasted the whole day. At the start of Day Three, the defense objected to the court opening each day with prayer, claiming this was prejudicial. To consider the question, court ended early. In Day Four, the judge began by rejecting the defense's motion to summarily void the law, and he overruled their objections to opening prayers. Then the actual trial began. On a charge of breaking the anti-evolution law, Scopes pleaded not guilty. Opening arguments were presented by both sides. The State then presented its factual case. In this prosecution, students testified that Scopes had indeed taught evolution, and the school textbook was submitted into evidence.

The State rested its case in the afternoon of Day Four, and the defense began. First, it accepted as true everything thus far admitted into evidence. (This effectively conceded the case and avoided the need to place Scopes on the witness stand.) The defense then hoped to present expert testimony about how best to understand the status of evolution with respect to the Bible. The prosecution objected, arguing the law was clear on what it thought evolution entailed and suggesting its interpretation required no expert testimony. Arguments on the admissibility of testimony from the scientists occupied all of Day Five, too.

Though he attracted most of the attention, Bryan stayed quiet in the proceedings until he rose to defend prosecution attempts to exclude scientists as expert witnesses. He saw this as an effort to broadcast the views of evolutionists around the world and was keen to prevent use of the Trial in this way. Bryan appealed to his populist tradition, explaining that the anti-evolution law expressed the wish of the common people of Tennessee, and outsiders should have no influence on the matter. He condemned evolution as a poison to the mind and a promoter of atheism. He said he could not understand how courts could prohibit teachers from using the Bible while at the same time allow the teaching of ideas condemning it. "Has it come to a time" he told the jury, "when the minority can take charge of a state like Tennessee and compel the majority to pay their teachers while they take religion out of the hearts of the children?" (Bryan in Rhea County Historical Society, 1978; p. 172).

When Dudley Field Malone rose for the defense to speak in favor of scientists' testimony, no one expected him to top Bryan. Malone had worked for the US State Department when Bryan was Secretary of State in Woodrow Wilson's cabinet, and he used respect for that relationship as a device to turn the argument around. Bryan was not the only interpreter of the Bible's meaning, Malone suggested. Many scholars disagreed with his approach and saw no contradiction between the Bible and modern science. This was what the defense hoped to show with expert testimony. Why was Bryan refusing these people a chance to be heard? Why was he closing out a genuine search for understanding? "I have never seen harm in learning, and understanding, in humility and open-mindedness," Malone told the crowd (Rhea County Historical Society, 1978; p. 183). Besides, he added with obvious reference to Bryan's role as his own mentor, teachers worked hard and took very seriously their role as moral guides. Malone then appealed to fairness and free inquiry. People have the right to think and to decide for themselves. Nothing good could come from attempts to close their minds. When Malone finished, the crowd erupted in such cheers the judge adjourned court for the day. Even Bryan congratulated Malone on his mastery of the argument. This speech persuaded many people the defense was raising legitimate issues at the Trial.

In Day Six, the judge rejected Malone's arguments and excluded all testimony of scientific experts. The language of the law was sufficiently clear and needed no elaboration, he ruled. Further arguments on this point—it was central to defense tactics—continued through the day, and written statements by scientists eventually were accepted into the record for use during the appeals process.

With scientific testimony excluded from the trial, the defense in Day Seven argued that the phrase "as taught in the *Bible*" in the anti-evolution law could be interpreted in many fashions. They moved to include expert testimony on what the *Bible* actually taught. In the discussion, events took an unexpected turn when the defense called Bryan, a member of the prosecution's team, as an expert witness on the *Bible*. This produced the most famous moment in the trial.

In questioning Bryan for the defense, Darrow sought to expose the anti-evolutionists as bullies and to show Bryan to be a simpleton. When called, Bryan agreed to stand as a witness against strenuous objections by fellow prosecutors. Bryan understood Darrow's tactic and knew the symbolic value of his decision. "I want the papers to know I am not afraid to get on the stand in front of [Darrow, who Bryan elsewhere called 'the greatest atheist or agnostic in the country'] and let him do his worst. I want the world to know" (Rhea County Historical Society, 1978; p. 299). For his part, Darrow knew the rules of cross-examination gave him the crucial advantage. He was going to press Bryan to the breaking point.

In the examination, Darrow questioned Bryan on a dozen old criticisms of Biblical literalism: where did Cain's wife come from, how did Jonah survive three days in the belly of a great fish, how did so much variety survive the flood, and so on. Bryan stood firm on his faith, accepting the Bible "as it was presented." Darrow ridiculed Bryan's answers as foolish and naive. He challenged Bryan on his rejection of science and directed questions to expose superficiality in Bryan's views on issues such as the age of the Earth, human evolution, and the history of languages. (To be fair, Bryan lacked preparation and was forced to follow Darrow's agenda.) As questioning continued, tempers boiled over. The crowd cheered each jab in what had become an entertaining joust. Bryan charged Darrow with attempting to undermine the Bible and promoting atheism. "We have the purpose," Darrow shouted in reply, "of preventing bigots and ignoramuses from controlling the education of the United States and you know it" (Rhea County Historical Society, 1978; p. 299). Lawyers on both tables made repeated objections in the proceedings as the discussion spun out of control. Frustrated with the farce this had become, the judge abruptly adjourned this session in the middle of one-too-many squabbles.

Darrow's cross-examination of Bryan (done with the jury excluded) provided a clash between the trial's two great symbols: God and faith personified in Bryan, and Darwin and atheism personified by Darrow. Still, Darrow's tactical purpose was to show the Bible had many interpretations and might not be contradicted by evolution. Thus, it was crucial for the defense when Bryan stated the Bible need not be interpreted *literally*: i.e., creation need not have been for 6 days of 24 hours, but could have involved 6 "periods" of indeterminate length. Though useful in publicity, this admission gained the defense no legal advantage in the trial because at the start of Day Eight, the judge deleted Bryan's testimony from the trial record. It had been a mistake to allow this testimony in the first place, he said. Bryan complained about sportsmanship. He wanted a chance to cross-examine Darrow. That, the judge finally ruled, would not take place in his courtroom.

Day Eight saw the trial's end. With nothing left to defend themselves with, the defense conceded. The facts about what Scopes taught were not in dispute, they said. Everyone understood Scopes would be found guilty at the trial stage, then the case would move to appeal. The judge instructed the jury on Day Eight, who deliberated briefly then returned a "guilty" verdict on the same day. In his only comment in the courtroom, Scopes spoke before sentencing "I feel I have been convicted of violating an unjust statute," he said. Not opposing the law would have been "in violation of my ideal of academic freedom—that is to teach the truth as guaranteed in our constitution—[and] of personal and religious freedom" (Rhea County Historical Society, 1978; p. 313). Scopes was fined $100. After numerous informal final speeches, the trial closed.

The defense assumed they'd lose at the trail stage. Their strategy was to move the case through the court system until the state or federal courts ruled on the law's constitutionality. This strategy was cut short in January 1927, however, when the Supreme Court of Tennessee ruled on Scopes' case in a manner that prevented further action. The judge had made a technical error. In Tennessee, fines above $50 had to be fixed by a jury. Scopes' $100 fine—the minimum allowed by the law—was fixed by the judge. This error by the judge meant the Scopes case needed to be returned to a jury to fix the fine. But the Supreme Court discouraged state prosecutors from trying the case again simply to reset the fine. "We see nothing to be gained by prolonging the life of this bizarre case."

Without further action by state prosecutors, Scopes' defense team were caught in a legal no man's land. Worse, the court indicated it would uphold the law if asked, and they signaled considerable sympathy with the prosecution's view of the case. States did have a right to control curriculum, they said, and the antievolution law did not threaten academic or religious freedom.

SCOPES TRIAL IN THE MEDIA

The Scopes Trial has two origins. First, it was a test case in law. Second, it was a promotional event, instigated by Dayton businessman George Rappleyea. He deliberately sought a way to place Dayton "on the map" and to improve business prospects in the town. Rappleyea and other town elders noticed the ACLU's advertisements for a test case against the anti-evolution law and thought a trial would attract attention within the state and region. When Bryan agreed to participate, then Darrow, then many others, the trial's publicity value sprinted well ahead of their expectations. Rappleyea and others rode this bandwagon, and as media attention grew, everyone seemed to gain something from stoking the coals and increasing attention even further.

Timing and technology were key to the sensationalism. The Scopes case was built into a media event over two months of anticipation. In this lead-time, coverage featured regular "developments," breaking stories, and many angles for comment. By the mid-July trial, attention was well and truly focused on this small Tennessee town. As a summer event the trial took place outside the annual schedules of government and business. In the years before air conditioning and improvements in public health, those who could afford to travel during summers decamped from cities. As a result, summer was a slow news period for cosmopolitan papers. In this "silly season," events were often promoted well beyond their real value by editors scrambling to fill pages and newscasts. Technology made heavy coverage easy. Special train services helped more than 200 reporters flock into Tennessee for the show. Telegraph and telephone companies laid hundreds of miles of new cables through Dayton anticipating heavy demand. (The trial produced more telegraph business than any previous event, with more than a million words sent out from Dayton, including daily transcripts.) Airplanes provided daily courier flights to rush photographs and newsreel film to the big cities. Chicago radio station WGN transmitted live reports from the trial across their national network, though this had more to do with their rush to claim status as a pioneer in live news than with their ideological interest in the issues of the case. The Scopes Trial was one of the first times in America that the means for exploiting the news value of events converged with sensational events.

Importantly, the Scopes Trial provided a story fitting many familiar narratives: divides between city and country, science and religion, traditional and modern. The clash of gladiators was a common story line, as was the defense of religion against atheism and the defense of science against fundamentalists. This made reporting easy. Press coverage exploited these stereotypes and showed predictable regional biases. In the North, papers represented the trial as a fight for intellectual freedom and science. They parodied Tennesseans as hillbillies, Bible thumping yokels and illiterates as a way to applaud their own modernity and sophistication. Papers in the South represented the Trial as a conflict between arrogant outsiders and traditional, law-abiding, respectful families. Bryan was represented as a defender of the Common Man, and critics complained about the tendency of outsiders to force their standards on local tradition. Overseas, the trial offered more evidence for the view that Americans lacked any sense of perspective and common sense. All reporters also reported on the desire of Dayton businessmen to gain attention for the town.

Based at the *Baltimore Sun*, H. L. Mencken (1880–1956) was one of America's most popular syndicated columnists in the 1920s. His cynical commentaries on the trial became famous in the North and created a lasting image of the trial. Mencken emphasized Dayton's carnival atmosphere, calling it "Monkey town." Though he had great respect for individuals, Mencken saw Tennesseans as a whole from an elitist Northern perspective. He enjoyed describing Dayton's population as naive sheep at best and at worst, inbred bumpkins. Mencken's commentary from the last day of the Trial was typical:

> Let no one mistake [the trial] for comedy, farcical though it may be in all its details. It serves notice on the country that Neanderthal man is organizing in these forlorn backwaters of the land, led by a fanatic, rid of sense and devoid of conscience. Tennessee, challenging him too [timidly] and too late, now sees its courts converted into camp meetings and its Bill of Rights made a mock of by its sworn officers of the law. There are other States that had better look to their arsenals before the Hun is at their gates. (Mencken, 1925 in Tompkins, 1965; p. 51)

Mencken directed his sharpest criticisms toward Bryan, whom he described as a fraud and opportunist—a power-hungry hypocrite who promised redemption and salvation without effecting real changes and who used people's genuine faith as a fraudulent means to gain popularity and profit. Though widely read, Mencken's depiction of Dayton provided only one perspective on events.

For a hungry media, the trial also provided easy opportunities for satirists and cartoonists. References to the Trial are widespread in every form of media. The 1925 song, "You can't make a monkey of me" provides one example. The chorus read,

> You can't make a monkey of me.
> There's not a monkey in my family tree.
> I've searched on each branch from Adam to me.
> I am inclined to believe
> The story of Adam and Eve.
> There's no chimpanzee
> In my pedigree.
> And you can't make a monkey of me.

SCOPES TRIAL IN MEMORY

Most later impressions of the Scopes Trial derive from the film *Inherit the Wind*. Originally a play (Lawrence and Lee, 1955), the 1960 film version was directed by Stanley Kramer and starred Gene Kelly and Spencer Tracy, who won an Academy Award for his performance. The play remains a common choice in schools and community theaters, and several remakes of the film have been released.

Inherit the Wind offers a Northern perspective on the trial, through the eyes of someone like Mencken. During promotions for the film, Scopes toured the country, speaking about the trial and the importance of free speech. This was the only time he used his celebrity status for publicity.

Importantly, *Inherit the Wind* does not present itself as an accurate reconstruction of the trial. Rather, the authors used events in Dayton as a vehicle for criticizing something else: the anti-Communist campaigns of the 1950s. Such a presentation emphasized only one dimension of the Trial: free speech and the right of citizens to hold controversial ideas. The film's dramatic scenes center on decisions to stand against the majority and to ally oneself with an unpopular, persecuted character. In this version of events, Bryan's character was a fear-mongering brute who hounded the innocent for his own glorification. (Bryan's real-life equivalent probably was meant to be Vice-President Richard Nixon, who built his career on pursuing alleged Communists.) Following other controversial films by Kramer, such as (1952) *High Noon*, the message of *Inherit the Wind* was clear: we must stand by those who dissent, even if we disagree; otherwise we may lose one of our most cherished principles.

ANTI-EVOLUTION CAMPAIGNS SINCE 1925

After Scopes' appeal failed in 1927, anti-evolution laws were not tested in court again until the 1960s. Studies of biology textbooks over the twentieth century identify a decrease in coverage of evolution following the Scopes Trial. This persisted through the early 1960s and was due, in part, to commercial pressures from textbook selection groups who were lobbied by anti-evolutionist campaigners. Seeking to produce non-controversial products also sellable to the widest markets, publishers complied with parents' demands to reduce coverage of evolution or to present it in diffuse form. (To be fair, the shift in content also reflected a shifting emphasis within science education toward practical and civic aspects of biology, such as nutrition, hygiene and ecology.)

Emphasis on evolution resurfaced following the centennial of Darwin's *Origin of Species* in 1959 and the 1960 release of *Inherit the Wind*. "One hundred years without Darwin is enough" became a rallying call for evolutionary biologists. These scientists took advantage of intense national interest in science education, following Soviet successes with Sputnik, to influence the federally sponsored Biological Sciences Curriculum Study (BSCS). This complete redesign of secondary biology curriculum placed evolution as one of three essential dimensions for the life sciences and gave genuine substance to the quip by Theodosius Dobzhansky that "nothing in biology makes sense except in light of evolution."

In 1966, Little Rock biology teacher Susan Epperson used the BSCS curriculum to campaign against a Scopes-like anti-evolution law enacted in Arkansas in 1929. Epperson's case went back and forth in state courts and finally was heard by the US Supreme Court in 1968. In *Epperson v. Arkansas*, the Court ruled the law unconstitutional on the grounds that the establishment clause in the First Amendment prevents states from protecting Biblical accounts of creation. Fearing adverse publicity during Epperson's campaign, the Tennessee legislature repealed, in 1967, their anti-evolution law.

Scientists claimed a major victory in this case. However, anti-evolution sentiment and the concern about teaching evolution in public schools were far from defeated. Anti-evolutionists adopted a new legal strategy in 1973, convincing the Tennessee legislature to pass a "Genesis bill," which required all science textbooks to carry a disclaimer ensuring evolutionary explanations about "the origin and creation of man and his world" were represented merely as theories and "not represented to be scientific fact." Science teacher organizations tested this law in court, which ruled the law unconstitutional on First Amendment grounds.

In parallel, several factions of anti-evolutionists have pursued a third legislative strategy since the 1970s. Representing their interests as promoting "creation science" and "scientific creationism," these factions argued that scientific research into Creation offered a legitimate, secular alternative to "evolution science" and thus merited "equal time" as a competing model in classroom discussions of origins. Organizations such as the Institute for Creation Research, the Creation Science Research Center, and the Bible Science Association publicized "scientific" research along these lines. This approach emphasized American values of democracy and evenhandedness—give people the facts and let them decide for themselves—and enjoyed legislative success when Arkansas (1981) and Louisiana (1982) passed "equal time" laws.

Civil liberties groups and science teachers organizations quickly tested "equal time" laws in federal court, where they were ruled unconstitutional (*McLean v. Arkansas* in 1982—text in Nelkin, 1982—and *Edwards v. Aguillard* in 1987). In his decision on the Arkansas law, Judge Overton criticized the two-model approach as artificial, relying on a false dichotomy of creation versus evolution. (Curiously, this was a key argument presented by Scopes' defense.) He saw no evidence that "creation science" had scientific merit, and he concluded it had no educational value as a secular "model" for origins. Overton's ruling was unambiguous: the two-model approach was a thinly veiled attempt to pursue "a religious crusade, coupled with a desire to conceal this fact" (Overton, 1982 in Nelkin, 1982; p. 208).

Since the 1960s, the establishment clause of the First Amendment has been central to judicial rejections of antievolutionist laws in the same way local authority was key to their acceptance in the 1920s. On such matters, courts rely heavily on a landmark 1971 US Supreme Court ruling, *Lemon v. Kurtzman*, which set a three-part test for interpreting the constitutional separation of church and state. First, laws must have a secular purpose. Second, a law's principal effect must neither advance nor inhibit religion. Third, a law must not foster "an excessive government entanglement with religion." Failing any of these tests was grounds for ruling a law unconstitutional.

With judicial barriers frustrating their efforts, antievolutionists have adopted other strategies. Successful campaigns of local protest have focused on changing administrative policies at the levels of state and local boards of education. These campaigns include lobbying individual teachers and principals, protesting at school board meetings, and influencing selection processes for textbooks and boards of education. Campaigning in populous states produces the greatest effects, as textbook preferences in these states heavily influence what publishers supply to the national market. For instance, in the early 1970s California schools comprised more than 10 percent of the textbook market and was targeted by anti-evolutionist campaigners as a way to influence the national curriculum. Using an "evolution is theory not fact" strategy, they convinced the curriculum committee of the state's board of education to approve only textbooks representing evolution as a speculation or tentative theory, as an idea subject to testing and of uncertain status. The claim was that this prevented dogmatism, but the explicit goal was to bolster anti-evolutionists in their promotion of Creation theories. Similar efforts in Texas, New York, and other states met with varied success in the 1970s and 1980s. (These are discussed in Nelkin, 1982.)

Two recent cases show anti-evolutionists continuing campaigns at the local level. In 1994, the Tangipahoa Parish Public Schools in Louisiana followed local protests and adopted a policy requiring teachers to read a disclaimer in class whenever evolutionary topics were presented. This disclaimer explained that evolution was "a scientific theory" and presenting it in school was "not intended to influence or dissuade [anyone from accepting] the Biblical version of Creation or any other concept." Supporters argued this was necessary to avoid offending those holding alternative theories of origins and to promote critical thinking within the sciences. Several parents challenged this policy in court on First Amendment grounds. In 1997, a US District Court in Louisiana agreed and ruled the policy unconstitutional; in June 2000, the US Supreme Court voted 6-3 to let this ruling stand.

In August 1999, the State Board of Education in Kansas voted 6-4 to de-emphasize evolution and to remove reference to the "big bang" theory of origins in their standards for science education across the state. This followed heavy lobbying from Christian fundamentalists and was spearheaded by strongly conservative members of the Board. Though these standards were not mandatory, opponents argued this decision would degrade the quality of science tests and textbooks in the state and thus reduce the competitiveness of Kansas students against national peers. How could Kansas not teach something so well accepted by the scientific community? (This was another argument at the Scopes Trial.) The policy remained in effect for a year. At their first opportunity—in primary elections where four members of the board stood for reelection—voters removed from office three of the board members who supported the policy. Observers (as of August 2000) expect the policy to be rescinded as soon as the new board of education meets.

In 2005 the town of Dover, Pennsylvania, became the scene of another anti-evolutionist court case surrounded by a media circus. In the previous year, the local school board made a policy decision about evolution: "Students will be made aware of gaps/problems in Darwin's theory and of other theories of evolution including, but not limited to, intelligent design. Note: Origins of Life is not taught." A legal challenge by parents followed in which both anti-evolution and pro-intelligent design aspects of the policy were challenged. In December 2005, US District Court Judge John Jones III ruled this policy was simply an alternative incarnation of the same creationism promoted since the 1920s and determined it unconstitutional. Voters in the next local election expressed their opinion of policy decision by re-electing none of the school board members who voted for the anti-evolution policy. Nevertheless, since the start of the Bush administration in 2001, more than 100 similar attempts to pass anti-evolution policies have taken place.

ANTI-EVOLUTIONISM IN A LARGER FRAME

Why are passions so strong against evolution? The answer has a great deal to do with a broader cultural clash in American society. Attacks on evolution and its teaching are surrogates in that larger clash. People who seek a clash present science and religion as polar opposites and deny their compatibility. Though this is only one approach to understanding that relation, it is popular among those evangelizing both for religious and scientific causes.

In biology, a small group of Victorian materialists—headed by Thomas Henry Huxley—promoted metaphors of science's "conflict" and "war" with religion as a way to raise the prominence of science. However, these metaphors generally are deceptive and historically inaccurate. Darwin struggled to make his views acceptable to the devout. He emphasized a clockwork universe governed by natural laws (such as natural selection) set down by a Creator who breathed life into the first beings, then set the laws to work. "There is grandeur in this view of life," Darwin argued. Some of Darwin's first supporters, such as the American botanist Asa Gray, took the Creator's involvement further by suggesting Divine guidance in the process of mutation: God steered the course of evolution by making some variation available and not others.

Scopes' defense team had no trouble locating highly reputable evolutionists who saw no conflict between their science and their faith. When William Jennings Bryan admitted at the Scopes Trial that the "days" of Genesis could be interpreted as "periods of indefinite length," he admitted a basic compatibility of science with scripture that many scientists agreed with at the time. Indeed, well before and since Darwin reconciliation has been far more common than generally is appreciated. Only those who insist on absolute literalism in their reading of the Bible (which no one prosecuting Scopes insisted upon) present views that cannot be reconciled easily with evolutionist thinking. In the much wider setting of American culture, antievolutionist sentiments seem to be tied only loosely to evolutionary ideas. Instead, they are grounded in a far deeper historical split between modernists and traditionalists.

This is a cultural divide that is neither exclusively religious nor scientific. Anti-evolutionism serves more as a rallying point for groups on either side of this divide. It is not, however, a fundamental issue. On one side of this divide, modernists see progress coming through new technology, better management, and system-building. They emphasize the urban and cosmopolitan. Culture comes in the form of the latest fashions, high art and the energy of youth. Knowledge, skill and merit should form the foundations for respect. On the other side, traditionalists celebrate the rural ideal of an agrarian society. Roots in the land and in craft build character. Culture comes from celebrations of heritage and the populist notion of the "common man." Respect for experience and authority are the foundations of society. Faith and charity—not greed and gluttony—offer moral foundations. Traditionalists criticize faith in technology and industrialization because they seem to do more harm than good, and they criticize the values celebrated by modernists in the fear they will undermine society. Each time twentieth century modernists dominated American culture, traditionalists urged a return to fundamentals, faith, family and tradition.

Anti-evolutionist campaigns in the 1920s grew out from a traditionalist backlash related to industrialization and the horrors of World War I. From their perspective, the assembly line degraded workers and eliminated craftsmanship. "Scientific" management to increase efficiency had the effect of reducing men to soulless machines. Urban lives of the working-class were filled with squalor, disease and hopelessness. World War I produced the inevitable result of mechanization, modernism, and their many mistakes: mass slaughter in the trenches, gas attacks, aerial bombing, machine guns, and single battles in which hundreds of thousands died. Over 8 million people were killed in the war; more than 21 million were wounded. Improvements in battlefield medicine meant more casualties than ever came home, allowing everyone to see firsthand the war's victims and their burns, amputations and shell shock.

With these experiences fresh, traditionalists feared modernism would continue to destroy everything they held dear. Other icons strengthened their alarm. Relativists rejected the notion of absolute truth, then criticized history as mere storytelling and propaganda. Academics began subjecting the *Bible* to standards of evidence rather than faith. The philosophy of Frederich Nietzsche, expressed in his (1901) *Will to Power*, celebrated force over justice in the principle that "might makes right." In this context science promoted skepticism and disrespect for authority. Evolutionists seemed to be preaching "survival of the fittest" and building an ethic in which the strong could justify destroying the weak in brutal fashion. From the perspective of traditionalists, defending modernism and its icons was tantamount to defending the horrors of total war and the hopelessness of urban masses. All this was packed into tags such as "atheist," "agnostic," and "evolutionist." In attacking evolution, traditionalists were striking a blow against modernism and its wider implications. Theirs was a fight for cultural survival.

Bryan's involvement with anti-evolutionists began after World War I and was grounded in his profound disgust with modernism. He grounded his criticisms of liberal education in surveys about faith in American universities, such as *The Belief in God and Immortality* published by Lueba in 1916. Lueba reported that over 50 percent of college teachers professed to atheism. For students, more time in college meant increasing loss of traditional values: 16 percent of first year college men claimed to be atheists. By the third year this rose to 30 percent; the final year, 40–50 percent. Traditionalists were outraged by such trends. Students arrived in college at a vulnerable stage of life, they argued. Young people were impressionable, and atheistic professors obviously were encouraging a rejection of traditional values, promoting skepticism, and leading their students into atheism, relativism and nihilism. Instead, traditionalists maintained, universities should endeavor to pass on heritage from one generation to the next. They should teach discipline, character, moral training, and reverence. They should guide students away from corruption. Why should parents pay for schools to hire modernists who mocked tradition and spread dangerous ideas? Campaigns to sack atheistic professors and prohibit objectionable curriculum were common elements in traditionalist backlashes. The prosecution team in the Scopes Trial regularly pressed this theme.

Bryan's antagonism to evolution went deeper than a presumed conflict with Genesis. "Darwin's guess gives students an excuse for rejecting [parental authority and] the authenticity of God," he wrote in a 1921 book, *The Bible and Its Enemies*. Such rejections lead to moral degradation: if people are taught they are animals, then they certainly will act like animals and tragedy would ensue. A culture based on Darwinism, he worried, would head directly toward the merciless, brutal Nietzsche-based militarism at the heart of Germany's machine in World War I.

To support his connection between evolutionism and the Kaiser, Bryan relied on writings of popular American biologists, such as Vernon Kellogg and Leon J. Cole. Kellogg's 1917 widely read first-hand account of conversations with German commanders, *Headquarters Nights*, described German thinking as grounded in a perversion of Darwinian principles of a great, violent, and fatal struggle for existence in the competition between nations in which industrial efficiency and brutality determined the outcome. The strong was expected to destroy and enslave the weak. "It is a point of view," Kellogg explained, "that justifies itself by a whole-hearted acceptance of the worst of Neo-Darwinism, the Allmacht of natural selection applied rigorously to human life and society and Kultur" (Kellogg, 1917; p. 13).

The contrast to Bryan's philosophy was absolute, and Bryan was no stranger to international struggles. He had served as Secretary of State (1913–1915) for Woodrow Wilson. As a pacifist, Bryan pressed for world law and international mechanisms for conflict resolution. He arbitrated 30 international treatises and pressed for cooling-off periods and commissions to investigate international grievances. (Bryan resigned over Wilson's aggressive response to the sinking of the *Lusitania*, though he supported the government after America entered the war.) To Bryan, the connection was simple: Darwinism was a philosophy underpinning the worst kinds of behavior. It had to be opposed.

Much anti-evolutionism in the twentieth century was grounded in challenges by traditionalists to modernism in its various incarnations (Numbers, 1992). Scientists who attack creationism on evidential grounds (e.g. Kitcher, 1982;

Futuyma, 1995; Pennock, 1999) miss this point and so miss an opportunity to create meaningful common ground. That suggests they have no real desire to reconcile a central division in our culture.

REFERENCES

Bryan, W. J. *The Bible and Its Enemies, an Address Delivered at the Moody Bible Institute of Chicago*. Chicago: The Bible Institute, 1921.

Futuyma, D. *Science on Trial: the Case for Evolution*. Sunderland, MA: Sinauer Associates, 1995.

Kellogg, V. *Headquarters Nights: a Record of the Conversation and Experiences of the German Army in France and Belgium*. Boston: Atlantic Monthly, 1917.

Kitcher, P. *Abusing Science: the Case against Creationism*. Cambridge, MA: MIT Press, 1982.

Lawrence, J., and R. Lee. *Inherit the Wind*. New York: Random House, 1955.

Leuba, J. H. *The Belief in God and Immortality: A Psychological, Anthropological and Statistical Study*. Boston: Sherman, French, and Co., 1916.

Nelkin, D. *The Creation Controversy: Science or Scripture in the Schools*. New York: W. W. Norton, 1982.

Numbers, R. *The Creationists*. New York: Knopf, 1992.

Pennock, R. T. *Tower of Babel: The Evidence Against the New Creationism*. Cambridge, MA: MIT Press, 1999.

Rhea County Historical Society. *The World's Most Famous Court: Tennessee Evolution Case*. Dayton, TN: Rhea County Historical Society, 1978.

Scopes, J. T. *Center of the Storm: Memoirs of John T. Scopes*. New York: Holt Rinehart & Winston, 1967.

Tompkins, J. R. *D-days at Dayton; Reflections on the Scopes Trial*. Baton Rouge, LA: Louisiana State University Press, 1965.

FURTHER READING

De Camp, L. S. *The Great Monkey Trial*. Garden City, NY: Doubleday, 1968.

Larson, E. J. *Summer for the Gods: The Scopes Trial and America's Continuing Debate Over Science and Religion*. New York: Basic Books, 1997.

Ruse, M. *Can a Darwinian Be a Christian? The Relation Between Science and Religion*. Cambridge: Cambridge University Press, 2004.

Toumey, C. *God's Own Scientists: Creationists in a Secular World*. New Brunswick, NJ: Rutgers University Press, 1994.

JOHN F. HAUGHT

DOES EVOLUTION RULE OUT GOD'S EXISTENCE?

In 1859 Charles Darwin published *On the Origin of Species*, his famous treatise on "evolution." It is one of the most important books of science ever written, and experts today still consider it to be largely accurate. Theologically speaking, it caused a fierce storm of controversy, and we are still wrestling with the question of what to make of it. Does Darwin's theory perhaps put the final nail in religion's coffin? Or can there be a fruitful encounter of religion with evolutionary thought?

For many scientists evolution means that the universe is fundamentally impersonal. In fact, the physicist Steven Weinberg asserts that evolution refutes the idea of an "interested" God much more decisively than physics does.[1] Only a brief look at Darwin's theory will show why it disturbs the traditional religious belief in a loving and powerful God.

Darwin observed that all living species produce more offspring than ever reach maturity. Nevertheless, the number of individuals in any given species remains fairly constant, which means that there must be a very high rate of mortality. To explain why some survive and others do not, Darwin noted that the individuals of any species are not all identical: some are better "adapted" to their environment than others. It appears that the most "fit" are the ones that survive to produce offspring. Most individuals and species lose out in the struggle for existence, but during the long journey of evolution there emerge the staggering diversity of life, millions of new species, and eventually the human race.

What, then, is so theologically disturbing about the theory? What is there about evolution that places in question even the very existence of God? It can be summarized in three propositions:

1. The variations that lead to differentiation of species are purely *random*, thus suggesting that the workings of nature are "accidental" and irrational. Today the source of these variations has been identified as genetic mutations. Most biologists today follow Darwin in attributing them to "chance."

---

2. The fact that individuals have to *struggle* for survival, and that most of them suffer and lose out in this contest, points to the basic cruelty of the universe, particularly toward the weak.
3. The mindless process of *natural selection* by which only the better adapted organisms survive suggests that the universe is essentially blind and indifferent to life and humanity.

These three ingredients—randomness, struggle, and blind natural selection—seem to confirm the strong impression of many scientific skeptics today that the universe is impersonal, utterly unrelated to any "interested" God. Darwin himself, reflecting on the "cruelty," randomness, and impersonality in evolution, could never again return to the benign theism of his ancestral Anglicanism. Though he did not casually forsake his religious faith, many of his scientific heirs have been much less hesitant to equate evolution with atheism.

From the middle of the last century up until today prominent thinkers have welcomed Darwinian ideas as the final victory of skepticism over religion. T. H. Huxley, Darwin's "bulldog" as he was known, thought evolution was antithetical to traditional theism. Ernst Haeckel, Karl Marx, Friedrich Nietzsche and Sigmund Freud all found Darwin's thought congenial to their atheism. And numerous others in our own time closely associate evolution with unbelief. Given this coalition of evolution and hostility to theism it is hardly surprising that the idea has encountered so much resistance from some religious groups.

Is the Darwinian—or now the "neo-Darwinian"—picture of evolution compatible with religion, and if it is, in what sense? Answers to this question fall into four distinct groups.

I. THE "CONFLICT" POSITION

Both scientific skeptics and biblically literalist "creationists" maintain that Darwinian evolution inevitably conflicts with religion. Skeptics find in evolution a most compelling basis for rejecting theism in particular. The three features of chance, struggle and blind natural selection seem so antithetical to any conceivable notion of divine providence or design that it is hard for them to understand how any scientifically educated person could still believe in God.

Richard Dawkins, the renowned British zoologist, presents this "conflict" position handily.[2] His thesis is that chance and natural selection, aided by immensely long periods of time, are enough to account for all the diverse species of life, including ourselves. Why would we need to invoke the idea of God if chance and natural selection alone can account for all of the creativity in the story of life? Before Darwin it may have been difficult to find definitive reasons for atheism. The order or patterning in nature seemed to beg for a supernatural explanation, and so the design argument for God's existence may have been plausible then. But this, Dawkins claims, is no longer the case. Evolutionary theory, brought up to date by the discoveries of molecular biology, has demolished the divine designer that most educated people believed in before the middle of the last century. Evolution has once and for all purged any remaining intellectual respectability from the idea of God.[3]

In his book *Natural Theology* which set forth the standard academic and theological wisdom of the early nineteenth century, William Paley had compared nature to a watch. If you chanced upon a watch lying alone on the ground, he wrote, and then examined its intricate structure, you could not help concluding that it had been made by an intelligent designer. It couldn't possibly be the product of mere chance. And yet, the natural world exhibits much more complex order than any watch. Thus, Paley concluded, there has to be an intelligent designer responsible for nature's fine arrangement. This designer, of course, would be none other than the Creator God of biblical religion.

But Dawkins argues that the divine designer is no longer needed:

> Paley's argument is made with passionate sincerity and is informed by the best biological scholarship of his day, but it is wrong, gloriously and utterly wrong. The analogy between . . . watch and living organism, is false. All appearances to the contrary, the only watchmaker in nature is the blind forces of physics, albeit deployed in a very special way. A true watchmaker has foresight: he designs his cogs and springs, and plans their interconnections, with a future purpose in the mind's eye. Natural selection, the blind, unconscious, automatic process which Darwin discovered, and which we now know is the explanation for the existence and apparent purposeful form of all life, has no purpose in mind. It has no mind and no mind's eye. It does not plan for the future. It has no vision, no foresight, no sight at all. If it can be said to play the role of watchmaker in nature, it is the *blind* watchmaker.[4]

Even though David Hume and other philosophers had already severely battered the design argument for God's existence, Dawkins thinks that only Darwin's theory of natural selection provided a fully convincing refutation of natural theology. "Darwin made it possible to be an intellectually fulfilled atheist."[5]

The order and design in nature may seem on the surface to point to a divine "watchmaker" who devised its intricate parts. But evolution, the skeptic will insist, has allowed us to look beneath the deceptive surface of nature's orderly arrangements. The pattern and design that seem so wonderfully miraculous to the scientifically illiterate can now be fully accounted for by Darwin's impersonal theory of evolution. The theory rules out any proper appeal to the God-hypothesis. If there is a watchmaker at all, it is not divine intelligence but blind natural selection that has put the parts of nature so wonderfully together over the course of billions of years of trial and error. The aimless forces of evolution are sufficient to explain all the marvels of life and mind.

Clearly it is this reading of evolution that leads so many religious opponents of Darwin to adopt the "creationist" position. Creationists agree with skeptics that evolution is incompatible with the idea of a Creator. One version of creationism known as "scientific creationism" or "creation science," rejects evolutionary theory as *scientifically* unsound, and offers the Bible as an alternative "scientific" theory.[6] On the surface scientific creationists seem to embrace scientific method. And they argue on the basis of the paucity of intermediary forms in the fossil record that the biblical account provides a "scientific" hypothesis more suited to the actual data of geology, biology, and paleontology than does Darwinian science.

Most scientists reply, however, that "creation science" is really not science at all. It does not seriously accept the self-revising method required by true science, nor does it acknowledge that the gaps in the fossil record might be compatible with other, revised versions of evolutionary theory, such as that of "punctuated equilibrium" proposed by Stephen Jay Gould and Niles Eldredge.[7] Creation science, they argue, would not even be worth discussing were it not for the fact that its devotees stir up so much public controversy in their attempts to keep evolutionary theory out of schools and textbooks.

II. THE "CONTRAST" RESPONSE

A second response to the question of whether theology is possible after Darwin, argues that since science and religion are such disparate or "contrasting" ways of looking at the world that they cannot meaningfully compete with each other. This means that evolution, which may be quite accurate as a scientific theory, bears not the slightest threat toward religion. This position rejects both "scientific creationism" *and* scientific skepticism, both of which posit a conflict between evolution and religion. It would argue as follows:

So-called "scientific creationism" is objectionable in the first place because from the point of view of good science it refuses to look at the relevant data. The scientific evidence in favor of evolution is overwhelming. Although evolutionary theory is certainly not unrevisable, this does not mean that the world and life did not evolve, at least approximately along Darwinian lines. In the second place, scientific creationism is *theologically* embarrassing. It trivializes religion by artificially imposing scientific expectations on a mythic-symbolic text. It completely misses the Bible's religious point by placing the text of *Genesis* in the same arena with science, as an alternative "scientific" account. Creationism thus subverts the deeper meaning of the biblical account of creation, its covenantal motifs, its fundamental message that the universe is a gift and that the appropriate human response to this gift is gratitude and trust. Creationism turns a sacred text into a mundane treatise to be placed in competition with scientific attempts to explain things.

If our primary question to the Bible is one of scientific curiosity about cosmic beginnings or the origins of life, we will surely miss its real intentions. Since the text was composed in a prescientific age, its primary meaning cannot be unfolded in the idiom of twentieth-century science. But that is exactly the demand put upon the Bible by scientific creationism. Needless to say, such an expectation ends up shriveling to prosaic dust a collection of deeply religious writings designed to open us to the ultimate mystery of the universe.

In the third place, scientific creationism is historically anachronistic. Creationists ironically situate the ancient biblical writings within the time-conditioned framework of modern science. They refuse to take into account the social, cultural and historical conditions in which the books of the Bible were fashioned over a period of two millennia. In doing so they close their eyes to modern historical awareness of the time-sensitive nature of all human consciousness, including that expressed in the sacred texts of religion. They are unable to discern the different types of literary genre—symbolic, mythic, devotional, poetic, legendary, historical, creedal, confessional etc.—that make up the Bible. And so they fail to read the scriptures in their proper context.

In spite of these problems, however, the contrast approach can entertain a certain empathy with the phenomenon of creationism. Creationism may be an unfortunate symptom of the much wider effort by traditionally religious people to cope with modernity. At heart creationists and other fundamentalists are sincerely and understandably troubled by the failings of the post-Enlightenment world. They deplore the breakdown of authority, the diminishment of "virtue," the absence of common purpose, the loss of a sense of absolute values, and the banishing of a sense of sacred mystery from our experience. For many creationists the notion of "evolution" sums up all the evils and emptiness of secularistic modernity. Creationism, in other words, is responding to something much more complex than the conflict between religion and science.

Moreover, according to our second approach, the phenomenon of creationism points to serious problems in the way science has been presented to the public by some of our most prominent scientific writers. These scientists also indulge in a conflation of science with belief: they unnecessarily fuse valuable scientific data with the ideology of "scientific materialism" which *is* antithetical to any religious perspective. Scientists of the stature of Carl Sagan, Stephen Jay Gould, E. O. Wilson, and Richard Dawkins, just to name a few, offer the theory of evolution already snugly wrapped up in the alternative "religion" of scientism and materialism. So, in a sense, creationism is an understandable, though ineffective, reaction to an alternative conflation of science and belief.

Scientific materialists, as the contrast position contends, generally write about evolution as though it were inherently anti-theistic. In doing so they uncritically accept the assumptions of secularistic ideology and culture. Stephen Jay Gould, for example, has stated that the reason so many people cannot accept Darwin's ideas is that, in his opinion, evolution is inseparable from a "philosophical" message, namely, materialism. He writes:

> . . . I believe that the stumbling block to [the acceptance of Darwin's theory] does not lie in any scientific difficulty, but rather in the philosophical content of Darwin's message—in its challenge to a set of entrenched Western attitudes that we are not yet ready to abandon. First, Darwin argues that evolution has no purpose. Individuals struggle to increase the representation of their genes in future generations, and that is all. . . . Second, Darwin maintained that evolution has no direction; it does not lead inevitably to higher things. Organisms become better adapted to their local environments, and that is all. The "degeneracy" of a parasite is as perfect as the gait of a gazelle. Third, Darwin applied a consistent philosophy of materialism to his interpretation of nature. Matter is the ground of all existence; mind, spirit and God as well, are just words that express the wondrous results of neuronal complexity.[8]

If one is to accept evolution, Gould implies, one must first embrace materialism. Like many other scientists today, and in spite of serious disagreements with fellow Darwinians, he clearly approves of the alliance of materialist assumptions and evolutionary theory. But this merger seems no less illustrative of the conflation of science with a belief system than is scientific creationism. In both instances it is the (con)fusion of science with ideology that paves the way to conflict.

In order to avoid this kind of confusion the "contrast" approach consistently maintains a clear distinction between science and belief systems, whether the latter are religious or materialist. A contrast approach seeks to liberate science from all ideology. Consequently it insists that evolutionary thinking is not in a position to tell us anything about God, nor can religious experience shed any significant light on evolution. Theology, moreover, should stick to its task of opening us up to religious experience, and scientists should stick to science, steering clear of the kind of ideological propaganda that Gould exemplifies. Evolution is a purely scientific theory that need not be cast in materialist terms. It is not evolution itself, but the materialist spin some scientists put on evolution, that is incompatible with religion. When it is stripped of its materialist covering evolutionary theory in no way contradicts theism.

This is how the contrast position seeks to make room for theology after Darwin. But, one might ask, can evolution really be extricated from materialist dogma? What about those theologically troubling aspects of Darwinian theory: chance, the struggle for survival, and impersonal natural selection? Do they not refute theism and require a materialist interpretation? Without getting bogged down in theological conjecture here, the contrast approach is content to show that none of these three items necessarily contradicts theism.

In the first place, the "chance" character of the variations which natural selection chooses for survival may just as easily be accounted for as the product of human ignorance. The apparent randomness of what we today call genetic mutations could be a mere illusion resulting from the limitedness of our perspective. Religions claim, after all, that any purely human angle of vision is always exceedingly narrow. Hence, what appears to be absurd chance from a purely scientific perspective could be quite rational and coherent from that of an infinite Wisdom.

Second, evolutionist complaints about the struggle, suffering, waste, and cruelty of natural process add absolutely nothing new to the basic problem of evil of which religion has always been quite fully apprised. The Bible, for example, has surely heard of Job and the crucifixion of Jesus, and yet it proclaims the paradoxical possibility of faith and hope in God in spite of all evil and suffering. One might even argue that faith has no intensity or depth unless it is a leap into the unknown in the face of such absurdity. Faith, according to this contrast theology—which habitually appeals to thinkers like Soren Kierkegaard—is always faith "in spite of" all the objective difficulties that defy reason and science.

Finally, there is no more theological difficulty in the remorseless law of natural selection, which is said to be impersonal and blind, than in the laws of inertia, gravity or any other impersonal aspects of science. Gravity, like natural selection, has no regard for our inherent personal dignity either. It pulls toward earth the weak and powerful alike—at times in a deadly way. But very few thinkers have ever insisted that gravity is an argument against God's existence. Perhaps natural selection should be viewed in the same way.

Moreover, the contrast approach also rejects Paley's narrowly conceived "natural theology" since it seeks to know God independently of God's self-revelation. Nature itself provides evidence neither for nor against God's existence. Something so momentous as the reality of God can hardly be decided by a superficial scientific deciphering of the natural world. Hence religion should in no way be troubled by evolutionary theory.

III. THE "CONTACT" APPROACH

A third repose to our question about the prospects for theology after Darwin is not content with the standoff endorsed by the contrast position just summarized. It allows that the contrast approach has the merit at least of shattering the facile fusion of faith and science that underlies most instances of apparent conflict. Its sharp portrayal of the ideological biases in both creationism and evolutionism is very helpful. Contrast may be an essential step in the process of thinking clearly and fruitfully about the relationship of evolution to religion.

But for many scientists and religious thinkers the contrast approach does not go nearly far enough. Evolution is more than just another innocuous scientific theory that theology can innocently ignore. Theologians need to do more than just show that evolution does not contradict theism. Evolution, according to what may be called the "contact" position, is a most appropriate framework in which to express the true meaning of theistic faith. The "contact" approach would go something like this:

Evolutionary science deepens not only our understanding of the cosmos but also of God. Unfortunately, many theologians have still not faced the fact that we live in a world after and not before Darwin, and that an evolving cosmos looks a lot different from the world-pictures in which most religious thought was born and nurtured. If it is to survive in the intellectual climate of today, therefore, our theology requires fresh expression in evolutionary terms. When we think about God in the post-Darwinian period we cannot have exactly the same thoughts that Augustine, Aquinas, or for that matter our grandparents and parents had. Today we need to recast all of theology in evolutionary terms.

In fact, evolution is an absolutely essential ingredient in our thinking about God today. As the Roman Catholic theologian Hans Küng puts it, evolutionary theory now makes possible: (1) a deeper understanding of God—not above or outside the world but in the midst of evolution; (2) a deeper understanding of creation—not as contrary to but as making evolution possible; and (3) a deeper understanding of humans as organically related to the entire cosmos.[9]

Skeptics, of course, will immediately ask how theology can reconcile the idea of God with the role of chance in life's evolution. This is a crucial question, and the contrast position's casual conjecture that chance may not really exist is unsatisfactory. In fact, chance is quite real. It is a concrete fact in evolution, but it is not one that contradicts the idea of God. On the contrary, an aspect of indeterminacy is just what we should expect if, as religion maintains, God is love. For love never coerces. It allows the beloved—in this case the entire created cosmos—to be or to become itself. If, as theistic religious tradition has always insisted, God really cares for the well-being of the world, then the world has to be something other than God. It has to have a certain amount of "freedom" or autonomy. If it did not somehow exist on its own it would be nothing more than an extension of God's own being, and hence it would not be a world unto itself. So there has to be room for indeterminacy in the universe, and the randomness in evolution is one instance of it.

In other words, if the world is to be something distinct from God it must have scope for meandering about, for experimenting with different ways of existing. In their relative freedom from divine coercion, some of the world's evolutionary experiments may work and others may not. But divine love does not crudely interfere. It risks allowing the cosmos to exist in relative liberty. In the unfolding of life, the world's inherent quality of being uncompelled manifests itself in the form of "contingent" occurrences in natural history (as Stephen Gould insightfully emphasizes), or in the random variations or genetic mutations that comprise the raw material of evolution. Thus a certain amount of chance is not at all opposed to the idea of God.

A God of love influences the world in a persuasive rather than coercive way, and this is why chance and evolution occur. It is because God is involved with the world in a loving rather than domineering way that the world evolves.[10] If God were a magician or a dictator, then we might expect the universe to be finished all at once and remain eternally unchanged. If God controlled the world rigidly instead of willing its independence, we might not expect the weird organisms of the Cambrian explosion, the later dinosaurs and reptiles, or the many other wild creatures that seem so alien to us. We would want our divine magician to build the world along the lines of our own narrowly human sense of clean perfection. But what a pallid and impoverished world that would be. It would lack all the drama, diversity, adventure, and intense beauty that evolution has produced. It might have a listless harmony to it, but it would have none of the novelty, contrast, danger, upheavals, and grandeur that evolution has in fact brought about over billions of years.

According to the contact position, God is not a magician but a creator.[11] And this God is much more interested in promoting freedom and adventure than in preserving the status quo. Since divine creative love has the character of letting things be, we should not be too surprised at evolution's strange and erratic pathways. The long struggle of the universe to arrive at life, consciousness, and culture is consonant with faith's conviction that love never forces but always allows for the play of freedom, risk and adventure.

Love even gives the beloved a share in the creative process. Might it not be because God wants the world to partake of the divine joy of creating novelty, that it is left unfinished, and that it is invited to be, at least to some degree, self-creative? And if it is self-creative can we be too disconcerted that it has experimented with the many different, delightful, baffling, and bizarre forms that we find in the fossil record and in the diversity of life that surrounds us now?

Ever since Darwin scientists have found out things about the natural world that may not be consistent with an innocent notion of divine design such as the one proposed by Paley and lampooned by Dawkins. But the new discoveries of an evolving cosmic story correspond very well with the self-giving humility of the God of religious experience, a God who wishes to share the divine creative life with all creatures, and not just humans. Such a God renounces any will to control the process of creation and gives to creatures a significant role, indeed a partnership, in the ongoing evolution of the world. Such a gracious self-giving love would be quite consistent with a world open to all the surprises that pertain to evolution.

In summary, the "hypothesis" of God, taken in consort with (and not as an alternative to) evolutionary theory, can help account for the complexity and consciousness that evolution has brought about. God may be thought of as the transcendent source not only of the order in the universe but also of the novelty and turbulence that evolution has brought with it. God creates by inviting (not forcing) the cosmos to express itself in increasingly more diverse ways. As novelty comes into the evolving world, the present order has to give way. And what we confusedly refer to as "chance" and "chaos" may be the result of the breakdown of present arrangements of order in the wake of novelty's coming into the world.

The ultimate origin of evolutionary novelty is God. God's will, in this version of the contact approach, is the maximization of novelty and diversity. And since the introduction of novelty and diversity is what turns the cosmos into a world of beauty, we may say that the God of evolution is a God who wants nothing less than the ongoing enhancement of cosmic beauty. Thus an evolutionary picture of the cosmos, with all of its craziness and serendipitous wanderings, corresponds quite well with the biblical understanding of an adventurous and loving God as the One "who makes all things new."[12]

However, God's role in evolution is not only that of being the stimulus that stirs the cosmos toward deeper novelty and beauty. Religious faith claims that the same God who creates also promises to save the world from suffering and death. This would mean that the whole history of cosmic evolution, in all its detail and incredible breadth, is permanently taken into God's loving memory. The suffering of the innocent and the weak, highlighted so clearly by evolutionary thought, becomes inseparable from the divine eternity. Theology cannot tolerate a deity who merely creates and then abandons the world. God is intimately *involved* in the evolutionary process and struggles along with all beings, participating in both their pain and enjoyment, ultimately redeeming the world so that nothing in its long evolution is ever completely forgotten or lost.[13]

This is only a brief sampling of how some contemporary theology is now being transformed by its encounter with evolutionary science. Many varieties of evolutionary theology exist today, and the "contact" position summarized in this section is just a small fragment of the rethinking going on in theology after Darwin. According to this third response, it is regrettable that so much contemporary religious thought goes the way of creationism or contrast. Although evolutionary theology is inevitably in need of constant revision—and prudence requires that theology avoid enshrining for all time any particular version of it—a number of theologians consider evolution to be at least provisionally the most appropriate and fruitful framework within which to think about God today.

IV. CONFIRMATION

A fourth approach goes even further than the contact position in establishing the close connection between theism and evolution. It argues that biblical religion with its distinctive notion of God provides much of the soil in which Darwinian ideas have taken root in the first place.[14] In this sense, religion can be said to support or "confirm" the evolutionary picture of nature, not by providing any additional scientific information—which is not religion's function anyway—but by providing part of the general picture of reality which has made evolutionary science historically possible.

For example, evolutionary theory could hardly have originated and thrived outside of a cultural context shaped by the specifically biblical picture of the nature of time rooted ultimately in a very particular understanding of God. The Bible understands time in terms of God's bringing about a new and surprising future. When through biblical faith some people became aware of a *promise* offered by a God who appears out of the future, they began to experience time in a new way. As the promised new creation beckoned them, they no longer felt the compulsion to return to a golden age in the past. Time became directional and irreversible at a very deep level of their awareness. And even when the idea of God dropped out of the intellectual picture of the cosmos in the modern period, the feeling of time as directional and irreversible remained deeply lodged in western sensibilities, including that of secular scientists. But it was an originally biblical perception of temporality that made it possible for science to embrace an evolutionary picture of the universe.

In contrast to this linear-historical sensitivity, the argument continues, most non-biblical religions and cultures have understood time as a repeating circle. Time's destiny, in both primal and Eastern religious traditions, is not something radically new, but instead a return to the purity and simplicity of cosmic origins. The Bible's emphasis on God as the source of a radically new future, on the other hand, breaks open the ancient cycle of time. It calls the whole cosmos, through the mediation of human hope, to look forward in a more linear way for the coming of God's kingdom, either in the indefinite future or at the end of time.

According to our fourth approach, therefore, it is only on the template of this stretched out view of irreversible duration that evolutionary ideas could ever have taken shape. Even though evolution does not have to imply a vulgar notion of "progress," it still seems to have required an irreversible, future-oriented understanding of time as its matrix. This view of time, the confirmation position claims, originally came out of a *religious* experience of reality as promise.

However, there may be an even deeper way in which faith in God nourishes the idea of evolution. The central idea of theistic religion, as the Catholic theologian Karl Rahner (among others) has clarified, is that the Infinite pours itself out in love to the finite universe. This is the fundamental meaning of "revelation." But if we think carefully about this central religious teaching it should lead us to conclude that any universe related to the inexhaustible self-giving love of God must be an evolving one. For if God is infinite love giving itself to the cosmos, then the finite world cannot possibly receive this limitless abundance of graciousness in any single instant. In response to the outpouring of God's boundless love the universe would be invited to undergo a process of self-transformation. In order to "adapt" to the divine infinity the finite cosmos would likely have to intensify its own capacity to receive such an abounding love. In other words, it might endure what we now know scientifically as an arduous, tortuous, and dramatic evolution.[15]

Viewed in this light, the evolution of the cosmos is more than just "compatible" with theism. Faith in a God of self-giving love, it would not be too much to say, actually anticipates an evolving universe. It may be very difficult to reconcile the religious teaching about God's infinite love with any other kind of cosmos.

NOTES

1. Weinberg, pp. 246–49.

2. Richard Dawkins, *The Blind Watchmaker* (New York: W.W. Norton & Co., 1986); *River Out of Eden* (London: Orion, 2001); and *Climbing Mount Improbable* (New York: Penquin, 2006).

3. Ibid., p. 6.

4. Ibid., p. 5.

5. Ibid., p. 6.

6. Duane Gish, *Evolution: The Challenge of the Fossil Record* (El Cajon: Creation-Life Publishers, 1985).

7. See Niles Eldredge, *Time Frames: The Rethinking of Darwinian Evolution and the Theory of Punctuated Equilibria* (London: Heinemann, 1986).

8. Stephen Jay Gould, *Ever Since Darwin* (New York: W. W. Norton, 1977), pp. 12–13.

9. Hans Küng, *Does God Exist*, trans. Edward Quinn (New York: Doubleday, 1980), p. 347.

10. For a discussion of this approach see John F. Haught, *The Promise of Nature* (New York: Paulist Press, 1993.)

11. See L. Charles Birch, *Nature and God* (Philadelphia: Westminster Press, 1965), p. 103.

12. For a development of these ideas, many of which are suggested by Alfred North Whitehead, see John F. Haught, *The Cosmic Adventure* (New York: Paulist Press, 1984).

13. These ideas are elucidated especially by what is called "process theology." See John B. Cobb Jr. and David Ray Griffin, *Process Theology: An Introductory Exposition* (Philadelphia: Westminster Press, 1976).

14. See, for example, Ernst Benz, *Evolution and Christian Hope* (Garden City, NY: Doubleday, 1966).

15. See Karl Rahner, *Hominization*, trans. W. J. O'Hara (New York: Herder & Herder, 1965).

FURTHER READING

There are several collections of readings in the philosophy of biology. Let me mention three in which I have had a hand, the first two co-edited with David Hull. *Readings in The Philosophy of Biology* (Oxford University Press, 1988) is a collection that covers the whole field, from Darwinism to Creationism, with articles by major philosophers and biologists, accessible to senior undergraduate and graduate students. *The Cambridge Companion to the Philosophy of Biology* (Cambridge University Press, 2007) is a collection of new pieces, primarily by philosophers, pitched for a general audience, but rather demanding. *The Oxford Handbook to the Philosophy of Biology* (Oxford University Press, 2007) edited by myself alone, is a large collection, moving into newer areas with a strong emphasis on some of the more value-laden aspects of the subject, including discussions on race, on the environment, on religion, and on feminism and biology. By far the best overall textbook is *Sex and Death: An Introduction to Philosophy of Biology* (University of Chicago Press, 1999) by Kim Sterelny and Paul Griffiths. For up-to-date, more technical discussions, you should look at the journal *Biology and Philosophy*, which has now been flourishing for twenty years.

INTRODUCTION. Mayr wrote many books in his long life, right up to the end. For his views on the philosophy of biology, go to his collection *Towards a New Philosophy of Biology: Observations of an Evolutionist* (Harvard University Press, 1988).

LIFE AND ITS ORIGINS. Most evolution textbooks now have a chapter on the origin of life. I recommend S. Freeman and J. C. Herron, *Evolutionary Analysis*, 3rd ed. (Prentice-Hall, 2004). Iris Fry, *The Emergence of Life on Earth* (Rutgers University Press, 2000) is a good historical-philosophical introduction to the topic.

EXPLAINING DESIGN. All of the modern books from which this section's pieces are taken are well worth reading. If you are at all serious about the subject, then you really must read Darwin's *Origin*, a surprisingly good read for a classic. There is a facsimile of the first edition published by Harvard University Press (with a foreword by Ernst Mayr). If you want some background to Darwin and his revolution, then look at my *The Darwinian Revolution: Science Red in Tooth and Claw*, 2nd ed. (University of Chicago Press, 1999).

DARWINISM. I recommend any and all of Stephen Jay Gould's collections of essays (first published in *Natural History*). The first collection *Ever Since Darwin* (Norton, 1977) is the classic. If you want to learn something about the characters in the history of evolution, then try my *Mystery of Mysteries: Is Evolution a Social Construction?* (Harvard University Press, 1999). My *Darwinism and its Discontents* (Cambridge University Press, 2006) will bring you right up to date on modern controversies.

MACROEVOLUTION. Just before he died, Gould brought out a truly massive work on his subject. I doubt anyone can read *The Structure of Evolutionary Theory* (Harvard University Press, 2002) straight through from beginning to end, but there is much to be found if you pick and choose. Much more fun is an earlier book, *Wonderful Life: The Burgess Shale and the Nature of History* (Norton, 1989). David Sepkoski and I have a collection, *Paleontology at the High Table* (University of Chicago Press, 2007) that tries to capture some of the excitement of the last thirty to forty years in what is now known as paleobiology.

CLASSIFICATION. The collections mentioned earlier all have pieces on the topics of this section. If you want a good read, telling about how scientists interact and try to impose their own views on others, read the history of the classification wars, *Science as a Process* (University of Chicago Press, 1988) by David Hull.

TELEOLOGY AND ADAPTATION. Read George Williams's book, *Adaptation and Natural Selection* (Princeton University Press, 1966) extracted in this section. Then go on to my *Darwin and Design: Does Evolution have a Purpose* (Harvard University Press, 2003).

HUMAN NATURE. Start with Wilson's *On Human Nature* (Harvard Univeristy Press, 1978) extracted here. More recent and more technical is Kim Sterelny's *Thought in a Hostile World* (Blackwell, 2003).

EVOLUTION AND ETHICS. Edward O. Wilson has just published a collection of his papers, *Nature Revealed: Selected Writings, 1949–2006* (Johns Hopkins University Press, 2006) where you will find many pieces on and around his views on morality. I give my views in *Taking Darwin Seriously: A Naturalistic Approach to Philosophy*, 2nd ed. (Prometheus, second edition 1998). Just about any conventional text on ethics will tell you why our positions are untenable.

GM FOODS. Go to the collection by me and David Castle, *Genetically Modified Foods* (Prometheus, 2002).

BIOLOGICAL METAPHORS. You should read the whole of James Lovelock's *Gaia: A New Look at Life on Earth* (Oxford University Press, 1979). The Internet carries a huge amount of stuff, for and against, on this topic. For more metaphor in science, look at my already-mentioned *Mystery of Mysteries*.

ENVIRONMENTAL ETHICS. Start with *Philosophy Gone Wild* (Prometheus, 1986) by Holmes Rolston III, and then go on to Edward O.

Wilson's *Biophilia* (Harvard University Press, 1984). There are lots of collections on this topic.

WHEN LIFE BEGINS. There is obviously a huge amount of stuff here. With Christopher Pynes I have a collection *The Stem Cell Controversy*, 2nd ed. (Prometheus, 2006). Really good is Jane Maienschein's *Whose View of Life? Embryos, Cloning, and Stem Cells* (Harvard University Press, 2003).

GOD AND BIOLOGY. Start with my overview history *The Evolution-Creation Struggle* (Harvard University Press, 2005). Robert Pennock's anti-ID book, *The Tower of Babel* (MIT Press, 1998) is essential reading. John Haught's recent books include *God After Darwin* (Westview, 2000). Pennock and I have an edited volume, *But is it Science: The Philosophical Question in the Creation-Evolution Controversy*, 2nd ed. (Prometheus, 2007). A more systematic survey of the issues is my *Can a Darwinian be a Christian? The Relationship between Science and Religion* (Cambridge University Press, 2000).